MW00844011

COOKING INNOVATIONS

USING HYDROCOLLOIDS FOR THICKENING, GELLING, AND EMULSIFICATION

COOKING INNOVATIONS

USING HYDROCOLLOIDS FOR THICKENING, GELLING, AND EMULSIFICATION

AMOS NUSSINOVITCH • MADOKA HIRASHIMA

CRC Press
Taylor & Francis Group
Boca Raton London New York

CRC Press is an imprint of the
Taylor & Francis Group, an **informa** business

CRC Press
Taylor & Francis Group
6000 Broken Sound Parkway NW, Suite 300
Boca Raton, FL 33487-2742

© 2014 by Taylor & Francis Group, LLC
CRC Press is an imprint of Taylor & Francis Group, an Informa business

No claim to original U.S. Government works
Version Date: 20130711

International Standard Book Number-13: 978-1-4665-3989-1 (eBook - Kindle)

This book contains information obtained from authentic and highly regarded sources. Reasonable efforts have been made to publish reliable data and information, but the author and publisher cannot assume responsibility for the validity of all materials or the consequences of their use. The authors and publishers have attempted to trace the copyright holders of all material reproduced in this publication and apologize to copyright holders if permission to publish in this form has not been obtained. If any copyright material has not been acknowledged please write and let us know so we may rectify in any future reprint.

Except as permitted under U.S. Copyright Law, no part of this book may be reprinted, reproduced, transmitted, or utilized in any form by any electronic, mechanical, or other means, now known or hereafter invented, including photocopying, microfilming, and recording, or in any information storage or retrieval system, without written permission from the publishers.

For permission to photocopy or use material electronically from this work, please access www.copyright.com (http://www.copyright.com/) or contact the Copyright Clearance Center, Inc. (CCC), 222 Rosewood Drive, Danvers, MA 01923, 978-750-8400. CCC is a not-for-profit organization that provides licenses and registration for a variety of users. For organizations that have been granted a photocopy license by the CCC, a separate system of payment has been arranged.

Trademark Notice: Product or corporate names may be trademarks or registered trademarks, and are used only for identification and explanation without intent to infringe.

Visit the Taylor & Francis Web site at
http://www.taylorandfrancis.com

and the CRC Press Web site at
http://www.crcpress.com

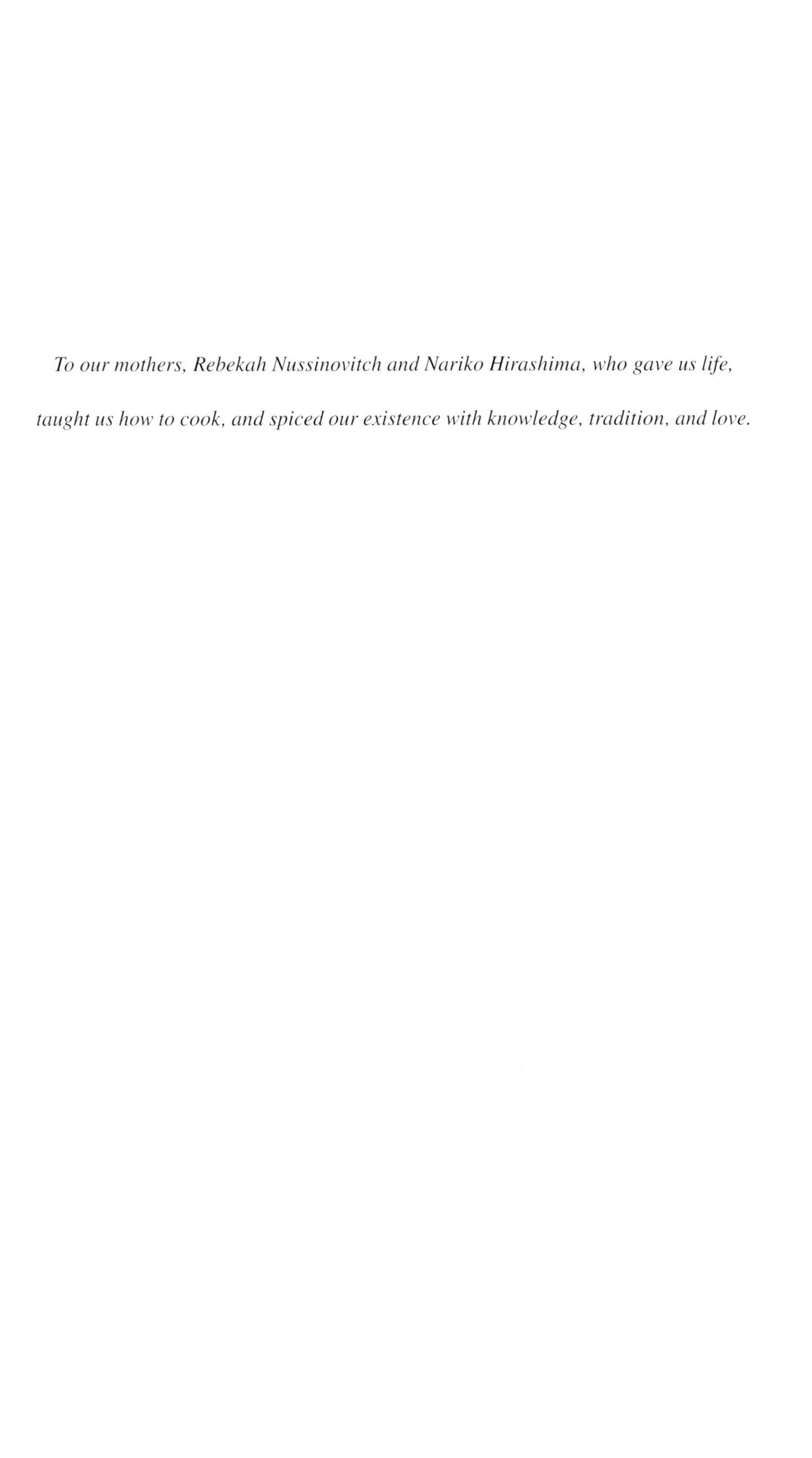

To our mothers, Rebekah Nussinovitch and Nariko Hirashima, who gave us life, taught us how to cook, and spiced our existence with knowledge, tradition, and love.

Contents

Preface xxi
Acknowledgments xxxi
The Authors xxxiii

1. Hydrocolloids—Where, Why, and When? 1
Introduction 1
Terminology 1
Classification 3
Economics 3
Gum Constituents and Their Effects on Processing 4
Functions of Hydrocolloids in Food Applications 5
Functions in Food Products 5
Viscosity Formation and Its Typical Food Applications 5
Gelation, Gel Types, and Linkages 6
Gel Textures 7
Gel-Enhancing Effects of Other Gums 7
Hydrocolloids in Emulsions, Suspensions and Foams, and in Crystallization Control 8
Other Unique Food Applications 9
Regulatory Aspects 10
References and Further Reading 10

2. Agar–Agar 13
Historical Background 13
Collection and Processing of Seaweed for Agar 13
Types of Agar Products 14
Regulatory Status and Toxicity 16
Structure of Agar 16
Agar–Agar Properties 16
Dissolution and Viscosity 16
Gelation and Melting 17
Gel Clarity 18
Gel Syneresis 18
Effect of Addition of Other Materials on Agar Properties 18
Commercial Food Applications 18
Agar in the Baking Industry 18
Agar in Confections 19
Agar in Wet and Dried Structured Fruit Products 19
Meat, Milk, Fish, and Other Products 20

Recipes with Agar–Agar 21
Mizu Yokan (Agar Jelly with Red Bean Paste) (Figure 2.3) 21
Ingredients 21
Preparation 22
Hints 22
Mitsu-Mame (Agar Jelly with Rice Dumplings) 23
Ingredients 23
Preparation 24
Hints 25
Orange Jelly (Figure 2.9) 26
Ingredients 26
Preparation 26
Hint 26
Awayuki-Kan (Agar Jelly with Egg White) (Figure 2.10) 27
Ingredients 27
Preparation 27
Hint 28
Xìngrén Jelly (Chinese Almond Pudding) (Figure 2.12) 28
Ingredients 28
Preparation 29
Hint 29
Agar Spaghetti (Figure 2.13) 29
Ingredients 29
Preparation 29
Tips for the Amateur Cook and Professional Chef 30
References and Further Reading 31

3. Alginates 35
Historical Background 35
Sources 35
Structure 36
Alginate Sources and Manufacture 36
Raw Materials 36
Alginate and PGA Production 36
Regulatory Status and Toxicity 36
Commercial Characteristics 37
Hydrocolloid Solution Preparation Procedures 37
Mechanism of Alginate Gelation 38
Degree of Conversion, Thixotropy, and Alginate–Pectin Gels 38
Diffusion Setting 39
Establishing Differences between Alginate Gels 39
Applications 40
Fruit-Like Products and Coatings on Fresh-Cut Fruit Pieces 40
Water Dessert Gels 41
Milk Puddings, Ice-Cream Stabilizers, and Other Dairy Products 41
Fish and Meat Preservation and Sausage Casings 41

Bakery Toppings, Fillings, Beverages, and Salad Dressings........................43
Further Applications 43
Recipes with Alginates...44
Yogurt Spheres (Figure 3.1) 44
Ingredients...44
Preparation..44
Artificial *Ikura* (Salmon Eggs) (Figure 3.2)...45
Ingredients...45
Preparation..45
Hints...46
Onion Rings (Figure 3.3) 47
Ingredients...47
Preparation..47
French Dressing (Figure 3.5) 49
Ingredients...49
Preparation..49
Hint..50
Tips for the Amateur Cook and Professional Chef...50
References and Further Reading..51

4. Carrageenan and Furcellaran 57
Introduction and Historical Background...57
Structure 58
Sources and Production...58
Accessible Types of Gum...59
Regulatory Aspects 60
Molecular Weight and Consistency...60
Solutions and Gels..60
Properties 60
Preparation..61
Viscosity...61
Effect of Molecular Weight...62
Gel Preparation and Mechanical Properties 62
Reactivity with Proteins 63
Applications...64
Milk Applications 65
Water Applications..66
Meat and Fish Applications...67
Recipes with Carrageenan and Furcellaran...68
Milk Pudding (Figure 4.1)...68
Ingredients...68
Preparation..69
Hints...69
Flan (Figure 4.2)...69
Ingredients...69
Preparation..69
Hint..70

Chocolate Drink (Figure 4.3) ... 71
Ingredients ... 71
Preparation ... 71
Hint ... 71
Ham (Figure 4.4) ... 71
Ingredients ... 71
Preparation ... 72
Hints ... 72
Tips for the Amateur Cook and Professional Chef ... 73
References and Further Reading ... 73

5. Cellulose Derivatives ... 77
Introduction ... 77
Manufacture ... 77
Properties of Methylcellulose (MC) and Methylhydroxypropylcellulose (MHPC) ... 78
Hydroxypropylcellulose (HPC) ... 79
Microcrystalline Cellulose (MCC) ... 80
Carboxymethylcellulose (CMC) ... 81
General Information ... 81
Chemical Nature and Manufacture ... 82
Chemical and Physical Properties ... 82
Solution Properties ... 82
Viscosity ... 83
Stability and Physical Data ... 84
Food Applications ... 85
Recipes with Cellulose Derivatives ... 88
Parsley Spaghetti (Figure 5.1) ... 88
Ingredients ... 88
Preparation ... 88
Hints ... 88
Takoyaki (Octopus Dumpling) (Figure 5.2) ... 89
Ingredients ... 89
Preparation ... 90
Hints ... 91
Soy Burger Patties (Figure 5.4) ... 91
Ingredients ... 91
Preparation ... 92
Hints ... 92
Pineapple Ice Cream (Figure 5.5) ... 93
Ingredients ... 93
Preparation ... 93
Hint ... 94
Sugarpaste (Figure 5.6) ... 94
Ingredients ... 94
Preparation ... 95
Hints ... 95

Low-Fat Whipped Cream (Figure 5.7) 95
Ingredients 95
Preparation 95
Hint 95
Tips for the Amateur Cook and Professional Chef 96
References and Further Reading 96

6. Curdlan 101
Historical Background 101
Production 101
Chemical Structure 101
Regulatory Status and Toxicity 102
Functional Properties 102
Solution Properties and Conformations 102
Aqueous Suspension Properties 102
Gelation 102
Commercial Food Applications 103
Recipes with Curdlan 105
Udon (Japanese Noodles) (Figure 6.3) 105
Ingredients 105
Preparation 106
Hint 106
Kamaboko (Fish Cake) (Figure 6.4) 106
Ingredients 106
Preparation 107
Hints 108
Sausages (Figure 6.5) 108
Ingredients 108
Preparation 109
Hints 110
Doughnuts (Figure 6.8) 111
Ingredients 111
Preparation 112
Hint 112
Kinugoshi Tofu (Soybean Curd) (Figure 6.9) 112
Ingredients 112
Preparation 113
Hint 113
Multilayered Jelly (Figure 6.10) 114
Ingredients 114
Preparation 114
Hints 115
Tips for the Amateur Cook and Professional Chef 115
References and Further Reading 116

7. Egg Proteins 119
Historical Background 119
The Structure of the Egg 119

The Composition of the Egg ... 119
Egg Yolk ... 120
Egg White ... 120
Essential Nutrients and Value of Eggs ... 121
Egg Yolk Emulsions ... 121
Egg White Foams ... 122
Gels ... 123
Egg Yolk Gels ... 123
Egg White Gels ... 123
Recipes with Eggs ... 124
Plain Omelette (Figure 7.2) ... 124
Ingredients ... 124
Preparation ... 124
Hints ... 125
Japanese Rolled Omelette (*Sushi* Egg) (Figure 7.3) ... 125
Ingredients ... 125
Preparation ... 125
Hints ... 126
Crabmeat and Egg Drop Soup ... 127
Ingredients ... 127
Preparation (Figure 7.5) ... 127
Hints ... 127
Chawan-Mushi (Savory Egg Custard) (Figure 7.6) ... 128
Ingredients ... 128
Preparation ... 129
Hint ... 129
Caramel Custard (Figure 7.7) ... 130
Ingredients ... 130
Preparation ... 130
Hints ... 131
Crème Brûlée (Figure 7.8) ... 131
Ingredients ... 131
Preparation ... 131
Hints ... 132
Cream Puffs (Figure 7.9) ... 132
Ingredients ... 132
Preparation ... 133
Hint ... 134
Meringues (Figure 7.10) ... 134
Ingredients ... 134
Preparation ... 134
Hints ... 135
Tips for the Amateur Cook and Professional Chef ... 135
References and Further Reading ... 135

8. Galactomannans ... 139
Introduction ... 139
Locust Bean Gum: Sources, Manufacturing, and Legislation ... 139

Guar Gum: Sources, Processing, and Regulatory Status 140
Tara Gum 140
Fenugreek Gum 141
Galactomannan Structure 141
Gum Solution Properties 142
Gelation and Interactions of Galactomannans 143
Stability 144
Food Applications 144
Recipes with Galactomannans 147
  Brown Sugar Sherbet (Figure 8.1) 147
    Ingredients 147
    Preparation 147
      Hints 148
  Raspberry Sauce (Figure 8.2) 148
    Ingredients 148
    Preparation 149
      Hint 149
  Banana and Nuts Ice Cream (Figure 8.3) 150
    Ingredients 150
    Preparation 150
      Hints 150
  Frozen Fruit Drink (Figure 8.4) 151
    Ingredients 151
    Preparation 151
      Hint 151
  Empanada Dough (Figure 8.5) 152
    Ingredients 152
    Preparation 152
      Hints 153
  Loukoums (Figure 8.6) 153
    Ingredients 153
    Preparation 153
      Hint 154
  Yoghurt Dressing (Figure 8.7) 155
    Ingredients 155
    Preparation 155
      Hints 155
  Sharbat Hilba (Fenugreek Soup) (Figure 8.8) 156
    Ingredients 156
    Preparation 156
  Pear and Fenugreek Jam (Figure 8.9) 157
    Ingredients 157
    Preparation 157
      Hint 158
Tips for the Amateur Cook and Professional Chef 158
References and Further Reading 158

9. Gelatin 161
Historical Background 161
Definitions 161
Manufacture and Sources 162
Acid Processing 162
Alkaline Processing 162
Final Products 163
Physical Properties 163
Technical Data 163
Bloom Strength 163
Chemical and Microbiological Properties 163
Food Uses and Applications 164
Gelatin Desserts 164
Confections 164
Ice-Cream Stabilizers, Dairy, Fish and Meat Products 165
Baked Goods, Carriers, and Miscellaneous 166
Regulations 167
Recipes with Gelatin 168
Fruit Jelly (Figure 9.1) 168
Ingredients 168
Preparation 168
Hints 169
Fruit Gummy Candy (Figure 9.2) 169
Ingredients 169
Preparation 170
Aspic Jelly Salad (Figure 9.3) 170
Ingredients 170
Preparation 171
Hint 172
Bavaroise (Bavarian Cream) (Figure 9.4) 172
Ingredients 172
Preparation 173
Hints 173
Marshmallows (*Guimauve*) (Figure 9.5) 173
Ingredients 173
Preparation 174
Hints 175
Tips for the Amateur Cook and Professional Chef 175
References and Further Reading 175

10. Gellan Gum 179
Historical Background 179
Structure and Chemical Composition 179
Source, Production Supply, and Regulatory Status 179
Manufacture 179
Nutritional Aspects 180

Regulatory Status ... 180
Functional Properties ... 180
Hydration ... 180
Low-Acyl (LA) Gellan Gum ... 180
High-Acyl (HA) Gellan Gum ... 181
Mechanism of Gelation and Gellan-Gum Gel Properties ... 181
Comparison to Other Hydrocolloids ... 182
Food and Other Applications ... 183
Recipes with Gellan Gum ... 186
Fruit-Juice Jelly (Figure 10.1) ... 186
Ingredients ... 186
Preparation ... 186
Hint ... 187
Low-Solids Jam (Figure 10.2) ... 187
Ingredients ... 187
Preparation ... 187
Hint ... 188
Liqueur Jelly (Figure 10.4) ... 188
Ingredients ... 188
Preparation ... 188
Pulp-Suspension Fluid Gel (Figure 10.5) ... 189
Ingredients ... 189
Preparation ... 189
Herb-Suspension Oil-Free Dressing (Figure 10.6) ... 190
Ingredients ... 190
Preparation ... 191
Hint ... 191
Tips for the Amateur Cook and Professional Chef ... 191
References and Further Reading ... 191

11. Gum Arabic ... 195
Introduction ... 195
Common Names, Economic Importance, and Distributional Range ... 195
Gum Arabic Production ... 195
Gum Arabic Properties ... 196
Gum Chemical Characteristics ... 196
Viscosity and Acid Stability ... 197
Applications of Gum Arabic ... 197
Recipes with Gum Arabic ... 199
Fruit Juice (Figure 11.1) ... 199
Ingredients ... 199
Preparation ... 200
Hint ... 200
Sugar-Free Candy (Figure 11.2) ... 200
Ingredients ... 200
Preparation ... 201
Hint ... 201

Brown Gravy (Figure 11.3) 202
Ingredients....... 202
Preparation 202
Hint....... 203
Tips for the Amateur Cook and Professional Chef....... 203
References and Further Reading....... 203

12. Konjac Mannan 207
Historical Background 207
The Plant and the Tuber 207
Manufacture 208
Structure....... 208
Technical Data....... 209
Food Applications 209
Recipes with Konjac Mannan 211
Kon-Nyaku (Konjac) (Figure 12.6) 211
Ingredients....... 211
Preparation 211
Hints....... 211
Konjac Jelly (Figure 12.9) 213
Ingredients....... 213
Preparation 213
Hint....... 213
Okara (Soybean Fiber) Konjac (Figure 12.10)....... 214
Ingredients....... 214
Preparation 214
Hint....... 215
Okara Konjac Recipes 215
BBQ *Okara* Konjac (Figure 12.11)....... 215
Ingredients....... 215
Preparation 216
Hint....... 216
Fried *Okara* Konjac (Figure 12.12)....... 216
Ingredients....... 216
Preparation 216
Hint....... 217
Regulatory Status 217
Tips for the Amateur Cook and Professional Chef....... 217
References and Further Reading....... 218

13. Pectin 221
Introduction....... 221
Nomenclature 221
Structure....... 222
Sources and Properties....... 222
Pectin Manufacture 223
Commercial Availability, Specifications, and Regulatory Status....... 224
Solution Properties 225
Viscosity....... 226

Pectin Gel Types and Properties 226
Applications 228
Recipes with Pectin 231
Apple Jam (Figure 13.1) 231
Ingredients 231
Preparation 231
Hints 232
How to Sterilize Jams and Preserves 232
Orange Marmalade (Figure 13.4) 233
Ingredients 233
Preparation 233
Hint 234
Milk Jam (Low Sugar) (Figure 13.5) 234
Ingredients 234
Preparation 234
Hints 235
Nappage Neutre (Clear Glaze) (Figure 13.6) 235
Ingredients 235
Preparation 235
Hint 236
Dessert Base (Figure 13.8) 236
Ingredients 236
Preparation 237
Hint 237
Tips for the Amateur Cook and Professional Chef 238
References and Further Reading 238

14. Starch 243
Introduction 243
Varieties of Starch 243
Structure and Composition 245
Functional Properties of Starch Suspensions 246
Starch Pastes and Gels 247
Effect of Food Ingredients on Starch Functionality 248
Properties of Available Starches 249
Properties of Dry Starch 249
Pregelatinized Starches 250
Modified Starches 250
Commercial Applications of Starches 252
Recipes with Starch 255
Blanc Manger (Figure 14.1) 255
Ingredients 255
Preparation (Figure 14.2) 256
Hints 257
Warabi-Mochi (Sweet Potato Starch Jelly) (Figure 14.3) 257
Ingredients 257
Preparation (Figure 14.4) 257
Hint 258

Creamy Corn Soup (Figure 14.5) ... 258
Ingredients ... 258
Preparation ... 259
Hints ... 259
Sardine Meatball Soup (Figure 14.6) ... 259
Ingredients ... 259
Preparation (Figure 14.7) ... 260
Hints ... 261
Squid Meatball Soup (Figure 14.8) ... 261
Ingredients ... 261
Preparation ... 261
Hints ... 262
Goo Lou Yok (Sweet and Sour Pork) (Figure 14.9) ... 262
Ingredients ... 262
Preparation ... 263
Hints ... 264
Mapo Tofu (Spicy *Tofu*) (Figure 14.10) ... 264
Ingredients ... 264
Preparation ... 265
Hints ... 265
Deep-Fried Mackerel with Vegetables (Figure 14.11) ... 266
Ingredients ... 266
Preparation (Figure 14.12) ... 267
Hints ... 268
Tips for the Amateur Cook and Professional Chef ... 268
References and Further Reading ... 269

15. Xanthan Gum ... 271
Introduction ... 271
Processing ... 271
Chemical Structure ... 272
Xanthan Gum Solutions ... 272
Stability in Different Media, under Different Technological Treatments ... 273
Solution Preparation ... 273
Xanthan Gum Interactions ... 274
Food Applications ... 275
Toxicity ... 277
Recipes with Xanthan Gum ... 277
Italian Dressing (Figure 15.1) ... 277
Ingredients ... 277
Preparation ... 278
Hints ... 278
Creamy Italian Dressing (Figure 15.2) ... 279
Ingredients ... 279
Preparation ... 279

Tempura (Figure 15.3) ... 280
Ingredients ... 280
Preparation ... 281
Hints ... 281
Chocolate Soufflé (Figure 15.4) ... 281
Ingredients ... 281
Preparation ... 282
Hints ... 282
Tips for the Amateur Cook and Professional Chef ... 283
References and Further Reading ... 283

16. The Use of Multiple Hydrocolloids in Recipes ... 287
Synergistic Combinations ... 287
Protein–Polysaccharide Interactions: Conjugates and Complexes ... 287
Applications ... 288
Locust Bean Gum (LBG) Interactions with Carrageenan and Xanthan ... 288
Carrageenan and Protein Synergy ... 289
Recipes with Multiple Hydrocolloids ... 290
Coffee Jelly (Figure 16.1) ... 290
Ingredients ... 290
Preparation ... 290
Hint ... 290
Hot Savory Jelly (Figure 16.2) ... 291
Ingredients ... 291
Preparation ... 291
Hints ... 291
Dulce de Batata (Sweet Potato Jam) (Figure 16.4) ... 292
Ingredients ... 292
Preparation ... 292
Hint ... 293
Low-Fat Ice Cream (Figure 16.5) ... 293
Ingredients ... 293
Preparation ... 294
Hint ... 294
Tips for the Amateur Cook and Professional Chef ... 295
References and Further Reading ... 295

Glossary ... 297

Alphabetical List of Hydrocolloid Manufacturers and Suppliers ... 301

Subject Index ... 319

Recipe Index ... 341

Preface

Hydrocolloids are among the most commonly used ingredients in the food industry. They function as thickeners, gelling agents, texturizers, stabilizers, and emulsifiers; in addition, they have applications in the areas of edible coatings and flavor release. Manufactured foods that are reformulated for reduced fat rely primarily on hydrocolloids to provide suitable sensory quality. Furthermore, hydrocolloids are currently finding increasing applications in the health arena: they provide low-calorie dietary fiber, among many other uses.

Many books have been devoted to a description of the different water-soluble polymers (hydrocolloids) and their uses. In 1969, a monograph by M. Glicksman, *Gum Technology in the Food Industry* (Academic Press), presented a technical compilation of information in the area of hydrocolloid technology as it pertains to the food industry. The need for such a book was apparent to most food technologists and scientists, particularly those engaged in the development of convenience foods. This book was followed by three more volumes by Glicksman (1982 to 1984) entitled *Food Hydrocolloids*, volumes I, II, and III (CRC Press). The first volume was composed of two parts, the first dealing with comparative properties of hydrocolloids and the second with biosynthetic gums. The second volume dealt with natural food exudates and seaweed extracts, and the third volume described cellulose gums, plant seed gums, and plant extracts. Those books were much more comprehensive than Glicksman's first monograph and were very useful for both food technologists and academics.

In 1980, an excellent book entitled *Handbook of Water-Soluble Gums and Resins* (McGraw Hill Company) was edited by R. L. Davidson. The book comprised 23 chapters written by advisors and contributors from universities and the industry. It contained information on where water-soluble gums and resins come from, how they are used, how they work, and their individual uses to obtain specific properties and performance. It gave an encyclopedic description of the major commercial varieties of both natural and synthetic gums and resins, each listing beginning with a concise overview, followed by full details on the chemistry, properties, handling uses, and other pertinent factors.

In 1997, a monograph by one of us (A. Nussinovitch) entitled *Hydrocolloid Applications: Gum Technology in the Food and Other Industries* (Blackie Academic & Professional) was published, composed of two parts. The first dealt briefly with a description of the known hydrocolloids. The second was devoted to information which is more difficult to locate, namely uses of hydrocolloids in ceramics, cosmetics, and explosives, for glues, for immobilization and encapsulation, in inks and paper, and for the creation of spongy matrices, textiles, and different texturized products. Another monograph by A. Nussinovitch entitled *Water-Soluble Polymer Application in Foods* (Blackwell Science) from 2003 was devoted to the uses of hydrocolloids in foods and in biotechnology, and discussed topics such hydrocolloid adhesives, hydrocolloid coatings, dry macro- and liquid-core capsules, multilayered products, flavor encapsulation, texturization, cellular solids, and hydrocolloids in the production of special textures. Yet another monograph by A. Nussinovitch from 2010, *Plant Gum Exudates of the World: Sources, Distribution, Properties, and Applications* (CRC), provided a description of the most extensive collection of plant gum exudates in print. The book included a chapter specifically devoted to food uses of plant exudates, including confectionery, salad dressings and sauces, frozen products, spray-dried products, wine, adhesives, baked products, and beverages, among many other industrial products and animal foods.

In 2009, the 2nd edition of *Handbook of Hydrocolloids*, edited by G. O. Phillips and P. A. Williams, was published. This excellent manuscript reviewed over 25 hydrocolloids, covering their structure and properties, processing, functionality, applications, and regulatory status. In addition to the traditional hydrocolloids, the book emphasized protein hydrocolloids and protein–polysaccharide complexes, expanded the coverage of microbial polysaccharides, and also discussed the role of hydrocolloids in emulsification and as dietary fibers.

These are just a few examples of the wealth of material existing in this field of science. Note that the inclusion of a book in this short list does not imply that it is any better than other published books on hydrocolloids or their widespread applications.

Although some food recipes can be located in a few of these many books, there are no scientific books fully devoted to the fascinating topic of hydrocolloids and their unique applications in the kitchen. A kitchen can be regarded as an experimental laboratory, with food preparation and cookery involving processes that are well described by the chemical or physical sciences. Finally, it is well established that an understanding of the chemistry and physics of cooking and the involvement of different ingredients (such as hydrocolloids) in these processes will lead to improved performance and increased innovation in this realm. Since the use of hydrocolloids is on the rise in many fields, the writing of a book that covers both past and future uses of hydrocolloids in the kitchen is both timely and of great interest.

General Approach and Aims

Each chapter in this book addresses a particular hydrocolloid, protein hydrocolloid, or protein–polysaccharide complex, in alphabetical order. The chapter starts with a brief description of the chemical and physical nature of the hydrocolloid, its manufacture, and its biological/toxicological properties. It is important to note that this book is not intended as a replacement for the already published books on hydrocolloid properties (some of which are mentioned above); our aim is not to compete with or repeat any of the information found in those books. In the present book, the emphasis is on practical information for the professional chef and amateur cook alike. Furthermore, such a volume may serve to inspire cooking students, and to introduce food technologists to the myriad uses of hydrocolloids, how they are used and for what specific purposes. Each chapter includes a few recipes demonstrating that particular hydrocolloid's unique abilities in cooking and those abilities are elaborated upon. Several formulations were chosen specifically for the food technologist, who will be able to manipulate them for large-scale use or as a starting point for novel industrial formulations. In summary, the volume is written such that chefs, food engineers, food science students, and other professionals will be able to cull ideas from the recipes and be initiated to the what, where, and why of a particular hydrocolloid's use. Each recipe/formulation is accompanied by color images showing the final food/product. Additional images illustrating some of the more important steps in the cooking/preparation or something that is crucial for an understanding of the cooking steps are provided, as needed, in the form of color photographs with explanations. We chose to discuss the most commonly used hydrocolloids in this book, but would like to refer you to supplemental material regarding additional, lesser used, hydrocolloids in the updates tab of the book's Web page at http://www.crcpress.com/product/isbn/9781439875889.

Chapter 1: Hydrocolloids—Where, Why, and When?

Hydrocolloids can be extracted from common or unique natural sources. These include cereal grains, wounded trees, seaweed, animal skin and bones and fermentation slime, among many others. Aside from those natural sources, synthetic gums are produced by skilled organic chemists. This introductory chapter deals with the terminology and classification of hydrocolloids, their market and economics, gum constituents and their effects on processing, functions of hydrocolloids in food applications and products, hydrocolloids as viscosity formers and their typical food applications, gelation, gel types and linkages, gel textures, gel-enhancing effects of other gums, the presence of hydrocolloids in emulsions, suspensions, and foams, and in crystallization control. Other unique food applications of hydrocolloids and regulatory aspects related to their use are also briefly discussed in this chapter.

Chapter 2: Agar–Agar

Agar applications in the kitchen are fundamentally based on its enormous gelling power, and total gel reversibility. Agar is unique among polysaccharides in that gelation occurs at a temperature that is far below the gel's melting temperature. Many uses of agar depend upon its high hysteresis. Moreover, as a gelling agent, agar is very useful in preparing strong water–gel systems for molds, or for creating the molds themselves. Typical recipes that take advantage of agar's properties are presented. Two examples are the preparation of *mizu yokan* (traditional Japanese agar jelly with sweet red-bean paste) and *mitsu-mame* (a traditional Japanese dessert with agar jelly and rice flour dumplings). Today, canned *mitsu-mame* has become quite popular in Japan, and it is less often prepared from scratch. A recipe for orange jelly is presented to demonstrate the influence of organic acids on the texture of the agar–agar product as well as to understand when this influence is at its strongest. Another important issue encountered in this recipe is the hysteresis between gel melting and setting. A recipe for condensed sweet agar jelly (*awayuki-kan*, which is agar jelly with egg white) is presented because the product has unique textural properties: it melts on the tongue like light snow (*awayuki* in Japanese means "light snow"). Another recipe is for *xìngrén* jelly (Chinese almond pudding), which originated in China. In fact, throughout this book we present recipes from all over the world to demonstrate the use of hydrocolloids in the kitchen, as well the universal importance of these ingredients.

Agar spaghetti has become quite popular, perhaps because its preparation is so much fun; this recipe is presented because it demonstrates the important steps required to prepare an agar–agar gel (or product) with a desired texture. It is important to note that within this formulation one can use any fruit juice, for example, orange, grape, apple, cranberry, strawberry, and so on, or even chicken stock, coffee or tea with milk to obtain any desired spaghetti-type product. As with all of the recipes in this book, explanations of its composition and the chemical/physical reasoning behind its preparation can be located in the recipe itself, along with a vast scientific background and the attached glossary.

Chapter 3: Alginates

Alginates' application in the kitchen relies on their gelling properties being essentially independent of temperature. Alginates from brown marine algae differ from most other gelling polysaccharides and proteins. In general, alginates are used to improve, modify, and stabilize food texture. They can be used as viscosity enhancers and gel formers, and in general for the stabilization of aqueous mixtures, dispersions, and emulsions. Therefore, the recipes chosen for this chapter are divided into food gels and recipes in which alginate serves as a thickener. Dropping of alginate solution into, for example, calcium chloride solution (among many others), produces alginate beads (gels), which, with different inclusions and under various conditions, have found their way into the kitchen and molecular gastronomy. Instead of the very common bead formation, here we present a recipe to produce reverse spheres. In the scientific literature, these are termed *liquid-core capsules*. In this case, the calcium (or other suitable) cations diffuse from the liquid core into the alginate solution followed by the spontaneous production of a membrane that holds the liquid core. The chapter includes a recipe for the preparation of the Japanese artificial *ikura* (salmon eggs). *Ikura* is a popular ingredient for *sushi*, and the use of artificial *ikura* reduces the cost of the *sushi*. Another recipe describes the use of sodium alginate for the texturization of onion rings (restructured food). This recipe is not normally prepared in the home, but consistently sized onion rings can be created for commercial purposes. The use of propylene glycol alginate as a thickener, stabilizer, and emulsifying agent at deliberately low pHs for the production of dressings is also demonstrated by a suitable recipe. Of course, one of the major properties of alginate is its unique texturizing ability.

Chapter 4: Carrageenan and Furcellaran

Red seaweed contains the naturally occurring polysaccharides carrageenan and furcellaran. Gels can be produced by heating and cooling solutions of these polysaccharides (with different suitable cations) to give soft, elastic gels with ι-carrageenan, and firm, brittle gels with κ-carrageenan and furcellaran. The recipes chosen for this chapter demonstrate the ability to use different carrageenans for characteristic formulations. The use of carrageenan and/or furcellaran in the kitchen is demonstrated by the preparation of milk pudding, and possible interactions between proteins and carrageenans, as well as their pH dependence, are discussed throughout the chapter. In addition, the preparation of an egg-free dessert jelly, that is, "flan," using a carrageenan-based product is described. A cold-process recipe for preparing chocolate milk with carrageenan is described, as is a preparation for ham including carrageenan. Adding carrageenan improves the texture of the ham because of the carrageenan–protein interaction that increases the ham's elasticity, and decreases its syneresis.

Chapter 5: Cellulose Derivatives

Cellulose derivatives, which have been approved for food applications include methyl cellulose (MC), hydroxypropyl cellulose (HPC), hydroxypropyl methylcellulose (HPMC), methyl ethylcellulose (MEC), and carboxymethylcellulose (CMC). The parsley spaghetti recipe is included to exemplify a formulation utilizing the thermogelation properties of MC. This is a modern gastronomy recipe in which other green leaves, or peas, can be used instead of parsley. The influence of cellulose derivatives on texture and technological features of foods is demonstrated by the preparation of *takoyaki* (octopus dumpling) in which the added MC leads to a softer, more liquid texture, with better resistance to freezing and thawing. A recipe for soy burger patties with HPMC is presented; this is a good recipe for vegetarian or macrobiotic diets, and can also be used to prepare soy burgers. Hydrocolloid inclusion and heating above a certain threshold temperature produce the gelation needed to bind product ingredients. A recipe for pineapple ice cream includes CMC, enabling the creation of an egg-free product. This recipe is also ideal for other fruits, such as mango, coconut, pear, orange, strawberry, and so forth. The addition of CMC to sugar paste leads to a thicker and more flexible product that can be utilized in cake decorations and sugar paste crafts.

Chapter 6: Curdlan

Curdlan is an extracellular microbial polysaccharide. It is a food additive that has the ability to form an elastic gel. Curdlan forms a gel at both relatively high and relatively low temperatures or upon neutralization or dialysis of an alkaline solution of curdlan. Curdlan is tasteless, odorless, and colorless, and possesses excellent gel-forming ability and water-retention properties, and it is therefore widely used in the food industry. One recipe chosen to demonstrate curdlan's unique properties is Japanese noodles—*udon*. Adding curdlan to this recipe makes the *udon* elastic, improves its resistance to freezing, and gives the noodles a better texture. In the recipe for *kamaboko* (fish cake), curdlan is included as a texturizing agent and a *surimi* mimetic. In a recipe for sausages, curdlan is used to improve water-holding capacity and as a fat mimetic. In doughnuts, curdlan is added to decrease oil uptake during deep-frying. In tofu (*kinugoshi* tofu, soybean curd), curdlan is added to modify the resistance to freezing and thawing, as well as heating. The gelling and adhesive properties of curdlan are demonstrated in a recipe in which different types of gels are glued together to create a multilayered jelly.

Chapter 7: Egg Proteins

Chicken eggs contain very high amounts of functional proteins, lipids, and lipoproteins. The chemical composition and structural characteristics of egg yolks and egg whites are related to their functional properties as emulsifiers, and as foaming and gelling agents. Since eggs and egg proteins are used in a huge number of recipes, we chose a few in which the proteins are used for some typical functional purpose. An omelette recipe is presented due to this food's vast popularity and the option to incorporate an unlimited variety of additions such as tomato, cheese, many kinds of vegetables, and many types of sauces. The Japanese rolled omelette (sushi egg) is also included to demonstrate differences between cultures; it is a very popular recipe in Japan. Crabmeat and egg drop soup (thickened with potato starch), *chawan-mushi* (savory egg custard, representative of typical Japanese cuisine), and caramel custard (which is widely eaten the world over, with many national variations) are also included in this chapter. A recipe for *crème brûlée* serves as a typical example of the use of egg proteins, and cream puffs are presented due to their popularity and their many possible variations. A recipe for traditional meringues is also included.

Chapter 8: Galactomannans

Different galactomannans include fenugreek, guar gum, tara gum, and locust bean gum (LBG), agents which are generally used for thickening purposes. Perhaps the most popular galactomannans in today's—and possibly also tomorrow's—kitchen are LBG and guar gum. The recipes we chose to present in this chapter include brown sugar sherbet, where LBG is added to prevent ice crystallization. Another interesting formulation is raspberry sauce. We chose this recipe since its texture lies somewhere between jelly and jam. It has a sour taste, which can be used as a sauce for sweets. In this case, LBG is used as a thickener and agar–agar as a gelling agent. Another recipe is for banana and nuts ice cream, in which guar gum is used both as a viscosity former and possible emulsifier. In the recipe for frozen fruit drink, guar gum is used as a thickener and in the recipe for empanada dough, both guar gum and starch serve as thickeners and guar gum additionally serves to prevent starch retrogradation. In the recipe for *loukoums*, more than one hydrocolloid is added to achieve a textural advantage. The included agar–agar, tara gum, and gum arabic influence consistency, increase viscosity, and offer possible synergism among the gums to achieve the desired texture. In the recipe for yoghurt dressing, tara gum is used to increase product creaminess, in *sharbat hilba* (a traditional Libyan soup), fenugreek extracted from seeds influences both thickness and taste. The thickening ability of fenugreek is further demonstrated in the preparation of pear and fenugreek jam.

Chapter 9: Gelatin

Gelatin is one of the most versatile biopolymers, with numerous applications in foods, confectionery, pharmaceuticals, and many other fields. Gelatins (~300,000 metric tons produced annually worldwide) are derived from the parent protein collagen; the origin of the parent collagen and the severity of the extraction procedures determine the properties of the final gelatin. Gelatins are mainly produced from bovine and porcine sources, but they may also be extracted from fish and poultry. The most important properties of gelatin are thermoreversible gel formation, texturizing and thickening, high water-binding capacity, emulsion formation and stabilization, foam formation, protective colloidal function, and adhesion/cohesion. Gelatin's unique gel-forming ability is emphasized in the recipes for fruit jelly, fruit gummy candy, and aspic jelly. A Bavarian cream recipe is included to demonstrate the smooth texture achieved in part by the use of gelatin. Another presented recipe is marshmallows, where the gelatin helps produce a texture that is softer than that of commercial marshmallows.

Chapter 10: Gellan Gum

Gellan gum, an extracellular polysaccharide, forms gels at low concentrations when a hot solution of it is cooled in the presence of gelation-promoting cations. Since there are functional differences between the two forms of gellan gum—low acyl (LA) and high acyl (HA)—with respect to hydration, gelation, stability, and texture, recipes based on both of these forms are presented. For example, we include a recipe for fruit juice jelly using LA gellan gum. This should be of great interest to the reader since it gives information on how to produce textures that are stable to freeze-thaw cycles. Another recipe is for low-solids jam, which owes its unique thermostability to the use of LA gellan gum. A recipe for liqueur jelly is presented since the inclusion of LA gellan gum enables the inclusion of alcohol in the formed product. In the pulp-suspension fluid gel, HA gellan gum's ability to suspend particles is demonstrated. The recipe for herb-suspension oil-free dressing manufactured with HA gellan gum is also included due to its ability to suit any vegetable.

Chapter 11: Gum Arabic

Gum arabic is unique in that it forms a solution at over 50% concentration. It is widely used in the industry. In this chapter we present recipes in which the gum is used for confections, and various other food products, beverages, and coatings. A recipe that includes fruit juice is presented. In this case, a combination of gum arabic and gum tragacanth is used not only because of the unique texture they confer on the beverage but also because of their ability to dissolve at over 75°C. This recipe produces imitation fruit juice. Gum arabic is not only used as an emulsifier but it reinforces the "fruit juice" taste; this recipe could serve as a basis for different imitation juices. Since the major application of gum arabic is in the confectionery industry, we present a recipe for sugar-free candy, which can be easily prepared in the home kitchen. The recipe was chosen because of the many alternatives, which can be selected in terms of favorite colors and flavors. Another interesting formulation can be found in the recipe for brown gravy where gum arabic is also used as an emulsifier.

Chapter 12: Konjac Mannan

Konjac mannan is a major component of konjac tubers; konjac is a perennial plant of the Araceae. Konjac mannan contains trace amounts of acetyl groups. Deacetylation occurs with alkali treatment, creating a chewy, irreversible gel. This gel has long been used as a traditional food in Japan. A recipe for the popular traditional Japanese food *kon-nyaku* (konjac) is described. Konjac jelly, one of the more popular jelly products in Japan, is also described. In this recipe, the texture is achieved using two hydrocolloids (konjac and carrageenan); interest in this recipe lies in the ability to replace carrageenan with other gelling agents such as agar–agar, gelatin, or gellan gum. *Okara* (soybean fiber) konjac is presented due to its ability to act as a meat substitute. The presented recipe contains a great deal of fiber and less fat and protein. Other recipes using *okara* konjac are also included in this chapter.

Chapter 13: Pectin

Pectins, belonging to the group of heteropolysaccharides, are present in all plant primary cell walls and are traditionally used in the food industry for the production of jams, low-calorie jams, and marmalades. First, we include two recipes for apple jam and orange marmalade that are based on natural sources of pectin, but for which further addition of pectin is an option. Apples contain a high concentration of high-methoxyl (HM) pectin and this might be sufficient for gelation; however, if more HM pectin is added to the recipe, the jam becomes harder. On the other hand, if low-methoxyl (LM) pectin is added, the jam becomes more thermostable and such jams can be prepared with less sugar. Other recipes that include the inherent addition of LM pectin, such as milk jam (low sugar), *nappage neutre* (clear glaze), and dessert base are included.

Chapter 14: Starch

No other single food ingredient compares with starch in terms of sheer versatility of applications in the food industry. Specialty starches have been tailored to build textural differentiation or enhance product aesthetics, and they are related to issues of nutrition, cost, shelf life, and the like. In this chapter, the question was not what to include, but rather what not to leave out. Here we provide a few recipes to demonstrate starch's versatility. The dessert recipes include *blanc manger* (cornstarch jelly) and *warabi-mochi* (sweet potato starch jelly). We include a recipe for creamy corn soup to illustrate starch's thickening ability. For holding meatballs together, sardine meatball soup and squid meatball soup (meatballs made from sardine or squid and potato starch) are included. In addition, recipes for *goo lou yok* (sweet and sour pork), *mapo tofu* (spicy tofu), and deep-fried mackerel with vegetables are presented to demonstrate the use of potato starch for thickening sauces.

Chapter 15: Xanthan Gum

Xanthan gum is manufactured by fermentation: it is an extracellular polysaccharide secreted by the microorganism *Xanthomonas campestris*. It is soluble in cold water and solutions exhibit highly pseudoplastic flow and synergistic interactions with galactomannans. The inclusion of xanthan gum appears to be beneficial in many recipes. Examples include salad dressings in which xanthan gum is used as a stabilizer and emulsifier at low pH. The use of xanthan gum as a stabilizer in *tempura* and chocolate soufflé is also described. In the latter recipe, other hydrocolloids, such as starches, are included to obtain a better consistency.

Chapter 16: The Use of Multiple Hydrocolloids in Recipes

Mixtures of hydrocolloids can be employed to obtain novel and improved characteristics in various food products. Such combinations can sometimes also lead to a reduction in cost. The synergistic effects might be due to the association of different hydrocolloid molecules or phase separation. In a few previous chapters, the reader can find recipes where more than one hydrocolloid is utilized. In this chapter, we add a few more unique recipes, such as coffee jelly, in which the interaction between xanthan gum and locust bean gum (LBG) contributes to improved texture, and hot savory jelly, in which κ-carrageenan, konjac, and xanthan gum are used. Here, the gum mixture contributes to the thermostability of the product. *Dulce de batata* (sweet potato jam) is based on agar–agar and LBG. The original South American recipe, based only on agar–agar, produces a rigid gel, but the inclusion of LBG makes the preparation more spreadable. In the recipe for low-fat ice cream, carrageenan, xanthan gum, LBG, and guar gum are combined to take advantage of possible interactions between the xanthan gum and the galactomannans as well as the added ability to thicken and hold the extra water in this unique ice cream.

Indices and Other Additions

After the last chapter, a glossary of terms used in the book appears. Such terms include the lesser known food ingredients, or those that are taken from different cultures. Also, an alphabetical listing of hydrocolloid manufacturers and suppliers is provided. Although it presents a vast collection of manufacturers, there are, of course, many more in each country. Inclusion in this list does not imply superiority over other companies, ports, or agents.

The book finishes with two comprehensive indices. The first index presents the food recipes according to categories, that is, soups, sauces, meat products, and so on. The second index is a general, alphabetical index, which should prove very useful for locating both theoretical information and commercial applications.

This book was written by two authors who are very much entranced by the field of hydrocolloids and their potential applications and by cooking, and we hope that this list of hydrocolloid manufacturers and suppliers will make it easier for readers to get samples or to buy hydrocolloids in order to try using them in their kitchens and/or labs.

What we had in mind when writing this book was much more than just a traditional cookbook. The useful, albeit not always purely theoretical background provided for the different hydrocolloids is extensive but to the point. We have made an attempt to include not only recipes that everyone can follow within the confines of their own kitchen—be they professional chefs or amateur cooks, but also products and scientific experiments that can be conducted and studied in university laboratories, such as those currently being performed by one of the authors (M. Hirashima) in Japan. As such, the book can be used as a textbook for cooking science and food-processing classes, and provides recipes that can be scaled up for industrial use, making it ideal for food technologists and engineers. We believe that such a book can bridge the gap between the scientist and the chef or in fact anyone who is interested in novel applications and textures in the kitchen. In addition, it is designed to serve as a guide for all those who want to introduce the fascinating world of hydrocolloids to the public.

We believe that this is going to be an extremely useful book, an essential purchase for personnel involved in food formulation, food science, and food technology, in particular: food scientists (chemists/microbiologists/technologists) working in product development, food engineers whose job typically involves figuring out how to make the products that the developers cook up, research chefs, for example, members of the Research Chefs Association, professional restaurant chefs who like to experiment with new creations, members of the American Culinary Federation and culinary education programs, and members of equivalent organizations in foreign countries. This book will be a useful addition to the traditional libraries of universities and research institutes where food science, chemistry, life sciences, and other practical industrial issues are taught and studied. In this sense, the book is quite unique, and we are confident that it will be a great success.

Amos Nussinovitch and Madoka Hirashima

Acknowledgments

This book was written over the course of 2 years. It describes cooking innovations, in particular the use of hydrocolloids in cooking for thickening, gelling, and emulsification. The book is fully devoted to the fascinating topic of hydrocolloids and their unique applications in the kitchen, with the belief that an understanding of the chemistry and physics of cooking and the involvement of hydrocolloids in these processes will lead to improved performance and increased innovation in this realm. The use of hydrocolloids is on the rise in numerous fields and consequently, the time is ripe for a book that covers both past and future uses of hydrocolloids in the kitchen. We have tried to include many traditional and nontraditional local and international uses of hydrocolloids in cooking, developed throughout the world. Our hope is that this book will assist readers who are in search of comprehensive knowledge about the fascinating field of hydrocolloids, as well as those seeking up-to-date information on the very different past and current uses and applications of hydrocolloids in cooking. The volume is written such that chefs and amateur cooks, food engineers, food science students and other professionals will be able to cull ideas from the recipes and be initiated to the what, where, and why of a particular hydrocolloid's use.

We wish to thank the publishers for giving us the opportunity to write this book. Special thanks to Stephen Zollo, the senior editor of Food Science and Technology at CRC Press, Taylor & Francis Group, for his efficient handling of the project, from its conception to the moment we received the green light to start cooking, photographing, and writing. Stephen's genuine interest, enthusiasm, and encouragement during the process were phenomenal and are deeply appreciated. We wish to thank our editor, Camille Vainstein, for working shoulder-to-shoulder with us when time was getting short and our editorial consultant, Yaara Nussinovitch, for her exceptional assistance in finalizing this manuscript. Adriana Szekely's help in locating and rectifying the many old or inaccurate references was above and beyond the call of duty. We are grateful to the illustrator of Kagemusha Art Club at Mie University, Risato Tano, for drawing the wonderful cover art and numerous illustrations.

During the writing and cooking sessions, we had to get numerous samples of hydrocolloids from all over the world. This stage can be complex, and we have therefore included an index of the names and addresses of manufacturers and suppliers, which will simplify the reader's access to obtaining/purchasing samples as well as gaining further information. These samples were used to prepare the recipes presented in the book and the supplemental material on the downloads/updates tab on the book's Web page at: http://www.crcpress.com/product/isbn/9781439875889.

In particular, we wish to thank Dr. Omri Ben-Zion of Nagum Company, Israel, for providing us with low-methoxyl pectin and gellan gum samples, both produced by CP Kelco Company. Special thanks to the technologist Maarit Saareleht from Est-Agar AS, Estonia, for providing us both with a furcellaran sample and additional data. The advice of Dr. Graham Sworn from Danisco is much appreciated.

We also wish to thank Takeda Chemical Industries (now Takeda Pharmaceutical Company), Japan, for providing us with curdlan samples, Daiwa Pharmaceutical Company for arabinoxylan samples, Hayashibara Company for pullulan samples, and Fuji Oil Company for soluble soybean polysaccharides. We are also thankful to Ms. Ayumi Shibamura from DSP Gokyo Food & Chemical Company for supplying us with xyloglucan samples, Dr. Makoto Nakauma from San-Ei-Gen F. F. I. for bacterial cellulose samples and carrageenans, Mr. Norio Matsuki from Shin-Etsu Chemical Company for methylcellulose and hydroxypropylmethylcellulose samples and suggested recipes, Mr. Kotaro Sasaki from Uenoya Company for konjac samples and photographs, and Dr. Kobi Meiri from the Tnuva Company, Israel, for contributing the different milk proteins.

We must also thank Prof. Katsuyoshi Nishinari for his great support in obtaining samples for us, Ms. Grace Forker for supplying kombucha samples from the United States, and students at Mie University in Japan for helping us make the dishes in our book and photograph them.

Last, but not least, we wish to thank the Hebrew University of Jerusalem in Israel and Mie University for being our home and refuge for many years of very extensive research and teaching.

Amos Nussinovitch and Madoka Hirashima
Israel and Japan, December 2012

The Authors

Professor Amos Nussinovitch was born in Kibbutz Megiddo, Israel. He studied Chemistry at the University of Tel Aviv, and Food Engineering and Biotechnology at the Technion-Israel Institute of Technology. He has worked as a food engineer at several companies and has been involved in a number of R&D projects in both the United States and Israel, focusing on the mechanical properties of liquids, semisolids, solids, and powders. He is currently in the Biochemistry and Food Science Department of the Robert H. Smith Faculty of Agriculture, Food and Environment of the Hebrew University of Jerusalem, where he leads a large group of researchers working on theoretical and practical aspects of hydrocolloids. Prof. Nussinovitch is the sole author of the books *Hydrocolloid Applications, Gum Technology in the Food and Other Industries*; *Water-Soluble Polymer Applications in Foods*; *Plant Gum Exudates of the World: Sources, Distribution, Properties and Applications*, and *Polymer Macro- and Micro-Gel Beads: Fundamentals and Applications.* He is the author or coauthor of numerous papers on hydrocolloids and on the physical properties of foods, and he has many related patent applications. This book is devoted to cooking innovations: specifically, using hydrocolloids for thickening, gelling, and emulsification. Prof. Nussinovitch has been working in this area for many years and has studied gel textures and structures, textures of hydrocolloid beads and texturized fruits, liquid-core hydrocolloid capsules, different hydrocolloid carriers for encapsulation, novel hydrocolloid cellular solids, and edible hydrocolloid coatings of foods, among many other applications.

Madoka Hirashima, Ph.D., was born in Kyoto, Japan. She studied the rheological properties of curdlan and cornstarch at the Graduate School of Human Life Science, Osaka City University. Dr. Hirashima worked at a food company as a new food developer, and then as a lecturer at several colleges. She is currently in Home Economics Education at the Faculty of Education, Mie University, where she teaches cooking as well as cooking science. She continues to study the rheological properties of polysaccharides, with a focus on the textures of starch and konjac products.

1

Hydrocolloids—Where, Why, and When?

Introduction

Hydrocolloids can be extracted from common or more unique natural sources, including cereal grains, wounded trees, seaweed, animal skin and bones, and fermentation slime, among numerous others. Aside from those natural sources, synthetic gums can be produced by skilled organic chemists. Hydrocolloids were used in biblical times and it is probable that either gum arabic or gum acacia was the "manna from heaven" that sustained the Israelites during their escape from Egypt as portrayed in Exodus. Seaweed has a very long history of use in foods and medicine. The *Chinese Book of Poetry* from ~800–600 B.C. described cooking with seaweed. Agar–agar (**Chapter 2**) served as a vital item in the diet of Orientals in the form of flavored and sweetened gels. Along with the many other examples of hydrocolloids used as foods or in foods (**Figure 1.1**), gums were also utilized for nonfood purposes, for instance, the utilization of gum arabic (**Chapter 11**), termed *kami* by the Egyptians, in textile glues, embalming fluids, and for pigment dispersal. The importance of hydrophilic colloids lies in their hydrophilic nature, making gums important textural components of most foods. Additional information, appearing on the book's Web site will be henceforth denoted as Supplemental, with the corresponding appendix letter.

Terminology

The classification of water-soluble gums is inconsistent, perhaps due to its random development over many centuries. In fact, this inconsistency reflects a compilation of expressions from broad geographical sources applied to a variety of impure natural substances. Initially, the word *gum* was presumably used to describe natural gum exudates oozing from trees. This term must have comprised every type of natural exudate containing numerous water-insoluble materials, for instance, resins, latex, and chicle, which would explain the incorrect usage of the word *gum* for many of the water-insoluble resins used in today's paint and chemical industries. The growth of trade led to uncertainty and a profusion of widespread names and trade names. For example, *gum arabic* (**Chapter 11**), which got its name from its introduction into Europe through Arabian ports, became "Turkey gum" during the Middle Ages when the shipment harbors were under the rule of the Turkish Empire; afterward, when export commerce built up in India, it received the name of *East Indian gum*. In different locations, analogous exudates were given local names that had little bearing in other areas. In addition, classification was sometimes based on mistaken information,

FIGURE 1.1 Examples of various food products in Japan containing hydrocolloids.

which led to additional confusion. For instance, seaweed gums (**Chapters 2, 3, and 4**), which were thought to be nitrogenous compounds, were referred to as *vegetable gelatin* until eventually it was noted that they do not include nitrogen; accepted terms then became *seaweed gum* and *seaweed mucilage.*

Some seaweed extracts were named according to their gelling properties or proclaimed abilities. For example, refined agar (**Chapter 2**) extract was termed *gelose* by Payen in 1859 to denote its gelling abilities. In 1879, Marchand proposed the name *phycocolle* to describe agar. In 1946, Tseng suggested the name *phycocolloid* for the algal polysaccharides obtained from brown and red seaweed. In 1958, Stoloff coined the abbreviation *polysacolloid* (polysaccharide hydrocolloid) to describe this cluster of substances. Another classification stressed the dissimilarity between "gums" and "mucilages." Accordingly, mucilages were portrayed as slimy materials extracted from plants, in contrast to gums, which have tacky properties. Since tackiness and sliminess are just two of the numerous physical properties manifested by hydrocolloids, and further, these properties can be eliminated or modified by suitable technological treatment, these categories were in effect meaningless. Moreover, given that they were not assigned on a chemical basis, it was proposed that they should be discarded. In 1947, Mantel plainly indicated the boundaries of the expression *water-soluble gums.* These substances do not dissolve in water in an accurate scientific sense but form colloidal dispersions. In 1953, Whistler and Smart proposed the generic phrase "glycan" for the word *polysaccharide*. The name was structured by replacing the suffix "-ose" in the basic sugar of the polymer (glucose) by the suffix "-an", which was usually found in names relating to sugar polymers. Publications from 1959 attempted to further achieve uniformity in this field. For example, instead of using the term *Danish agar* for gum extracted from *Furcellaria fastigiata* it was named *furcellaran* (**Chapter 4**). Larch gum (**Supplemental H**) was generically termed *arabinogalactan* and was sold under the trade name "Stractan"—both names conforming to the -an suffix. In 1963, Glicksman proposed to define a gum as any polymeric material that can be dissolved or dispersed in water to give a viscous solution or dispersion. This definition was based on the characteristic functional properties of gums as used in the industry. Furthermore, it included synthetic polymers and proteins such as gelatin, casein, and others that exhibit viscosity and/or gel-forming properties in water. This definition differentiated between water-soluble polymers and oil-soluble resins (which are also called *gums*), such as gum kauri and gum copal. It also distinguished them from other polymers, such as rubber, gum chicle, and other incorrectly named gums.

Classification

Traditionally, most of the gums were categorized as polysaccharides and grouped in accordance with their plant source. Consequently, agar–agar, alginate, and carrageenan (**Chapters 2, 3, and 4**) were assembled under the seaweed group. Gum arabic (**Chapter 11**), gum karaya, and gum tragacanth (**Supplemental H**), among others, were categorized in the tree exudate group. Additional gum-like substances, such as pectin (**Chapter 13**) and starch (**Chapter 14**), were classified as separate groups, while proteins, such as gelatin, were not included at all. In addition, there was no room for synthetic gums such as cellulose derivatives (**Chapter 5**) and vinyl polymers, which required a new category. A practical approach to classification would be the use of botanical origin to classify important plant gums. For example, locust bean gum (LBG) and guar gum (**Chapter 8**) are derived from similar plant seed sources and have similarities in their chemical structure; moreover, they can sometimes be employed for similar purposes. To enable wide-range sorting, all types of gums that are employed in the food industry would be included and room should be left for novel gums that might be developed in the future. As a result, Glicksman proposed an all-inclusive classification composed of three categories: natural gums found in nature, modified gums (semisynthetic) that are based on chemical modifications of natural gums, and synthetic gums that are manufactured by chemical synthesis.

Economics

The world hydrocolloid market is valued at ~$4.4 billion (composed of ~70% starches (**Chapter 14**), ~12% gelatin (**Chapter 9**), ~5% carrageenan (**Chapter 4**), ~5% pectin (**Chapter 13**), and ~4% xanthan gum (**Chapter 15**), followed by LBG (**Chapter 8**), alginates (**Chapter 3**), carboxymethylcellulose (CMC) (**Chapter 5**), guar gum (**Chapter 8**), and many others), with a total volume of ~260,000 tons (which can be broken down into 46.3% gelatin, 11.8% guar gum, 8.5% carrageenan, 7.7% pectin, 7.7% gum arabic (**Chapter 11),** 4.6% xanthan gum, 3.9% LBG, 3.9% CMC, 3.1% alginates and 2.7% agar). These represent small quantities since these substances are required in only very minute amounts in foods to impart the desired functional properties to the products. In several foods, the maximum quantities of gums allowed are much higher than those used in actual practice. The hydrocolloid market has been growing at a rate of 2–3% in recent years. The selection of a particular hydrocolloid depends on its functional characteristics as well as its price and supply security. Therefore, it is not surprising that the low-cost starches are the most frequently used viscosity formers. The less expensive gums are the plant exudates and seed flours because less processing is required. More costly are seaweed extracts since they require both expensive collection and drying. Modified natural products are fairly expensive, with prices determined by the cost of the initial substances plus the degree and complexity of the modification. Most expensive are the synthetic polymers if they are produced via sophisticated synthetic processes that make use of costly reagents and complex equipment. Nevertheless, a more expensive thickener such as xanthan gum may still be the first choice due to its unique rheological properties. It is important to note that selection of gums to attain common functional properties is frequently based exclusively on cost comparisons. There is no absolute association between cost per weight and effective use level. Gelatin is the most commonly used gelling agent but due to outbreaks of bovine spongiform encephalopathy (mad cow disease), there has been substantial interest in substitute sources of gelatin, such as fish skin. The carrageenan market has had to face the introduction of cheaper grades, which can be used in cases where gel clarity is not important. It should be noted that nearly all hydrocolloids are used in nonfood industrial applications. The main customers are usually the textile and paper industries, which use the gums as sizing and coating agents. Nevertheless, there are a few gums (e.g., gelatin, pectin, carrageenan, and some other natural gums), which find their major applications in foods.

Gum Constituents and Their Effects on Processing

Gum constituents are included in almost every natural food and are mostly responsible for the structure and textural properties of the plant. In prepared foods, hydrocolloids as food additives impart textural and functional properties. In addition, convenience foods almost always include hydrocolloids within their ingredients. The use of gums to obtain higher quality foods can be exemplified by ice cream: homemade ice cream generally has poor textural qualities, for instance the presence of ice crystals, a sandy texture, and lack of smooth meltdown. Today's industrially manufactured ice creams contain a variety of hydrocolloids as stabilizers and emulsifiers to eliminate these quality defects. In more or less all food processing, there is a modification in the moisture content or the physical shape of water. Both residual gum constituents and addition of hydrocolloids control the type of physical transformation and rate of migration of the water component. In general, the included gums might serve to manipulate processing conditions via modifications in water retention, lessening of evaporation rates, modifications in freezing rates, alterations in ice-crystal formation, and participation in chemical reactions. These practical effects are manifested in the textural qualities of the product. These useful effects are taken into account along with aspects such as cost, accessibility, simplicity of usage, and legal restrictions related to the hydrocolloid's use.

Functions of Hydrocolloids in Food Applications

Functions in Food Products

Gums are employed for an extensive array of applications. A few examples are: as adhesives in bakery glazes; binding agents in sausage; bulking agents in dietetic foods; crystallization inhibitors in ice creams and sugar syrups; clarifying agents in beer and wine; clouding agents in fruit juices; coating agents in confectionery; emulsifiers in salad dressings; encapsulating agents in powdered fixed flavors; film formers in sausage casings and other protective coatings; flocculating agents in wine; foam stabilizers in whipped toppings and beer; gelling agents in puddings, desserts, aspics, and mousses; mold-release agents in gum drops and jelly candies; protective colloids in flavor emulsions; stabilizers in beer and mayonnaise; suspending agents in chocolate milk; swelling agents in processed meats; syneresis inhibitors in cheeses and frozen foods; thickening agents in jams, pie fillings, sauces, and gravies; and, whipping agents in toppings and icings. It is important to note that the broad applications of gums are confined to their two main properties—to serve as thickening and gelling agents. The thickening ability, that is, viscosity production, is the key feature in the use of hydrocolloids as bodying, stabilizing, and emulsifying agents in foods. A few gums that have gelling abilities are helpful in foods where shape retention is needed before any application of pressure. The most widespread, gelled food item for consumption is gelatin dessert gel; additional recognized food gels are starch-based milk puddings (**Chapter 14**), gelatin aspics (**Chapter 9**), and pectin-gelled cranberry sauce (**Chapter 13**).

Viscosity Formation and Its Typical Food Applications

While hydrocolloids are utilized to impart viscosity to a food item, the main issue is product stability. This can be controlled by proper selection of the hydrocolloid. It is vital to avoid degradation of the polymer, which will result in viscosity reduction and thus product deterioration. Nearly all gums are long-chain polymers; as a result of their molecular breakdown, their viscosities tend to decrease in solution. Such degradation can be the result of high shearing and/or high processing temperatures. Natural, low-viscosity gums are more stable than high-viscosity types. Proportional stabilities should be evaluated based on equal viscosity rather than equal concentration. Viscosities of hydrocolloid systems are dependent on concentration, temperature, degree of dispersion, solvation, electrical charge, previous thermal and mechanical treatment and the presence or absence of electrolytes and nonelectrolytes. Hydrocolloids are frequently used exclusively for thickening in products such as dry beverage mixes and pie fillings. In the latter, if the filling is based on fruits, gums are used to increase the viscosity of the fruit juice in order to decrease the flow of the filling once it is in the pie shell. In beverages, the gums are used as body generators, particularly for sugar-free dietetic drinks. In soup mixes or prepared soups, body and consistency are achieved with starches (**Chapter 14**) and other gums. Similarly, sauces and sauce mixes include hydrocolloids to obtain the desired consistency (**Figure 1.2**). If low-pH sauces are produced, gum tragacanth (**Supplemental H**) and propylene glycol alginate (**Chapter 3**) can provide resistance against acid degradation. Today, gums are finding novel uses in dried pet foods. In dry dog food, the

FIGURE 1.2 Japanese sauces and dressings containing xanthan gum, starch, guar gum, and tamarind gum.

addition of water can hydrate and thicken the meat-like pieces, resulting in a thick gravy-like sauce surrounding the meat-like chunks.

Gelation, Gel Types, and Linkages

Gels from hydrocolloids can maintain their firm structural form under pressure. Gels have a continuous network of solid material enclosing a continuous or delicately separated liquid phase. The solid usually consists of long-chain molecules in a structure of fibrils interlinked by primary or secondary bonds at broadly divided locations all along the molecule. Gels demonstrate properties of both solids and liquids. The resemblance to solids is reflected in their structural stiffness and characteristic elastic response, and to liquids in their compressibility, electrical conductivity, and vapor pressure. Gelation starts with a continuous decrease in the Brownian movement of colloidal particles occluded within the gel. This decrease originates from the application of long-range forces among the molecules, which in sequence results in hydration and coherence of the particles. Then an increase in viscosity parallels the advancing gelation; the liquid is absorbed by the swelling solid and is then slowly immobilized. With the progress in gelation, a 3D network that contains liquid is progressively developed. Further progress creates a large continuous structure with an apparent rigidity. Fractions of the large molecular chains in the network can also react with other parts by cross-linking to further increase the rigidity of the whole structure. Gelation can be induced from either the sol- or solid-state condition. From the sol state, the gel can be structured by generating forces between solute molecules; this can be achieved by adding a nonsolvent, evaporating the system's concurrent solvent, adding a cross-linking agent, reducing solute solubility by chemical reaction, changing the temperature, or adjusting the pH. In the solid state, gels are formed when sufficient liquid is imbibed by the solid phase. In this case, the gel is considered to be in an intermediate state of hydration between sol and solid. The gel is composed of a continuous network that entraps the liquid phase consisting of solvent and solutes, some of them comprising noncross-linked polymeric materials. Cross-linking mechanisms consist of hydrogen bonding, electrovalent linkage, and direct covalent linkage.

Gel Textures

Gel textures vary extensively, from smooth, elastic gelatin–water gels to brittle carrageenan–water gels (**Chapter 4**). In gelatin gels, elasticity can be achieved if long-chain molecules possessing freely rotating links exist, there are weak secondary forces between the molecules, and the molecules are interlocked in a small number of places along their length to form a 3D network. On the one hand, elastic gel properties prevail in a polymeric system with a low degree of cross-linking; on the other, a more rigid gel structure is achieved when a high degree of closely spaced cross-linking exists. Brittle gels can also be obtained by some form of precipitation rather than true gel formation. Such gels might occur when poorly formulated pectin (**Chapter 13**) and alginate (**Chapter 3**) gels cross-linked by calcium ions are involved. With some gels, when constituent particles come into close contact and create a clot, and then shrink, fluid is exuded from the gel; this phenomenon is termed *syneresis*.

The use of proteins in the manufacture of simulated meat products is an important food application. A chewy gel is required for such products. Chewy gels can be obtained under specific conditions of protein concentration, pH, and heating. The major concern is to produce gels that contain enough water to be enjoyably moist and yet have sufficient firmness of consistency to provide just the "right" resistance to bite. When soy proteins were used to form simulated meat fibers, the desired texture was achieved by changing the pH. More often than not, food gels contain considerable quantities of sugar, which contribute to flavor but also have a vital functional effect on the gel. Sugar acts as a plasticizer to enable greater separation of the polymer chains and it competes with the polymer for water, or in other words, decreases the solubility of the polymer. These effects can be complementary when they increase the elastic properties of the gel, or they can increase the concentration of the cross-linkages, resulting in a broken gel structure.

Gel-Enhancing Effects of Other Gums

Upon addition of nongelling hydrocolloids to systems that include gelling agents, special textures can be achieved. Synergy occurs when carrageenan and LBG are combined. Carrageenan gels (**Chapter 4**) have a brittle, crumbly-type texture. The addition of LBG (**Chapter 8**) converts that texture to an elastic and tender one, which is sometimes stronger than carrageenan alone. The same effect is also observed when LBG is added to pectin (**Chapter 13**) or agar (**Chapter 2**) gels. The texture of products such as meringues, pie fillings, and marshmallows can be modified by the addition of nongelling hydrocolloids. An example is the addition of gum arabic (**Chapter 11**) to agar gels to soften their texture. When agar-based marshmallows are produced, addition of gum arabic to the composition produces a tender consistency. With pectin gels, addition of CMC (**Chapter 5**) or sodium alginate (**Chapter 3**) imparts a smooth texture to the product. The addition of CMC to typical alginate dessert gels improves their resistance to textural deterioration as a result of freezing and thawing cycles.

Hydrocolloids in Emulsions, Suspensions and Foams, and in Crystallization Control

In most cases, hydrocolloids perform as emulsion stabilizers by increasing the viscosity of the aqueous phase so that it approximates or even surpasses the viscosity of the oil phase. In this method, the tendency of the dispersed phase to coalesce is reduced, and the emulsion is stabilized. For example, in the case of French salad dressing, that is, an oil-in-water emulsion, gum tragacanth (**Supplemental H**) and propylene glycol alginate (**Chapter 3**) can be used to stabilize the emulsion. Other salad dressings can be stabilized with starches (**Chapter 14**). The inclusion of minute amounts of supplementary gums to the starch will also enhance the body and shelf life of the product. Hydrocolloid stabilizers can be utilized to stabilize other emulsions. In those cases, they serve to promote smoothness, improve body and meltdown, and enhance resistance to heat shock. In many applications, the use of a mixture of a few gums is preferred due to synergistic effects.

A stable suspension of solids in a liquid phase can be achieved in chocolate-flavored products. In products such as chocolate milk or syrups, the cocoa solids are suspended by increasing the viscosity of the liquid phase. For this purpose, carrageenans (**Chapter 4**) that can both increase viscosity and react directly with small amounts of protein to form stable colloidal suspensions have the advantage. Other products, which include milk protein solids, such as buttermilk and yoghurt, can benefit from efficient suspension ability. Hydrocolloids are used to coat particles and change their surface properties. The coating has an affinity for the continuous phase and thus connects the phases and stabilizes the suspension. Sometimes surfactants are also included to better displace the air on the particles with the protective gum. In addition, plasticizers such as glycerol and propylene glycol are added to obtain smooth spreading.

The formation of stable foams can be enhanced by adding different proteins. Egg albumin (**Chapter 7**), gelatin (**Chapter 9**), milk (**Supplemental G**), and soy proteins can be used in confectionary foams. The proteins are dispersed in a sugar solution and the suspension is whipped. The foam is stabilized by heat-denaturing the egg white protein. If gelatin is used, then a stable foam is formed upon cooling. In addition to heat and pressure, chemical means can also be used for such purposes. Hydrocolloids such as sodium alginate (**Chapter 3**), carrageenan, and LBG (**Chapter 8**) can be added to react with the protein and produce stable foams. In the case of nougat confections, sugar crystals form a grainy structure that holds the whipped egg or vegetable protein foam in a delicate state, preventing its breakdown (**Figure 1.3**). Icings (i.e., aerated protein systems) on baked goods can be stabilized by thermostable gel stabilizers such as agar (**Chapter 2**), carrageenan, or gelatin. If the process does not include a heating stage, then CMC (**Chapter 5**) and cold-water-soluble carrageenans are used. When whipped toppings include 20% butterfat, the addition of LBG, carrageenan, or gum karaya (**Supplemental H**) is beneficial. If the butterfat is replaced with vegetable oil and nonmilk proteins, then an emulsifier is needed to stabilize

FIGURE 1.3 Japanese sweets containing hydrocolloids.

the dispersed fat and a stabilizer such as carrageenan is added to stabilize the foam. When stable citrus juice foams are required, modified soy albumin can be used as a whipping agent in addition to methylcellulose (MC), which is used as a stabilizer.

Hydrocolloids can affect crystallization by three mechanisms. The hydrocolloid can attach itself to a growing crystal surface via hydrogen bonds or negative- and positive-charge bonding. The hydrocolloid can compete for the crystal's building blocks. For example, hydrocolloids will compete with ice crystals for water molecules. The third mechanism is the combination of hydrocolloids with impurities that affect crystal growth.

In foods, the principal crystalline materials are ice (water) and sugar. Good-quality ice cream, for example, has ice crystals of minimal size.

Other Unique Food Applications

An efficient flavor-fixation process involves molecular absorption, where flavor molecules are attracted to the hydrocolloid molecules (gum arabic). The porous gum particles take in and hold the volatile liquid flavors and protect them. The industrial process includes spray drying. Other gums may also be suited for such an operation. Powdered flavors are used to a large extent in the food industry. Except for the gum, the encapsulating process must consider various factors, including the percentage of solids in the initial formulation, the ratio of gum to oil, the choice of an appropriate emulsifier, the viscosity of the preparation, and drying temperatures, among other factors. Spray-dried gum-based products suffer from several disadvantages: the exposed surface of the particles is large; in gelatin media, the addition of an unsuitable gum at some concentrations can cause unattractive cloudiness. In addition, in cold-water beverages, the gum is more gradually dissolved and in some cases the smallest gum–flavor powder particles tend to aggregate and settle in the package. Aside from spray drying, flavors are also fixed in gelatin and the formed slabs of gelled material are broken into fine particles. Such slab fixation is effective in chewing gums and desserts. Additional techniques for fixing flavors are freeze drying and drum drying. Microencapsulation of flavor oils has also been performed using the basic technique of coacervation with gum arabic (**Chapter 11**) and gelatin (**Chapter 9**). Another technique includes encapsulation of flavor oils within insoluble calcium-alginate films.

The use of edible protective films prepared from different hydrocolloids is gaining in popularity. Alginate films have been used to coat meat and fish. Other edible films have been produced from gelatin (**Chapter 9**), pectin (**Chapter 13**), and CMC (**Chapter 5**), among many others. These films are thin and inexpensive, and can be used not only to coat fresh produce for shelf-life extension but also to produce pouches for soluble coffee, tea, soups, and sugar. Such coatings can also be used as wrapping for sticky candies and confections. Another function of gums in food processing is the prevention or minimization of syneresis. It is possible that gum addition can be of benefit not only in preventing syneresis, but also in improving the texture of the treated product. Such an effect has been achieved in processed cheeses and cheese spreads.

Regulatory Aspects

Hydrocolloids are regulated as either food additives or food ingredients. In every chapter of this book, the relevant hydrocolloid's regulatory status is discussed. Food additives are authorized if a reasonable technological need for their addition can be demonstrated by the user of the additive, rather than by its supplier or manufacturer. In addition, additives can be used if they present no hazard to health at the levels used and the consumer is not misled in the process.

REFERENCES AND FURTHER READING

Bikerman, J. J. 1953. *Foams*. New York: Reinhold Publishing Corp.

Davidson, R. L. 1980. *Handbook of water-soluble gums and resins*. New York: McGraw-Hill.

Dickinson, E. and P. Walstra. 1993. *Food colloids and polymers, stability and mechanical properties*. Cambridge: Royal Society of Chemistry.

Glicksman, M. 1969. *Gum technology in the food industry*. New York and London: Academic Press.

Glicksman, M. 1983. *Food hydrocolloids*, vols. 1, 2, 3. Boca Raton: CRC Press Inc.

Harris, P. 1990. *Food gels*. New York: Elsevier Science Publishing Co.

Hoefler, A. C. 2004. *Hydrocolloids: Practical guides for the food industry*. St. Paul, MN: American Association of Cereal Chemists.

Hollingworth, C. S. 2010. *Food hydrocolloids: Characteristics, properties and structures*. New York: Nova Science Publishers, Inc.

Holy Bible, New International Version. Exodus 16:5–13.

Howes, F. N. 1949. *Vegetable gums and resins*. Waltham, MA: The Chronica Botanica Co.

Imeson, A. 1992. *Thickeners and gelling agents for food*. London: Blackie Academic and Professional.

Laaman, T. R. 2010. *Hydrocolloids in food processing*. Oxford: Willey-Blackwell (IFT Press).

Mantel, C. L. 1947. *The water soluble gums*. New York: Reinhold Publishing Corp.

Marchand, L. 1879. Note sur la phycocolle ou gelatine produite par les algues. *Bull. Soc. Botan. France* 26:264–87.
Nishinari, K. and E. Doi. 1994. *Food hydrocolloids: Structure, properties and functions*. New York: Plenum Press.
Nussinovitch, A. 1997. *Hydrocolloid applications, gum technology in the food and other industries*. London: Blackie Academic and Professional.
Nussinovitch, A. 2003. *Water soluble polymer applications in foods*. Oxford, UK: Blackwell Publishing.
Nussinovitch, A. 2010. *Plant gum exudates of the world: Sources, distribution, properties, and applications*. Boca Raton: CRC Press, Taylor and Francis Group.
Payen, M. 1859. Sur la gelose et les nids de salangane. *Compt. Rend. Paris* 49:521–30.
Phillips, G. O. and P. A. Williams. 2009. *Handbook of hydrocolloids*. Cambridge: Woodhead Publishing Limited.
Smith, F. and R. Montgomery. 1959. *The chemistry of plant gums and mucilages*. New York: Reinhold Publishing Corp.
Stephen, A. M. 1995. *Food polysaccharides and their application*. New York: Marcel Dekker Inc.
Whistler, R. L. and C. L. Smart. 1953. *Polysaccharide chemistry*. New York: Academic Press.
Whistler, R. L. and J. N. BeMiller. 1993. *Industrial gums: Polysaccharides and their derivatives*, 3rd ed. San Diego: Academic Press.

2

Agar–Agar

Historical Background

Agar–agar, more commonly referred to as, simply, *agar*, was the first phycocolloid to be used as a food additive. Phycocolloids, that is, agar, alginates (**Chapter 3**), and carrageenans (**Chapter 4**), are defined as gelling agents extracted from marine algae, which are traded and utilized in many ways exclusively for their colloidal characteristics. Agar was discovered in Japan in the mid-17th century; Tarozaemon Minoya is most commonly credited with this discovery, in 1658. *Kanten*, the Japanese term for agar, means "cold sky," and relates to the cold weather conditions in the mountains where it was produced.

From Japan, its use as a food was introduced to natives in the Far East by Chinese settlers (Buddhist priests). In the coastal areas of Japan, use of a variety of agar-gel-like seaweed extracts almost undoubtedly dates back to prehistoric times. The use of agar for the manufacture of fruit and vegetable jellies was introduced to Europe by Dutch people living in Indonesia. In 1882, Robert Koch initiated the use of agar as a culture medium and it remains the main bacteriological medium in use today.

Collection and Processing of Seaweed for Agar

Japan and Spain currently manufacture ~70% of the world supply of agar. *Gelidium* is one of the algal sources for the traditional production of agar in Japan, with as many as 24 local species. Another source is algae of the genus *Gracilaria*. In Japan, seaweed is gathered by divers equipped only with masks at water depths of up to ~9 m, or at below 18 m by divers equipped with diving gear. The seaweed is collected, stored in tubs or rafts, and towed to shore, where it is dried and partially bleached, then taken for final processing. Deepwater seaweed is regarded as yielding the best gelling extracts, and the most favorable harvesting period is April to September.

The extraction of agar from seaweed was initially based on straightforward boiling to obtain a jelly mass. Progress in the techniques of preparation and purification, according to lore, came about unintentionally: an innkeeper who had prepared a dish of seaweed jelly for noble guests threw the leftovers outside, where they froze during the night and thawed out the next day. This produced a dry, semitransparent material which, when reboiled in water, yielded a clearer jelly with improved features. As a consequence, the purification of agar using the low-cost process of freezing and thawing became the preferred practice, and is still in use today. This traditional method was later modified. American Agar & Co. (San Diego, California) started to manufacture agar industrially in 1939. The technology included the use of freezing containers comparable to those employed for the manufacture of popsicles. After the Second World War, the technique was applied in Japan, Spain, Portugal, and Morocco. The resultant seaweed extract is then concentrated tenfold to contain 10 to 12% agar.

Agar extraction from seaweed begins with mechanical cleanup of the red algae followed by washing with water. It is then boiled in 15 to 20 times its volume of water. Addition of 0.01 to 0.05% sulfuric acid or 0.05% acetic acid enhances extraction. The alga is boiled for about 2 h and then further cooked for 8 to 14 h. Sodium bisulfite (i.e., the common reducing agent sodium hydrogen sulfite, $NaHSO_3$) or the bleaching agent calcium hypochlorite [$Ca(ClO)_2$] can be introduced to remove the color of the agar, yielding a product of the highest possible value. The hot extract is passed through a filter and the remaining seaweed can be re-extracted. The filtrate is then cooled, gelled, and cut. This step is followed by freeze drying or pressing dehydration to yield the dried product. The agar is marketed in dry, ground, packed form (powdered agar), or in bars or strings.

The best quality agar is manufactured with a carefully chosen assortment of raw materials. *Gelidium amansii* (a rigid seaweed) should be a main component in the seaweed extract blend, along with a small amount of soft seaweed (e.g., *Ceramium*). For rigid-type seaweed, extraction under pressure (gauge pressure of 1–2 kg/cm^2 for 2–4 h) increases yields and decreases processing time. Precise conditions have to be established to prevent degradation of the extracted agar. Treatment of *Gracilaria* with alkali results in enhanced agarose content and reduced agaropectin (see "**Structure of Agar**" in **Chapter 2**), and sulfate contents in the agarose fraction. Lab experimentation has shown that pretreating seaweed with enzymes or gamma irradiation, or extraction in ammonia medium, enhances extraction; however, none of these pretreatments are applied in its manufacture.

Confirmation of a new, simplified agar-extraction method by "acid pretreatment" of agarophytes (i.e., seaweed, typically red algae, that produce the hydrocolloid agar in their cell walls) can be located in the relevant literature. Following extraction, the hot sol must be filtered and the gel dehydrated. The gel is commonly frozen prior to dehydration, leading to the elimination of 48% of the water by sublimation, 40% in the process of drip defrosting, and 12% by vaporization. Freeze dehydration can be either mechanical or natural, and serves, for example, for *Gelidium* extracts, for which pressing dehydration is not appropriate. When gelled extracts are packed in cloth with a closed mesh, the water can be expelled from the gel by application of pressure. This process, termed *syneresis*, can be achieved by semi-automated methods. Agar purity is increased in syneresis, since the process eliminates more of the soluble impurities. Therefore, these agars contain lower ash content, which is beneficial to the consumer.

Types of Agar Products

More than a few types of manufactured agar exist. *Kaku-kanten*, agar in bar form (**Figure 2.1**), comes in rectangular pieces weighing an average 7.5 g each. It is sold for home use in bags of one or two pieces. Agar is also sold in string form, termed

FIGURE 2.1 *Kaku-kanten*, a bar-type agar.

hoso-kanten or *ito-kanten* (**Figure 2.2**). These strings are 28 to 36 cm in length, with a commercial unit having a net weight of 15 to 30 kg. Smaller quantities are sold to the public, and compactly packed strings are used for overseas delivery. The major seller, however, is powdered (fine) agar, although agar flakes are also in demand. Agar flakes are produced from *Gelidium* species via freezing process, whereas the powder is produced from alkali-treated *Gracilaria* using a pressing-dehydration (nonfreezing) method.

FIGURE 2.2 *Hoso-kanten* or *ito-kanten*, a string-type agar.

Regulatory Status and Toxicity

Agar is extracted from red algae of the class Rhodophyceae. Agar is sold commercially in a variety of nearly odorless forms, including strips, flakes, and powders. It comes in a range of colors from white to pale yellow. It is only soluble in boiling water. The permissible limits for impurities are: arsenic, not more than 3 ppm as As; ash, not more than 6.5% on a dry weight basis; acid-insoluble ash, not more than 0.5% on a dry weight basis, residual gelatin (**Chapter 9**) that must pass a purity test; heavy metals such as Pb, not more than 10 ppm; insoluble matter, not more than 1%; loss on drying, not more than 20%, and starch levels and water absorption that must pass a test.

Agar is classified as generally recognized as safe (GRAS) by the U.S. Food and Drug Administration (FDA) and is consequently permitted for use as an additive for human consumption in the United States. Permission was based on the positive and long (over three centuries) experience with this product in the Far East, as well as additional controls for its toxicological, teratological (i.e., the study of abnormalities in physiological development), and mutagenic aspects. Agar is not degraded in the human digestive tract and as a result, it can be found in the feces of animals and humans. It is nontoxic, even when consumed in high quantities. Food-grade agar was found to be noncarcinogenic to mice and rats fed up to 50,000 ppm orally for 103 weeks. Moreover, no correlated histopathological effects were observed. Upon daily consumption of 22.5 g of food-grade agar by human subjects, which is quite a bit higher than the normal daily consumption of this hydrocolloid, no effect on calcium or copper uptake was observed.

Structure of Agar

Agar is extracted from a number of red algal species. Its structure has been studied for decades, first and foremost by Japanese scientists. Agar extract is composed of two groups of polysaccharides: agarose, a neutral (nonionic) polysaccharide that serves as the gelling component, and agaropectin, a nongelling ionic (charged) polysaccharide.

Agaropectin has no marketable value and is in principle discarded during the commercial production of agar. Agarose and agaropectin contents differ in different types of commercial agar. The proportion of agarose in agar-bearing seaweed can range from 50 to 90%. The two polymer constituents can be commercially fractionated. The structure of agaropectin is more complex than that of agarose and is less well understood. More details on the chemical structure of agar can be located in the scientific literature.

Agar–Agar Properties

Dissolution and Viscosity

The physical properties of agar solutions and gels are important to the producer, the scientist, and the cook. Even though agar is soluble in boiling water, when used in bar, flake, or string form, overnight soaking in cold water helps achieve full dissolution.

Even soluble-type agar can benefit from a short soak (on the order of minutes) for rapid and good dissolution. The pH throughout soaking and boiling should be kept neutral.

Measurements of agar solution viscosity should be carried out at temperatures that are higher than the gelling temperature (> 40°C). A linear relationship was found between the logarithm of relative viscosity (i.e., the ratio of the viscosity of a solution to the viscosity of the solvent) and agar concentration. Moreover, a correlation between degree of viscosity and firmness of agar gels was confirmed. The presence of ions tends to reduce the viscosity of agar solutions.

Gelation and Melting

One of the most important characteristics of agar, which makes it unique among gelling agents, is that gelation occurs at temperatures far below the gel's melting temperature. Agar makes rigid gels at a concentration of ~1% (w/w). The sol sets to a gel at about 30 to 40°C. This gel is rigid and sustains its shape. In general, all gels are surprising in that they are self-supporting forms that can be shaped at low concentrations. For agar gel, as little as 0.1% agar (the rest being water) is sometimes sufficient; nevertheless, for practical uses, higher concentrations are required. Long ago, the word *brittle* best described the properties of an agar gel. Today, elasticity and rigidity can be achieved using different agars. Agar gelation is completely reliant on its agarose content. Agarose produces "physical gels," that is, a structure in which polymer molecules are connected solely by hydrogen bonding. The gel is melted by heating to ~85–95°C. There is a relationship between gel strength and melting point: both increase in the same direction, although some exceptions can be found. Setting temperatures have been found to increase with agar concentration, as have the gelation temperatures of agarose. At a predetermined (constant) temperature, the higher the setting point of the sol, the higher the rigidity coefficient (i.e., the ratio of the force acting on a linear mechanical system, such as a spring, to its displacement from equilibrium) of the gel. The gelling temperature of agarose increases with increasing methoxyl (the monovalent group, $-OCH_3$) content; moreover, the slower the cooling rate, the higher the temperature of agarose gelation. An important property of agar gels is their high gelling hysteresis, that is, the difference between gelling (~38°C) and melting (~85°C) temperatures.

The definition of gel strength varies with corporation, country, and individual scientist. This may be due to different methodologies, instruments, or testing conditions. Details on gel strength as defined by Marine Colloids, the Meer Corporation, and Japanese companies can be found elsewhere. Furthermore, different gel testers have been invented to measure this gel property. The mechanical behavior of agar, carrageenan (**Chapter 4**), and alginate (**Chapter 3**) gels was evaluated for a wide range of gum and setting-agent concentrations, as well as for different gel-preparation methods. It was found that the higher the gum content, the stronger the gel. A linear increase in strength of agar gels was reported when the gum concentration was raised from 1 to 2%. It should be noted that the preparation conditions might influence the measured mechanical properties, and the consequentially perceived texture. Thus, particular attention should be paid to these conditions. It is important to note that dissolution of agar in boiling acid solutions causes significant degradation. Gel stability is best achieved at pHs slightly over 7.0. The addition of 10 mg/l sodium carbonate to slightly acidic agar improves gel strength.

Gel Clarity

Agarose sols are transparent and colorless. They are much clearer than agar gels, which are usually opaque and hazy. Agar sols are hazy and sometimes yellowish. When agar gels are frozen (at around 0°C), they collapse upon thawing and do not recover their gel phase. Nevertheless, the gel can be remelted and regelled, producing a new gel with nearly identical properties to the original one.

Gel Syneresis

Upon gelification and solidification, the agar gel shrinks and expels water, a phenomenon known as *syneresis* (tearing). The quantity of syneresed water is inversely proportional to the square of the concentration of the hydrocolloid for nearly all useful concentrations. Syneresis is not only influenced by the concentration of the gel, but also by other factors, such as holding time, apparent gel strength, rigidity coefficient, pressurization, and total sulfate contents. While agar gel or agar-based food is being masticated, mechanically induced syneresis occurs, resulting in a feeling of "juiciness" in the mouth. The extent of this syneresis diminishes with increasing gum concentration.

Effect of Addition of Other Materials on Agar Properties

The effect of adding inorganic salts on the time required to set agar sols has been methodically studied. Potassium sulfate is best at accelerating gelation, whereas iodine is the least effective. The transparency of a 1% agar gel can be reduced by inclusion of ferrous salts (at 1 ppm). Gel strength can be slightly increased by inclusion of sodium salt (table salt). The addition of sugar also contributes to agar strength. Effects of sugars have been ascribed to hydrogen bonding between the hydroxyl groups in the polymers and sugars, and to structural changes in the solvent water. Agar gel is also strengthened by the addition of locust bean gum (LBG) (**Chapter 8**). If sodium alginate (**Chapter 3**) is added to an agar gel before setting, the alginate must then be cross-linked by an appropriate cation diffusing into the gel from an external source. Agarose–gelatin (**Chapter 9**) gels at high concentrations tend to interface with one another.

Commercial Food Applications

Agar is a food additive. As already noted, it is classified as GRAS by the FDA. In Europe it is considered an E406 additive. In the registration service of the *Chemical Abstracts*, it has been assigned the number 9002-18-0.

Agar in the Baking Industry

The functions of agar depend on its features. By varying seaweed source and/or processing method, a vast range of agar textures, from the traditional brittle to very elastic, can be attained. Agar gels are used in the baking industry for their durability at high temperatures. Agar is employed as a stabilizer in chiffon pies, icings, toppings, meringues, pie fillings, and other similar items for consumption. An icing is a coating or topping for cakes and other sweet goods, which contains shortening, milk solids, and stabilizers, whipping agents, salt and flavoring. To minimize problems with flat icings, which include melting, disappearing, sticking to a cellophane wrapper, cracking and peeling, and in general decreased visual appeal and limited stability, the water content of the icings should be decreased, and a gum added to bind the free water. Agar or viscosity-formers such as LBG (**Chapter 8**), alginate (**Chapter 3**), carrageenan (**Chapter 4**), pectin (**Chapter 13**), or karaya gum (**Supplemental H**) can be used for this purpose. Doughnut glazes have a tendency to fracture as a consequence of sugar crystallization: the inclusion of agar as a stabilizer at levels of ~0.5 to 1.0%, depending on the sugar content, increases the viscosity of the glaze and, in a blend with glucose, an inverted sugar, alters its crystallization characteristics. A superior icing stabilizer is obtained by blending the agar with surface-active agents such as sorbitan esters (lipophilic nonionic surfactants that are used as emulsifying agents) of higher fatty acids.

Agar in Confections

Agar can be used at concentrations of ~0.3 to 1.8% in confections. Another use of agar in the sweets industry is as a filler in candy bars. Agar in its powdered form is widely used in jelly candies due to its solubility and the strength it confers upon the product. High-quality results are achieved if the agar is immersed for 1 or 2 h before being cooked with sugar, corn syrup, invert sugar, and monosodium citrate as an acidulant, with added color and flavoring. Subsequent to mixing, the jelly is deposited into starch- or paper-lined slabs. Very high levels of sugar (about 80%) are used in the confectionery industry. At these levels, the increasing stability breaks down for agarose and *k*-carrageenan (**Chapter 4**), a result that has also been detailed for deacylated gellan (**Chapter 10**). It has been argued that an increasing scarcity of water molecules at high sugar levels and their relative hydrogen bonding with sugar gradually deprive the polysaccharide helices of the required hydration layer for thermodynamic stability. This effect is not seen for gelatin (**Chapter 9**) samples, which do not demonstrate a drop in network strength at high concentrations of sugar. Other sweet confections based on agar and methylcellulose (MC) (**Chapter 5**) or invert sugar and sugar-heated solutions have been suggested. Once they are combined, the two solutions are boiled and cooled.

Agar in Wet and Dried Structured Fruit Products

Novel structured fruit products made of pulp, a wide range of hydrocolloids, and other additives, have been the focus of numerous patents and commercial applications. These products are generally composite materials in which particulates are embedded in a polymeric gel matrix. Numerous patents discuss the option of using a mixture of alginates (**Chapter 3**) with supplementary hydrocolloids, for instance agar or carrageenan (**Chapter 4**), together with fruit pulp and other traditional food additives, to create simulated fruit products. The accessible technological information on such products focuses mainly on the processes of producing different gel systems containing pulp, sugar, and acid. Broad research into the dependence of composite texturized fruit products on pulp properties has been carried out. In addition to studying the combined effect of fruit pulp, sugar, and gum on texturized fruit products, the ability to contain high proportions of highly acidic fruit pulp and the juice contained therein to manufacture succulent, hydrocolloid-based, texturized fruit products and multilayered texturized fruits has also been discussed. Drying either gels or texturized hydrocolloid fruit products produces a cellular material with a typical porous structure. Dried agar-texturized fruit gels are not widespread. Nevertheless, their fabrication by freeze-dehydrating gels consisting of fruit concentrate or purée could serve as a basis for novel fruit bars or food snacks. Such items are dry and crunchy with a high content of fruit ingredients. A dried texturized fruit product is typified by its porosity and pore-size distribution, among other parameters. Such products might be low-calorie natural cellular-based products or high-calorie natural fruit products with the inclusion of fatty substances within their structure by infusion or additional processing methodology.

Meat, Milk, Fish, and Other Products

Soft-gelled meat and fish products can be prepared with agar at levels of 0.5 to 2.0%. This method of "packaging" diminishes transit damage to brittle tissues and thus reduces textural loss. For such aims, agar is preferred to carrageenan (**Chapter 4**) or gelatin (**Chapter 9**) due to its higher gel strength and melting temperature. An example of such a methodology is the production (for export purposes) of large chunks of canned tuna preserved in agar jelly to Western Europe by the Japanese in the 1960s.

Canned corned beef contains gelatin and some other natural thickening agents. Agar and carrageenan are also used to create films (coatings) that include water-soluble antibiotics for the shelf-life extension of coated poultry. The technical feasibility and optimal conditions for the preparation of rice coated with type-4 resistant starch using LBG (**Chapter 8**), agar, and a mixture of LBG and agar were studied. Coating was performed through a simple soaking and drying method. This process could be readily transferable to industrial manufacture of low-calorie rice with high consumer acceptance. The best coating solution for coated rice was a 0.1% mixture of LBG and agar. Rice coated with this mixture exhibited lower carbohydrate digestibility compared to rice prepared using LBG or agar alone. This coating also resulted in significantly lower blood glucose levels than consumption of uncoated rice.

Agar has also been used in the manufacture of dehydrated powdered fish extract, which is usually found in soups and flavoring preparations. Another usage not specifically limited to agar is its addition to cheeses and cream cheese at levels of 0.05 to 0.85% to improve stability and sensory properties. An interesting use of agar was for gelification of cream or milk to form a solid material that can be dissolved in hot coffee or tea. Special use of agar at levels of ~0.05 to 0.15% for fining of wines, and numerous other uses of agar in vegetarian or health products, can be located in the literature, for example, as a bulking agent in breakfast cereals, in the preparation of starchless desserts, in aspic salads, and in puddings, fruit butter, jams, and preserves.

In recent years, the use of microparticles prepared from hydrocolloids, for example, agarose, has been explored for a variety of applications in the food industry. Characteristically, the microparticles are structured by a water-in-oil emulsion route, and these spherical hydrogel beads have garnered much interest as encapsulation matrices for an extensive variety of food supplements such as probiotics and antioxidants. Despite these promising functions, microspheres have proven difficult and expensive to manufacture due to the quantity of oil required. Very recently, some patents have described the use of crushed or sheared hydrocolloid microparticles in semisolid foods and beverages, for example: crushed agarose, forming microparticles, was used in fruit juices to disperse fruit pulp, and crushed agarose was used for a gel-based condiment that does not flow off the surface of semisolid food products. Another source portrays the encapsulation of additives in a crushed agar gel as part of a fluid gel beverage.

Recipes with Agar–Agar

Mizu Yokan (Agar Jelly with Red Bean Paste) (Figure 2.3)

Ingredients

(Serves 5)
1.5 g agar–agar (*kaku-kanten*; bar-type agar)[†]
250 mL (1 cup) water
25 g ($2^3/_4$ Tbsp) white sugar
100 g red bean paste (**Hint 1**)

[†]ANY FORM OF AGAR CAN BE USED.

FIGURE 2.3 *Mizu yokan* (agar jelly with red bean paste).

FIGURE 2.4 Preparation of *mizu yokan* (agar jelly with red bean paste). (**A**) Put water and squeezed agar–agar in a pan. (**B**) Heat until agar is fully dissolved. (**C**) Add red bean paste and stir. (**D**) Pour mixture into a mold and chill until it sets. (**E**) Cut pieces of *mizu yokan*.

Preparation (see also Figure 2.4)

1. Soak agar–agar in water for at least 1 h, and then squeeze out the water (**Hint 2**). Put water and squeezed agar–agar in a pan (**Figure 2.4A**) and heat at 90–100°C until completely dissolved (**Figure 2.4B**) (**Hint 3**).
2. Add white sugar to the pan. Then add red bean paste gradually with stirring (**Figure 2.4C**), and boil the mixture down to 350 g.
3. Remove mixture from heat and cool until thickened (at around 40°C) (**Hint 4**).
4. Pour cooled mixture into molds and cool to less than 20°C (in a refrigerator or in an ice-water bath) until it sets (**Figure 2.4D**).
5. Remove *mizu yokan* from molds and slice (**Figure 2.4E**).

Formulation pH = 6.59

Hints

1. If you have red beans, you can make red bean paste (**Figure 2.5**). Wash 300 g red beans and strain the water. Add 400 mL (1⅗ cups) water to the washed beans in a pan (**Figure 2.5A**), and bring to a boil. Add another 400 mL water, and return to a boil, then strain water (**Figure 2.5B**). Add 1200 mL (4⅘ cups) water to the boiled red beans and heat for 40–50 min until beans can be easily broken with your fingers (**Figure 2.5C**). Put a colander over a bowl, pour the boiled red beans and water into it, and crush the red beans with a pestle under a steady flow of water (**Figure 2.5D**). Add a large amount of water to the sediment in the bowl (**Figure 2.5E**), let it sit, then decant the clear layer of water. Repeat this process 4–5 times (**Figure 2.5F**). Put the sediment in a cotton bag or cotton cloth (**Figure 2.5G**) and squeeze out the water (**Figure 2.5H,I**).
2. When using agar–agar in powdered form, you do not have to presoak in water.

FIGURE 2.5 Preparation of red bean paste. (**A**) Put water and the washed red beans in a pan. (**B**) Strain water. (**C**) Cook until beans can be easily broken with your fingers. (**D**) Put a colander over a bowl, pour the boiled red beans and water into it, and crush the red beans with a pestle under a steady flow of water. (**E**) Add a large amount of water to the sediment in the bowl and let it sit. (**F**) Decant the clear layer of water. Repeat 4–5 times. (**G**) Put the sediment in a cotton bag or cloth. (**H**) and (**I**) Squeeze out the water through the cloth.

3. Agar–agar should be completely dissolved before adding white sugar. If sugar dissolves first, the agar–agar will never dissolve, and the *mizu yokan* will not set.
4. If you do not cool this mixture until thickened, the agar–agar solution and red bean paste will separate and the paste will set under the agar–agar jelly, because the specific gravity of the paste is higher than that of the agar–agar solution.

This recipe reminds Japanese people of summer. Eating something cold is good on hot summer days.

Mitsu-Mame (Agar Jelly with Rice Dumplings)

Ingredients

(Serves 5)

3.5 g agar–agar (*kaku-kanten*)

300 mL (1⅕ cups) water

50 g boiled red peas

FIGURE 2.6 *Shiratama* (waxy) rice flour.

50 g *shiratama* rice flour†
40 mL (a little under ⅕ cup) water for dumplings
150 g canned fruit‡
100 g canned fruit syrup
50 g (a little under ⅓ cup) white sugar
60 mL (a little over ¼ cup) water

†WAXY RICE FLOUR (FIGURE 2.6).

‡ORANGE, PEACH, PINEAPPLE, AND SO FORTH.

Preparation (see also Figure 2.7)

1. Soak agar–agar in water for at least 1 h, and then squeeze out water. Put water and squeezed agar–agar in a pan, heat at 90–100°C until completely dissolved (see **Figure 2.4A,B**), and boil down to 250 g.
2. Pour the mixture into a square mold, and cool to less than 20°C.
3. Put *shiratama* rice flour in a bowl, and add water gradually with stirring to make a dough (**Figure 2.7A**) (**Hint 1**), then shape dumplings by hand (**Figure 2.7B**).
4. Boil the rice dumplings for 1–2 min. (**Figure 2.7C**), and then put them in cold water (**Hint 2**).
5. Mix canned fruit syrup, white sugar, and water in a pan, heat to a boil, and then cool in the refrigerator or an ice-water bath.
6. Cut canned fruit into bite-size pieces.
7. Remove the solidified agar jelly from the mold and cut into 1-cm cubes.
8. Arrange the cut agar jelly, rice dumplings, cut canned fruit (**6**), and boiled red peas in glass serving bowls (**Figure 2.7D**), and pour the cold syrup (**5**) over it.

Shiratama rice dumpling pH = 6.69

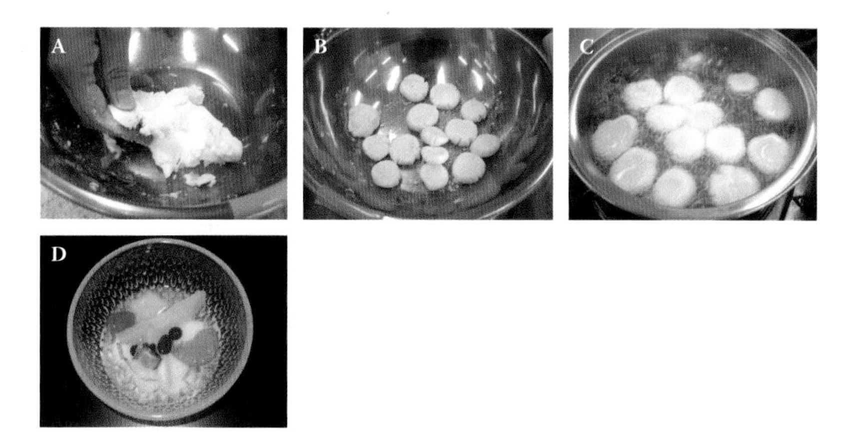

FIGURE 2.7 Preparation of *mitsu-mame* (agar jelly with rice dumplings). (**A**) Prepared dough. (**B**) Shape dough into dumpling balls. (**C**) Boil rice dumplings. (**D**) Serving suggestion for *mitsu-mame*.

FIGURE 2.8 Canned *mitsu-mame*, a popular Japanese product.

Hints

1. Add water very gradually to *shiratama* rice flour for complete absorption of water. If you add it too quickly, the flour will only take up part of the water.
2. Cold water firms up the *shiratama* dumplings.

Mitsu-mame is a traditional Japanese dessert. Today, it is less often prepared from scratch, and canned *mitsu-mame* has become quite popular in Japan (**Figure 2.8**).

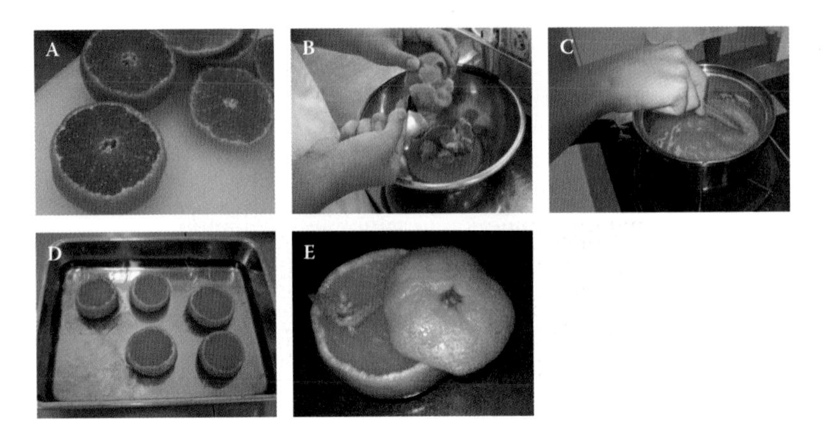

FIGURE 2.9 Preparation of fruit juice jelly. (**A**) Cut off the top of each orange. (**B**) Core with a spoon, separate pulp from skin. (**C**) Boil the puréed orange and sugar dispersion. (**D**) Cool down until set. (**E**) Garnish with mint leaves and serve.

Orange Jelly (Figure 2.9)

Ingredients

(Serves 5)

4 g agar–agar powder

100 mL (⅖ cup) water

25 g (2¾ Tbsp) white sugar

5 oranges

Mint leaves for garnish

Preparation

1. Cut off the top of each orange (**Figure 2.9A**), core with a spoon (or melon baller), and separate pulp (about 400 g) from skin (**Figure 2.9B**).
2. Purée the pulp in a blender.
3. Put water and agar–agar in a pan, and heat at 90–100°C until completely dissolved.
4. Add white sugar to the pan, boil the mixture until sugar dissolves (**Figure 2.9C**), and cool it to 60°C (**Hint 1**).
5. Add puréed orange pulp to the cooled mixture, pour into the five orange skin "cups," and cool to less than 20°C (in a refrigerator or in an ice-water bath) until it sets (**Figure 2.9D**).
6. Add mint leaves as garnish (**Figure 2.9E**).

Hint

1. The organic acids in fruit juice weaken the gel strength of agar–agar jelly. When the temperature of the agar–agar solution is below 70°C, organic acids no longer affect the agar–agar molecules. This is why the agar–agar solution should be cooled down before mixing with fruit juice.

FIGURE 2.10 *Awayuki-kan* (agar jelly with egg white).

Awayuki-Kan (Agar Jelly with Egg White) (Figure 2.10)

Ingredients

(Serves 5)

5 g agar–agar (*kaku-kanten*)

300 mL (1⅕ cups) water

70 g (a little over ⅖ cup) white sugar

30 g egg white (1 egg white)

Vanilla flavoring

Preparation

1. Soak agar–agar in water for at least 1 h, and then squeeze out the water. Put water and squeezed agar–agar in a pan, and heat at 90–100°C until completely dissolved (**Figure 2.4A, B**).
2. Add white sugar to the pan and boil the mixture down to 300 g.
3. Remove from heat and cool the mixture to 40–50°C (**Hint 1**).
4. Whip egg white until stiff peaks form (**Figure 2.11A**); gradually fold the egg white into the cooled mixture (**Figure 2.11B**), then add vanilla flavoring.
5. Pour the mixture into moistened molds (**Figure 2.11C**), and cool to less than 20°C (in a refrigerator or in an ice-water bath) until it sets.

pH = 8.69

FIGURE 2.11 Preparation of *awayuki-kan* (agar jelly with egg white). (**A**) Whip egg white until stiff peaks form. (**B**) Add whipped egg white to cooled solution of dissolved agar–sugar. (**C**) Pour mixture into moistened molds and cool until set.

Hint

1. In contrast to *mizu yokan*, the specific gravity of the whipped egg white is lower than that of the agar–agar solution. Thus, if the agar–agar solution is not cooled before mixing with the whipped egg white, the agar–agar jelly will set under the egg white.

Awayuki is Japanese for "light snow." *Awayuki-kan* will melt on your tongue like light snow.

Xìngrén Jelly (Chinese Almond Pudding) (Figure 2.12)

Ingredients

(Serves 5)

4 g agar–agar powder

300 mL (1⅕ cups) water

30 g (3⅓ Tbsp) white sugar

40 g almond powder for xìngrén jelly

200 mL (⅘ cup) milk

A { 40 g white sugar
100 mL (⅖ cup) water }

Grenadine

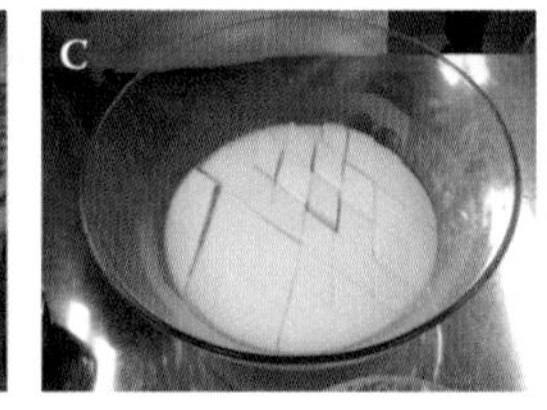

FIGURE 2.12 Preparation of *Xìngrén* jelly (Chinese almond pudding). (**A**) Cool until milk and almond powder solution mixture sets. (**B**) Cut almond "pudding" into diamond shapes. (**C**) Pour syrup over "pudding" pieces and shake the bowl gently.

Preparation

1. Put water and agar–agar in a pan, and heat at 90–100°C until completely dissolved (**Figure 2.4A,B**).
2. Add white sugar to the pan, and boil mixture down to 250 g.
3. Put 150 g lukewarm milk (at 30–40°C) in the pan. In a bowl, mix 50 g milk and almond powder and add it to the mixture in the pan (**Hint 1**).
4. Pour the mixture into a glass bowl, and cool to 10–20°C until it sets (**Figure 2.12A**).
5. In a separate pan, heat 40 g white sugar and 100 mL water (**ingredients**), boil down to 100 g, and add a little grenadine.
6. Cut the set almond mixture into diamond shapes (**Figure 2.12B**), pour the syrup (**A**) over them and shake the bowl gently (**Figure 2.12C**).

pH = 6.11

Hint

1. The lipids, casein, and lactose in the milk counteract agar–agar gelation. The strength of the jelly weakens with increasing amounts of milk.

This recipe comes from China. Originally, apricot seeds were used for their therapeutic effect according to Chinese medicine. The almond powder can be omitted.

Agar Spaghetti (Figure 2.13)

Ingredients

(Serves 5)
3 g agar–agar powder
250 g (1 cup) fruit juice†
40 g white sugar
Syringe, silicone tube

†YOU CAN USE ANY FRUIT JUICE, FOR EXAMPLE, ORANGE, GRAPE, APPLE, CRANBERRY, STRAWBERRY, AND SO FORTH. YOU CAN ALSO USE CHICKEN STOCK, COFFEE OR TEA WITH MILK.

Preparation

1. Mix fruit juice and agar–agar well in a pan (**Figure 2.13A**), and heat at 90–100°C until completely dissolved (**Figure 2.13B**).
2. Add white sugar to the mixture in the pan and heat. When the sugar is dissolved, remove pan from heat.

FIGURE 2.13 Preparation of agar spaghetti. (**A**) Mix fruit juice and agar. (**B**) Heat the mixture. (**C**) Prepare the syringe-silicone tubing. (**D**) Fill tubing with hot product. (**E**) Cool product to set in tube. (**F**) Extrude the gelled product.

3. Attach silicone tube to a syringe (**Figure 2.13C**), and fill it with the agar solution (**Figure 2.13D**).
4. Remove the silicone tube from the syringe, and put it in a bowl of cold water for 15 min. (**Figure 2.13E**).
5. Fill the syringe with air, attach the cold tube with the set agar jelly and extrude the jelly (**Figure 2.13F**).

pH of grape juice agar spaghetti = 3.04

pH of orange juice agar spaghetti = 3.80

Agar spaghetti is a modern gastronomical creation. Extruding the agar jelly from the tube is fun and an activity that can be shared by the whole family.

Tips for the Amateur Cook and Professional Chef

- Agar is defined as a strongly gelling hydrocolloid from marine algae.
- Agar is unique for commercial purposes because it forms firm gels at concentrations as low as 1%.
- Once the agar solution starts to boil, it should be boiled (without foaming) for exactly 20 seconds.
- If foaming occurs, disperse it by swirling and stirring for a few minutes.
- The melting temperature of an agar gel is a function of the agar's concentration and molecular weight.
- Around neutral pH, agar is compatible with most other polysaccharide gums and proteins, in the sense that flocculation or marked degradation does not occur when their dispersions are mixed.

- A good level of agar for use in icings will range from 0.2 to 0.5%.
- At concentrations of 0.1 to 1.0%, agar is a useful antistaling agent in breads and cakes.
- Addition of LBG (**Chapter 8**) to agar enhances the gel's elasticity and reduces its brittleness.
- Tannic acid (found, e.g., in squash, apple, and prune) may inhibit agar gelation. This can be avoided by adding small quantities of glycerol.
- To avoid hydrolysis of the polymer when the pH is lowered and the temperature is high, keep the gel under these conditions for the shortest time possible.

REFERENCES AND FURTHER READING

Araki, C. H. 1937. Acetylation of agar like substance of *Gelidium amansii. J. Chem. Soc. Japan* 58:1338–50.

Araki, C. H. 1956. Structure of agarose constituent of agar–agar. *Bull. Chem. Soc. Japan* 29:43–44.

Armisen, R. 1999. Agar. In *Thickening and gelling agents for food*, 2nd ed., ed. Alan Imeson, 1–21. London: Blackie Academic and Professional.

Armisen, R. and F. Galatas. 1987. Production, properties and uses of agar. In *Production and utilization of products from commercial seaweed*, ed. D. J. McHugh, 1–57. Rome: FAO Fisheries Technical Paper No. 288.

Armisen, R. and F. Galatas. 2009. Agar. In *Handbook of hydrocolloids*, ed. G. O. Phillips and P. A. Williams, 82–107. Oxford, UK: Woodhead Publishing Limited.

Carr, J. M., Sufferling, K., and J. Poppe. 1995. Hydrocolloids and their use in the confectionery industry. *Food Technol.* 49:41–2, 44.

Choi H. J., Lee, C. J., Cho, E. J., Choi, S. J., and T. W. Moon. 2010. Preparation, digestibility, and glucose response in mice of rice coated with resistant starch type 4 using locust bean gum and agar. *Int. J. Food Sci. Technol.* 45:2612–21.

Davidson, C. J. 1906. The seaweed industry of Japan. *Bull. Imp. Inst. Japan* 4:125–49.

Deszczynski, M., Kasapis, S., MacNaughton, W., and J. R. Mitchell. 2003. Effect of sugars on the mechanical and thermal properties of agarose gels. *Food Hydrocolloids* 17:793–9.

Ellis, A. and J. C. Jacquier. 2009. Manufacture and characterisation of agarose microparticles. *J. Food Eng.* 90:141–5.

FDA. 1972. *GRAS (generally recognized as safe) food ingredients: Agar–agar*, PB 221-225, NTIS. U.S. Washington, DC: U.S. Department of Commerce.

FDA. 1973. *Evaluation of health aspects of agar–agar as a food ingredient,* Food and Drug Administration PB-265502. Bethesda, MD: Federation of American Societies for Experimental Biology.

FDA. 1973. *Mutagenic evaluation of FDA 71-53 (agar–agar)*, PB-245-443. Washington, DC: U.S. Department of Commerce.

FDA. 1973. *Tetratologic evaluation of FDA 71-53 (agar–agar)*, PB-223820. Washington, DC: U.S. Department of Commerce.

Food Chemicals Codex III. 1981. In *Agar*, 11–2. Washington, DC: National Academy Press.

Glicksman, M. 1969. *Gum technology in the food industry*, 199–266. New York and London: Academic Press.

Glicksman, M. 1983. *Food hydrocolloids*, vol. 2, *Seaweed extracts*, 63–73. Boca Raton: CRC Press Inc.

Guiseley, K. B. 1968. Seaweed colloids. In *Encyclopedia of chemical technology*, vol. 17, 2nd ed., 763. New York: Kirk-Othmer, John Wiley and Sons Inc.
Laaman, T. R. 2011. *Hydrocolloids in food processing*. Ames, IA: Wiley-Blackwell Publishing & Institute of Food Technologists.
Lawrence, A. A. 1973. *Edible gums and related substances*, 165–74. Park Ridge, NJ: Noyes Data Corporation.
Lawrence, A. A. 1976. *Natural gums for edible purposes*, 238–49. Park Ridge, NJ: Noyes Data Corporation.
Mantel, C. L. 1965. *The water-soluble gums*. New York: Hafner Publishing Company.
Matsuhashi, T. 1990. Agar. In *Food gels*, ed. P. Harris, 1–51. London and New York: Elsevier Applied Science.
Meer, W. 1980. Agar. *Handbook of water-soluble gums and resins*, ed. R. L. Davidson, 7.2–7.14. New York: McGraw-Hill.
Moirano, A. L. 1977. Sulfated seaweed polysaccharides. In *Food colloids*, ed. H. D. Graham, 347–81. Westport, CT: Avi Publishing.
Morris, V. J. 1986. Gelation of polysaccharides. In *Functional properties of food macromolecules*, ed. J. R. Mitchell and D. A. Ledward, 121–70. Amsterdam: Elsevier.
Newton, L. 1951. *Seaweed utilization*, 107–8. London: Sampson Low.
Nussinovitch, A. 1997. *Hydrocolloid applications: Gum technology in the food and other industries*. London: Blackie Academic & Professional.
Nussinovitch, A. 2003. *Water soluble polymer applications in foods*. Oxford, UK: Blackwell Publishing.
Nussinovitch, A. 2010. *Polymer macro- and micro-gel beads: Fundamentals and applications*. New York: Springer.
Nussinovitch, A., Corradini, M. G., Normand, M. D., and M. Peleg. 2000. Effect of sucrose on the mechanical and acoustic properties of freeze-dried agar, *k*-carrageenan and gellan gels. *J. Texture Studies* 31:205–23.
Nussinovitch, A., Gershon, Z., and L. Peleg. 1998. Characteristics of enzymatically produced agar-starch sponges. *Food Hydrocolloids* 12:105–10.
Nussinovitch, A., Jaffe, N., and M. Gillilov. 2004. Fractal pore-size distribution on freeze-dried agar-texturized fruit surfaces. *Food Hydrocolloids* 18:825–35.
Nussinovitch, A., Kopelman, I. J., and S. Mizrahi. 1991. Modeling of the combined effect of fruit pulp, sugar and gum on some mechanical parameters of agar and alginate gels. *Lebensm.-Wiss. U.-Technol.* 24:513–7.
Nussinovitch, A., Peleg, L., and Z. Gershon. 1995. Properties of agar-starch sponges. In *The 8th International Conference and Industrial Exhibition on Gums and Stabilizers for the Food Industry*, July 1–14, The North East Wales Institute, Cartrefle College, Wrexham, CLWYD, UK.
Nussinovitch, A., Velez-Silvestre, R., and M. Peleg. 1993. Compressive characteristics of freeze-dried agar and alginate gel sponges. *Biotechnol. Progr.* 9:101–4.
Rees, D. A. 1969. Structure, conformation and mechanism in formation of polysaccharide gels and networks. In *Advances in carbohydrate chemistry and biochemistry*, ed. M. L. Wolform and R. S. Tipson, 267. New York: Academic Press.
Selby, H. H. and T. A. Selby. 1966. Agar. In *Industrial gums*, ed. R. L. Whistler, 15–49 (translated by Yabuki and S. Suzuki in NKRL Report, No. 4). New York: Academic Press.
Smith, F. and R. Montgomery. 1959. The structure of polysaccharides from seaweed. In *The chemistry of plant gums and mucilage*, 426. New York: Reinhold Publishing Corporation.
Stanley, N. F. 1995 In *Food polysaccharides and their applications*, ed. A. M. Stephen, 187–204. New York: Marcel Dekker.

Tseng, C. K. 1946. Phycolloids: Useful seaweed polysaccharides. In *Colloid chemistry*, vol. VI, ed. J. Alexander, 630. New York: Reinhold.
Weiner, G. and A. Nussinovitch. 1994. Succulent, hydrocolloid-based, texturized grapefruit products. *Lebensm.-Wiss. u.-Technol.* 27:394–9.
Whistler, R. L. and J. N. BeMiller. 1993. *Industrial gums: Polysaccharides and their derivatives*, 3rd ed. New York: Academic Press.

3

Alginates

Historical Background

Alginate was discovered in 1881 by E. C. C. Stanford, who was searching for useful products from kelp. He developed an alkali extraction method for a thick substance termed "algin," that is, from algae, and afterward precipitated it by means of mineral acid. Algin was isolated by A. Krefting 15 years later. In 1929, Kelco Co. began the profitable manufacture of algin in California. The extracted substance was initially utilized for can-sealing purposes and as a boiler compound. Utilization of alginate as an ice-cream stabilizer became significant in 1934. Ten years after that, propylene glycol alginate (PGA) was developed and manufactured commercially, leading to the establishment of alginate-production plants in the United States, Europe, and Japan.

Sources

Alginates are naturally occurring polysaccharides. They are extracted from brown seaweed, in contrast to agar (**Chapter 2**) and carrageenan (**Chapter 4**), which are extracted from red seaweed. The most widely used brown seaweeds are *Laminaria hyperborea, Macrocystis pyrifera*, and *Ascophyllum nodosum*. Most alginates in the United States are extracted from *M. pyrifera*, a massive kelp found in sea beds.

The kelp fastens itself to the rocky ocean bottom by means of a root-like structure termed *holdfast* against the physically powerful marine currents. The kelp is cropped ~1 meter below the surface of the water by mechanical means. This allows access to the sun's rays, supporting algal expansion. The harvested kelp is later processed. In North America, more than a few additional *Laminaria* species are employed for domestic manufacture including *Laminaria digitata, Laminaria cloustoni*, and *Laminaria saccharina*. In Europe, *Laminaria* species and *A. nodosum* are utilized for production, while the Japanese crop *Eklonia cava* and the South African, *Eklonia maxima*. Alginate is located as a mixed calcium/sodium/potassium salt of alginic acid in the cell wall and intercellular spaces. The alginate molecules provide the strength and elasticity required for algal expansion in the ocean. The gum, usually sold as sodium alginate, is water-soluble and used as a viscosity former, and to form gels in the presence of calcium and/or other polyvalent metal ions.

Structure

Stanford suggested that alginate is a nitrogenous material, with the formula $C_{76}H_{76}O_{22}(NH_2)_2$. However, later separation techniques proved that the pure product is nitrogen-free. Alginic acid is a linear copolymer (a polymer derived from two or more monomeric species, also termed *heteropolymer*) composed of D-mannuronic acid (M) and L-guluronic acid (G). Regions can consist of one unit or the other, or both monomers in alternating sequence, that is, M blocks, G blocks, or heteropolymeric MG blocks, respectively. Further information on the chemical composition of alginate can be located elsewhere.

The molecular weight, as well as the proportion and arrangement of M and G units, influence the behavior of a particular alginate. The proportion of M blocks varies from 61% in *M. pyrifera* to 31% in *L. hyperborea*. The content of alternating-block segments ranges from 26.8% in *Laminaria* to 41.7% in *Macrocystis*. Dissimilar relative amounts of monomer have been found in alginates. Molecular weights of commercial alginates are in the range of 32,000–200,000, and are related to a degree of polymerization of 180–930.

Alginate Sources and Manufacture

Raw Materials

Commercial production of alginate is based on *L. hyperborea* from North Atlantic coastal regions, *M. pyrifera* from the west coast of the United States, and *A. nodosum* from Northern Europe and Canada. These are cropped by coastal gathering, mechanical collection, or cutting by hand. The harvested seaweed is then transported to industrial units where it can be processed, either wet or after drying.

Alginate and PGA Production

Today, alginate is produced by modifications of processes invented in 1936 by H. C. Green, and in 1938 by V. C. E. LeGloahec and J. R. Herter. These processes are fully detailed elsewhere. Even in the numerous extraction plants where advances in processing can be noted, Stanford remains accountable for the fundamental process. Alginate extraction includes several steps. The first is milling of the seaweed. This is followed by washing and treating it with strong alkali and heating to extract the alginate. The hydrocolloid is precipitated with calcium chloride and then treated with acid to form alginic acid. The alginic acid can be reacted with sodium carbonate (or other bases) to manufacture alginate salts, or treated with propylene oxide to produce PGA. Following the reaction, up to 90% of the carboxyl groups are esterified, and the remaining groups either remain unbound or are neutralized with sodium or calcium. There are various PGAs, aimed at specific applications, on the market.

Regulatory Status and Toxicity

Alginates are regarded as very safe. They are generally recognized as safe (GRAS) under the Code of Federal Regulations (CFR) (Substances GRAS in food, 21CFR182).

PGA is approved as a food additive under 21 CFR 172.858, for use as a thickener, stabilizer, or emulsifier in foods as per that regulation. The utilization of edible salts of alginic acid is approved by all relevant selected sections of Title 21, Code of Federal Regulations (21 CFR) and the *Federal Register* containing the regulations and rules appropriate for Food and Color Additive Petitions and Food Ingredient and Food Packaging Notifications for the United States and the European Economic Council (EEC). The Food *Chemical Codex* contains monographs on alginic acid, ammonium alginate, calcium alginate, potassium alginate, and PGA. Alginic acid, its edible salts or PGA are on the lists of approved emulsifiers/stabilizers published by the EEC and the Council of Europe. The FAO/WHO joint expert committee recommends an acceptable daily intake of 25 mg/kg body weight for alginic acid and its edible salts, and 25 mg/kg body weight for PGA.

Commercial Characteristics

Alginates have specific uses. The contents and proportion of M and G residues are not quantified by the producer. However, they can be estimated from information on the seaweed source and the mixed amounts employed to achieve the desired gel strength and solution-thickening ability. Sodium alginate is an accessible product that is sold in huge amounts; in addition to alginic acid, calcium, potassium, and ammonium alginates can be procured as well. Alginic acid—the free acid form of alginate—has limited stability: to stabilize it and make it water-soluble, the acid is converted into a marketable alginate by the incorporation of salts such as sodium carbonate, potassium carbonate, ammonium hydroxide, magnesium hydroxide, calcium chloride, or propylene oxide.

Hydrocolloid Solution Preparation Procedures

Prior to its dissolution in water, it is recommended that the gum be dry-mixed with other formulation ingredients such as sugar or starch (**Chapter 14**), or that it be dispersed in glycerol or vegetable oil, to better separate the gum particles and hence facilitate their dissolution, possibly using high-shear stirring. If dispersion is not carried out correctly, swollen clumps will form, negating contact between the water molecules and their center. In this case, very high-shear mixing must be performed to resolve the problem. Overheating must be avoided during the mixing so as not to degrade the alginate, which will result in decreased viscosity or gel strength. The use of soft water is advised to prevent undesired cross-linking of the alginate by calcium.

A smooth alginate solution with long flow properties is produced upon dissolution in water. The flow characteristics of the alginate solution depend on the type of salt, sequestering agent (i.e., a chemical that combines with metal ions and removes them from their sphere of activity) and polyvalent cation, polymer size, temperature, and shear rate, the concentration of the hydrocolloid in solution, and the presence of other miscible solvents. The viscosity of the solution is related to the gum's molecular weight but it is also influenced by the level of residual calcium from the manufacturing process. The influence of pH on alginate solutions varies, depending on the type of alginate used. Sodium-alginate solutions are not stable above pH 10. PGA is more stable at acidic pH, and sodium alginate precipitates at pH < 3.5. Sequestering agents prevent the calcium which is inherent in the alginate from reacting with it, or preventing the alginate's reaction with polyvalent ions in the solution. The addition of sequestering agents reduces viscosity relative to that without these agents. This is true for both M- and G-type alginates. Monovalent salts also reduce the viscosity of dilute sodium alginate. The effect of salt is pronounced after long storage and is dependent upon the alginate source, the degree of polymerization, gum concentration in the solution, and the character of the monovalent salt used.

Mechanism of Alginate Gelation

Aside from hydrocolloids, a number of divalent cations can be involved in the formation of an alginate gel. Calcium is most suitable for food purposes, owing to its nontoxicity. Borax can be used to manufacture gels for nonfood purposes. Gels are usually produced by ionic bridging between the calcium ions of two carboxyl groups on adjacent polymer chains. Gel formation and the consequent gel strength obtained from an alginate are directly related to the amount of G blocks and the average G-block length. High G content and long G blocks confer high calcium reactivity to alginates and the strongest gel-forming ability. The degree of polymerization should be > 200 to achieve optimal gel strength. The mechanism involves a cavity that operates as a binding position for calcium ions between two diaxially linked G residues, creating a 3D structure termed *egg-box*; the calcium interacts with the carboxyls and with the electronegative oxygen atoms of the hydroxyl groups. The alginate gel is regarded as a semisolid material: after gelation, the water molecules are actually entrapped by the alginate network, but retain their ability to migrate. The reaction of soluble sodium alginate with calcium results in gels with a range of consistencies. To obtain a stiff gel, a minute quantity of calcium is adequate. The stoichiometric reaction is ~0.75 mg calcium per g of algin. The most favorable alginate–calcium complexing is achieved with about 40% of the stoichiometric amount, but varies with the presence of other soluble solids, calcium source, and pH, among other things.

Degree of Conversion, Thixotropy, and Alginate–Pectin Gels

Alginate gel properties depend on the type of alginate used, the extent of conversion to calcium alginate, the calcium ion source and the processing. High-G alginates are required to structure strong brittle gels. High-M alginates form weaker, more elastic gels, which are not as prone to syneresis. The degree to which calcium replaces sodium (calcium conversion) to obtain calcium alginate governs gel texture. A small degree of conversion produces enhanced viscosity. Higher conversion results in the advance of gel structures. Yet further conversion produces increased gel strength and thixotropic performance (i.e., the constructed gel changes into a fluid upon application of shear, and reforms when the shearing is discontinued). Irreversible gels are fabricated through the use of supplementary calcium. These gels do not reform upon cessation of shearing. Alginate–high-methoxy pectin (**Chapter 13**) blends (revealing synergy) can create a gel at small solid contents and a broad range of pHs, quite the opposite of high-methoxy pectins which produce gels with high solid (sugar) contents.

Because fruits are rich in pectin, the addition of alginate to cooked fruits (e.g., apples) results in a rigid gel, and alginates with a high G content are employed.

Diffusion Setting

One method of preparing alginate gels is diffusion setting. Alginate beads can be prepared by merely dropping the gum solution into a calcium-chloride bath. Beads have small diameters and therefore the time needed for calcium to diffuse into them is not limiting, particularly if its concentration is intentionally increased. High-calcium salt concentrations (~15%) can be used as cross-linking agents (to induce gelation) for a few seconds. Less bitter calcium salts, such as calcium lactate or calcium acetate, are utilized for food purposes. Acidic baths can also be employed when a calcium salt, which is only soluble under acidic conditions, is solubilized in the alginate solution prior to extrusion. Diffusion occurs both in and out of a gel piece. This must be taken into account when a product including components with low molecular weights, such as sucrose, acids, and so forth, is prepared. Internal setting takes place while calcium is released under controlled conditions from inside the product. The rate of calcium release is controlled by pH and the solubility of the calcium salt. To avoid an excessively rapid reaction between the alginate and the calcium (or other cross-linking agent), a sequestering agent is added to the system and, depending on pH, the system's ability to entrap or liberate the ions is modified. Therefore, it is feasible to compose a preparation at neutral pH and to adjust the pH rapidly to an acidic one, inducing the suitable sequestering agent to release its entrapped component, which then reacts with the alginate. Calcium release should be monitored using a particular ion-selective calcium-ion electrode. Full information about the calcium concentration at any specific stage of the procedure allows the development of improved structured products.

Establishing Differences between Alginate Gels

Gel strength is defined as the maximal force in Newtons required to break a gel, or the stress at failure (i.e., stress that fractures a gel). It can be assessed by means of a universal testing machine (UTM). After setting, alginate gels are thermostable, that is, they do not melt upon reheating. At the same time, as all ingredients are mixed into the internally set alginate system, it is basically ready for gelation. Gelation can be delayed by increasing the solution temperature. Under such conditions, the alginate chains have additional thermal energy, which allows alignment, and associations then occur when the solution is chilled. Owing to calcium's accessibility to all alginate molecules, a smaller degree of syneresis is detected in these systems. Gelation by internal setting is different from systems in which gelation occurs by diffusion: there the molecules closest to the calcium source react first, resulting in greater gel shrinkage and syneresis.

The higher the concentrations of the alginate and calcium, the higher the temperature required to prevent gelation. At 0.6% sodium alginate, and sufficient calcium cations to give 60% conversion, a temperature of 80°C is needed to prevent gelation. A gel made of alginic acid (without calcium ions) can be manufactured by maintaining the pH at below 3.5 (acid setting bath). Nevertheless, the resultant gels are grainy and undergo syneresis. In calcium setting, reducing pH lowers the amount of calcium required to obtain a gel. Under acidic conditions, protein is also capable of interacting with sodium alginate by electrostatic interaction. A decrease in pH enhances the strength of this interaction. At acidic pH, proteins are destabilized and denatured in the presence of alginate, and high-molecular-weight complexes are also formed.

Applications

There are numerous applications for alginates in foods. A few examples are fruit-pulp texturization, diffusion-set gels, protein extrusion, storage-life extension of potatoes, entrapment of enzymes, inclusion in fish patties, and use in structured beef products. The use of alginates is expected to increase in the near and distant future.

Fruit-Like Products and Coatings on Fresh-Cut Fruit Pieces

Fruit and other texturized products can be prepared from alginates. In 1946, W. J. S. Peschardt developed a method to prepare analog (artificial) cherries. He dropped a colored and flavored alginate–sugar solution into a soluble calcium-salt solution. Following the immediate creation of an external calcium-alginate membrane, gelation of the inner part of the "cherries" was induced by slow diffusion of calcium cations into the spherical particle and cross-linkage with the alginate within. The produced artificial cherries were thermostable, enabling their inclusion in baked goods. In fact, ~60 years ago in England, this product could be found in fruitcake. The marketable product was composed of natural cherry purée, corn syrup, alginate, flavoring, and color. The inclusion of cherry purée enabled the manufacturer to call the product *cherry balls* instead of *imitation cherries*.

Simulated fruits can be effectively produced by means of a two-step mixing internal-setting procedure. The first phase consists of mixing together alginate, anhydrous dicalcium phosphate, disodium hydrogen ortho-phosphate, glucose, sucrose, and water. In parallel, fruit purée, sucrose, glucose, citric acid, and sodium citrate dihydrate are also blended. At neutral pH, the anhydrous dicalcium phosphate is insoluble. Upon pumping the two mixtures, first under high-shear mixing, then under no-shear conditions, the lowered pH causes calcium ion release from the anhydrous dicalcium phosphate leading to a gelling reaction. This process can also be executed in a continuous manner.

There are many recipes and patents that deal with the formation of texturized fruits. One example is the Unilever patent (BP 1,484,562), which describes a co-extrusion system for preparing reformed black currants. In general, fruit purées offer new outlets for visually imperfect or excessively small fruit. To restructure the purée, a texturizing agent such as alginate is required to control the functional properties of the product. Novel gel systems were formed from alginates and peach purée, without supplementary calcium or sugar. These gels were shown to be the product of interactions between pectin (**Chapter 13**) and alginate. Novel alginate–peach gels involved minimal preparation and could be scaled up in a very straightforward manner and employed by the food industry for a variety of end products: final fruit products that contain up to 99% fruit and are highly nutritious, and have commercial potential as ready-to-eat snack foods or as constituents of baked, frozen, and/or canned foods.

Coatings based on alginate or gellan gum (**Chapter 10**) were used on fresh-cut papaya pieces to preserve the quality of the minimally processed fruit. Formulations including glycerol and ascorbic acid revealed somewhat improved water-barrier properties as compared to the uncoated samples. Inclusion of sunflower oil in the formulations increased the water-vapor resistance of the coated samples. The coatings also improved the firmness of the fresh-cut product, and the inclusion of ascorbic acid as an antioxidant in the coatings helped maintain the cut fruit's nutritional quality throughout storage.

Water Dessert Gels

Eye-catching edible gels can be manufactured with alginates. Calcium is frequently the cation of choice for cross-linking; nevertheless, divalent or trivalent metal ions can also be employed. It is suggested that the calcium ion concentration and pH be properly selected to gain more control over reaction rates and the texture of the formed product. Gel formation that proceeds too rapidly creates a grainy gel, whereas extremely slow gelation results in very soft gels. Controlled release of calcium ions can be achieved by using a suitable salt to manage its solubility and by using sequestering or retarding agents. Such systems have been the subject of many patents and have been engaged for the manufacture of dessert gels and candied jellies, fruit jams and jellies, and jellied salads and broths.

Milk Puddings, Ice-Cream Stabilizers, and Other Dairy Products

The manufacture of milk pudding based on a specifically treated mix of water-soluble, alkali-metal alginate, a mild alkali, and a minute amount of calcium salt has been reported. A simple mix was used to form a good gel in water, which could be admixed with powdered skim milk to structure a milk pudding when dissolved in cold water. Ice-crystal growth in ice creams can be slowed by alginate to achieve a smooth texture. An identical addition can also contribute to delaying product melting and controlling overrun. Minute quantities (i.e., 0.1 to 0.5%) of sodium alginate are most frequently used as ice-cream stabilizers. Given that the stabilizers comprise water-holding properties and dispersive capabilities, they add to high-quality body properties and texture protection. Sodium alginate may react with the calcium in the ice-cream mix, thereby decreasing the concentration of calcium ions in the water. This is also useful for getting rid of the clumping of fatty globules in the manufactured goods.

A minute addition of sodium alginate to soft cheese spreads can prevent separation of water and oil, diminish surface hardening, and produce a better textured processed cheese. Inclusion of ~0.15% sodium alginate is adequate to thicken whipped cream, that is, to act as a stabilizer upon whipping. The manufacture of synthetic creams can be facilitated by the addition of alginate to achieve a quicker whip, greater tolerance to overwhipping, stabilized overrun, and prevention of syneresis.

Fish and Meat Preservation and Sausage Casings

Block-freezing fatty fish, such as herring or mackerel, in alginate jelly might prevent oxidative rancidity. An alginate film that excludes air is formed around each fish piece, making rancidity almost impossible. Storage with the coating also assists in decreasing the off-flavors and unpleasant odors correlated with fish. This procedure has been extended to other fish, as well as shellfish. Herring can also be dusted with alginates to assist in its canning. The film-forming sodium alginate and tamarind kernel powder, guar gum (**Chapter 8**), and agar–agar (**Chapter 2**) may also be beneficial for shelf-life extension and curing preservation of salted and dried mackerel.

Highly water-soluble fish-meat protein can be stably manufactured by conjugation with alginate oligosaccharides. Frozen mature salmon meat can be used as the raw material for a highly stable meat–alginate oligosaccharide conjugate. In addition, inhibition of protein denaturation in the processing line is vital to manufacturing a meat–alginate oligosaccharide conjugate with high water solubility. The water-soluble fish meat might be used to develop new food items such as protein-rich beverages, nutritional emulsifiers, and a source of protein for people who cannot chew foods.

Poultry parts have been coated with calcium-alginate films. Whole beef cuts were also coated with calcium-alginate films prior to freezing, by dipping in a 10–15% sodium-alginate solution, then 3.5–5% calcium chloride, followed by 10–20% glycerol (as a plasticizer). An edible sodium alginate–cornstarch (**Chapter 14**) meat coating was used to coat beef steaks, pork chops, and skinned chicken drumsticks to improve color, appearance, odor, texture, and juiciness. Raw poultry items can serve as a source of human pathogens that may cross-contaminate additional foods. Despite efforts to control them during rearing, shipping, and processing, elimination of poultry contamination by these organisms presents a significant challenge. To improve effectiveness without changing processing speeds at the plant, edible gels based on pea starch and calcium alginate containing antimicrobials can be applied to chicken surfaces. Theoretically, the antimicrobial agents will slowly diffuse from the gels or coating material into skin irregularities, and application after defeathering will provide increased contact time with target microorganisms and yield improved effectiveness.

An edible alginate coating was reported that improves the quality of frozen pork. The coating decreased meat loss during thawing, and was very useful in maintaining the functional properties of the frozen meat. In addition, no significant differences in the structures of the control and coated pork were observed. Alginates were also used as part of restructured meat products and to produce synthetic sausage casings. In 1955, the Visking Corporation pioneered alginate casings under the name of Tasti-Jax. The casings were not successful in the United States but were extensively used in Germany. A characteristic casing was made of 6% sodium alginate extruded into an acidic solution of 10% calcium chloride, which was then plasticized by washing with glycerol and calcium lactate. On the one hand the casings were elastic, hygienic, and easier to handle than natural casings. On the other, likely due to interaction with sodium salts, the casings became swollen, weaker, and less attractive, and did not succeed in contracting during cooking. Attempts to eliminate these shortcomings were suggested. Extruded edible casings manufactured from blends of sodium alginate with pectin (**Chapter 13**) or gelatin (**Chapter 9**) for use with breakfast pork sausage were assessed. Sensory analysis demonstrated a preference for the pectin casings over the gelatin ones.

The quality of low-salt sausages enriched with inulin (**Supplemental F**) was found to vary when some of the pork's back fat was replaced with olive oil. The addition of an olive oil–alginate emulsion and inulin resulted in a low-salt, reduced-fat product that was richer in monounsaturated fatty acids, while retaining sensory notes similar to those of the traditional sausage used as a control; this product achieved a good acceptability rating.

Bakery Toppings, Fillings, Beverages, and Salad Dressings

Alginates have been used to produce nonsticky icings, which do not crack. Alginates also successfully maintain the stability of foam and aerated structures and improve the texture of whipped sugar toppings. Alginates can minimize syneresis in baking jellies and enhance their stability at elevated baking temperatures. Pie fillers are stabilized with alginates. Moreover, better clarity of frozen and canned cherry and peach pastry fillings has been achieved by combining PGA with a modified waxy cornstarch (**Chapter 14**).

Fruit-pulp precipitation in drinks can be delayed by inclusion of PGA or sodium alginate within the formulation. PGA is also an efficient stabilizer in fermented fruit-milk drinks. An alginate–phosphate blend is useful as a stabilizer in chocolate-milk drinks. Another combination of pectin (**Chapter 13**), polyol alginate and a fatty acid ester stabilizes acidic drinks composed of an aqueous blend of vegetable extract, juice, and milk products. PGA might help in stabilizing beer foam, and sodium alginate can be utilized to clarify wine and eliminate tannins, nitrogenous substances, and coloring material.

Dressings are in essence vegetable-oil emulsions, which also include salt, sugar, spices, and flavorings. Sodium alginate and PGA slow separation of the oil and water phases, thus stabilizing the dressings or sauces (see "**Tips the Amateur Cook and Professional Chef**" in **Chapter 3**). Gum tragacanth (**Supplemental H**) can also be used for this purpose. Inclusion of PGA with cornstarch (**Chapter 14**) in salad dressings improves their stability, producing a final soft product in which the oil does not separate out upon standing. PGA and xanthan gum (**Chapter 15**) stabilize oil-in-water emulsions in the presence of salt as a dissolved electrolyte. Low-salt concentrations are particularly effective when PGA–xanthan gum combinations are used in the presence of polysorbate-60.

Further Applications

Alginate gels could be useful in numerous other applications. Yeast cells have been immobilized in alginate beads for alcoholic drinks and ethanol production and in the second fermentation of champagne. Alginate thermostability could be beneficial for microwaving alginate-based products. In addition, alginates enable the fabrication of novel fried foodstuffs that have better taste and texture and are more easily adapted for mass production. PGA is beneficial in the manufacture of dietetic dressings. Alginates have also been used in macaroni, spaghetti, and dough manufactured goods to allow a higher rate of extrusion.

Recipes with Alginates

Yogurt Spheres (Figure 3.1)

Ingredients

(Makes 10–15)
1000 mL (4 cups) water
5 g sodium alginate powder
100 g yogurt†
Sugar

†YOU CAN ALSO USE CERTAIN TYPES OF PURÉE WHICH CONTAIN RELATIVELY HIGH CONCENTRATIONS OF CALCIUM, GREEN LEAVES SUCH AS BASIL, JEW'S MARROW, KALE OR WATERCRESS‡, OLIVE, PEAR, STRAWBERRY, LEMON, MOZZARELLA, TOFU WITH CALCIUM, AND SO FORTH.

‡TO PREPARE 100 G GREEN LEAF PURÉE, USE 120 G FRESH GREEN LEAVES. BOIL LEAVES IN SALTED WATER (100°C) FOR 5 MIN, THEN COOL IN COLD WATER, AND MASH IN A FOOD PROCESSOR OR BLENDER.

Preparation

1. Pour water in a bowl. Add sodium alginate and mix with a hand mixer or a blender.
2. To remove bubbles or lumps in the sodium-alginate solution, place it in the refrigerator until they disappear, usually overnight (**Figure 3.1A**).
3. Scoop yogurt, place scoops in the sodium-alginate solution (**Figure 3.1B**).
4. Once the jelly has set (about 2 min.), remove the yogurt spheres from the solution (**3**) (**Figure 3.1C**).

pH of sodium-alginate solution = 7.82

pH of yoghurt = 4.06

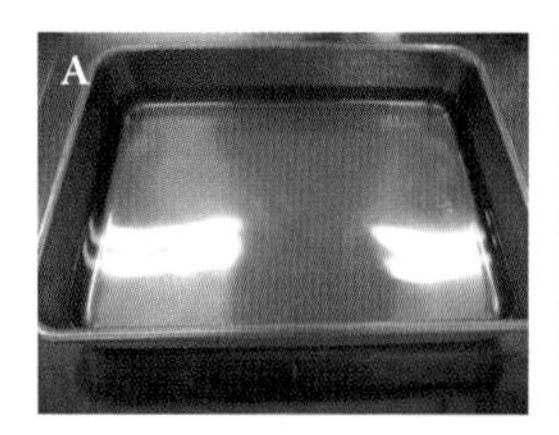

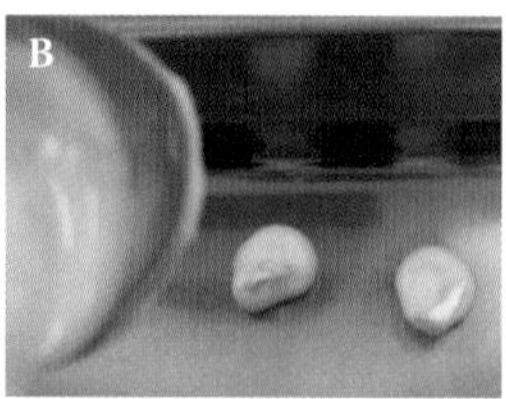

FIGURE 3.1 Reverse sphere. (**A**) Dissolve alginate. (**B**) Scoop yoghurt into sodium-alginate solution. (**C**) Remove finished reverse sphere.

Artificial *Ikura* (Salmon Eggs) (Figure 3.2)

Ingredients

(Makes ~20 artificial salmon eggs)
1 g sodium-alginate powder
1 g gum arabic powder
1 g κ-carrageenan powder
22 g calcium chloride
1 g guar gum powder
16 g (20 mL) vegetable oil
Water, β-carotene†, red or yellow food coloring‡
Straw (7 mm diameter), syringe (1.0–2.5 mL) with needle

†YOU CAN USE A β-CAROTENE SUPPLEMENT.

‡IF YOU DON'T HAVE THE FOOD COLORING, YOU CAN USE TEA OR COFFEE.

Preparation

1. Make 100 g 1 wt% sodium-alginate solution (**Hint 1**), 100 g 1 wt% gum arabic solution (**Hint 1**), 100 g 1 wt% κ-carrageenan solution (**Hint 1**), 100 g 2 wt% calcium-chloride solution, and 100 g 20 wt% calcium-chloride solution (**Hint 2**).
2. Add red or yellow food coloring to the 2 wt% calcium-chloride solution to obtain a light-orange color, then dissolve guar gum in it.
3. Dissolve β-carotene (about 0.5 g) in the vegetable oil for a deep-orange color (**Figure 3.2A**).

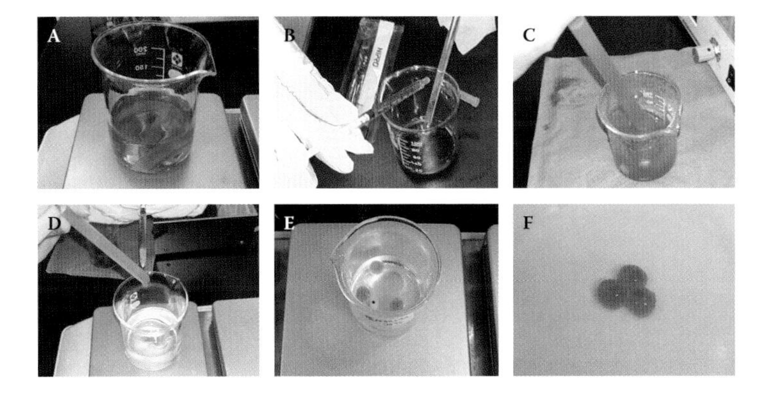

FIGURE 3.2 Artificial *ikura* (salmon eggs). (**A**) Add B-carotene to vegetable oil to obtain a deep-orange color. (**B**) Draw colored vegetable oil into a syringe, and affix to a needle that has been inserted in a straw. (**C**) Suck the guar gum–calcium-chloride solution into the straw. (**D**) Drop the mixture in the straw into the sodium alginate, gum arabic, κ-carrageenan, and water solution, injecting the colored vegetable oil from the syringe into the center of the drop. (**E**) Soak the drop in calcium-chloride solution. (**F**) Remove cross-linked product.

4. Pour 30 g 1 wt% sodium-alginate solution, 20 g 1 wt% gum arabic solution, 20 g 1 wt% κ-carrageenan solution **(1),** and 20 mL water into a cup. Combine the solutions by mixing well with a stirring rod.
5. Pour 100–150 g water into another cup.
6. Pour 50 g 20 wt% calcium-chloride solution into a third cup.
7. Poke syringe needle into a straw at about 2 mm from one end.
8. Draw colored vegetable oil (**3**) into the syringe (**Figure 3.2B**), and suck the guar gum–calcium-chloride solution (**2**) into the straw, with the syringe needle attached at the bottom (**Figure 3.2C**).
9. Attach the syringe to the needle in the straw **(8)**.
10. Drop the mixture in the straw (**9**) into the mixed solution of sodium alginate, gum arabic, κ-carrageenan and water (**4**), injecting the colored vegetable oil in the syringe into the center of drop **(9)** (**Figure 3.2D**).
11. Move the drop (**10**) to the cup with water (**5**) using a spoon, and then move it into the cup with the calcium-chloride solution (**6**). Soak for 30 min. (**Figure 3.2E**) then remove (**Figure 3.2F**).

pH of 1 wt% sodium-alginate solution = 8.00

pH of 1 wt% gum arabic solution = 6.16

pH of 1 wt% κ-carrageenan solution = 8.96

pH of 2 wt% calcium-chloride solution = 5.93

pH of 20 wt% calcium-chloride solution = 6.69

pH of the calcium chloride–guar gum mixture = 5.06

Hints

1. The solute dissolves easily in hot water.
2. One g solute plus 99 g water becomes 1 wt% solution, and 2 g solute plus 98 g water and 20 g solute plus 80 g water become 100 g 2 wt% and 20 wt% solution, respectively.

Japanese do not eat salmon eggs (*ikura*) on a daily basis; they are considered a delicacy. *Ikura* is a popular ingredient for *sushi*, and use of artificial *ikura* reduces the cost of the *sushi*.

Onion Rings (Figure 3.3)

Ingredients

(Makes 10–20 rings)
100 g onion
3 g sodium alginate
35 g (a little under ⅓ cup) starch
180 mL (a little under $^3/_4$ cup) soft water (low in calcium)†

A {
16 g calcium chloride
184 mL water
75 g (a little over ⅖ cup) soft wheat flour or all-purpose flour
}

10 g starch‡
2 g ($^1/_2$ tsp) baking powder
100 mL (⅖ cup) milk
Water, vegetable oil for deep-frying, salt, sugar and ketchup for serving.

†IF TAP WATER IS RICH IN CALCIUM (HARD WATER), ADD 0.5 G SODIUM CITRATE TO IT.

‡YOU CAN USE ANY TYPE OF STARCH: WHEAT, CORN, POTATO, TAPIOCA, AND SO FORTH.

Preparation

1. Make 200 g 8 wt% calcium-chloride solution using **ingredients A**.
2. Chop onion into small pieces.
3. Mix starch, sodium alginate, salt and sugar, add water and mix with a hand mixer or a blender (**Figure 3.3A**).
4. Add the chopped onion (**2**) to the sodium-alginate dispersion, and mix thoroughly with a spoon (**Figure 3.3B**).
5. Form the onion mixture into rings (**Figure 3.3C**), soak them in the 8 wt% calcium-chloride solution for a few minutes (**Figure 3.3D**). Remove onion rings from the solution.
6. Mix wheat flour, starch, baking powder, and milk; then use to coat the onion rings (**5**) (**Figure 3.3E**).
7. Heat vegetable oil to 170°C. Deep-fry the coated onion rings (**6**) in hot oil for 5 min. (**Figure 3.3F**).
8. Garnish onion rings with salt, serve with ketchup (**Figure 3.3G**).

pH of starch and sodium-alginate mixture = 6.33
pH of 8 wt% calcium-chloride solution = 6.77
pH of onion rings before dipping in batter (**5**) = 5.71

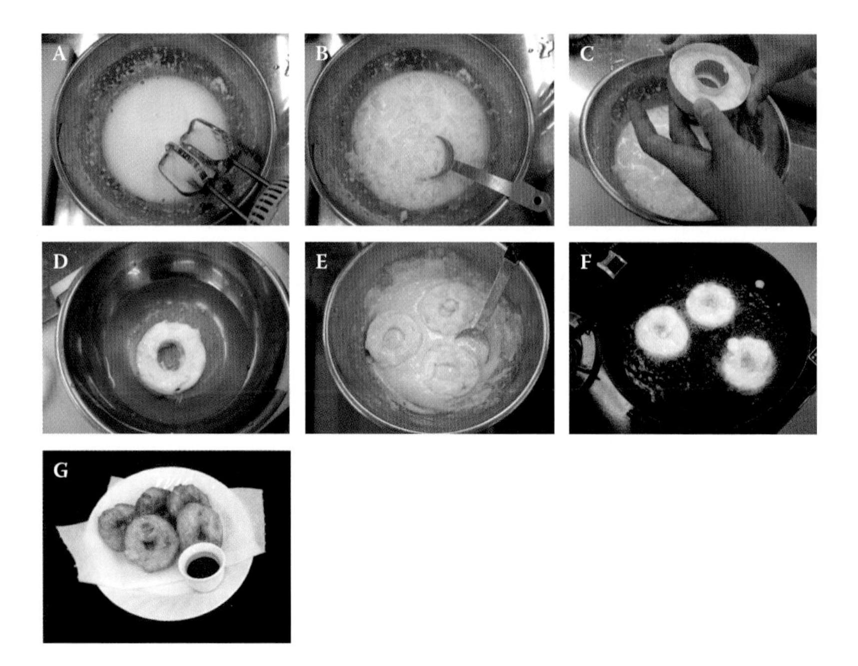

FIGURE 3.3 Onion rings. (**A**) Mix starch, sodium alginate, salt, sugar, and water. (**B**) Add chopped onion to the alginate dispersion. (**C**) Form onion rings. (**D**) Soak onion rings in calcium-chloride solution. (**E**) Coat onion rings with the powder mixture and milk. (**F**) Deep-fry onion rings. (**G**) Garnish and serve.

This recipe is less commonly used when cooking at home. When onions are sliced into rings, the size is nonhomogeneous. For commercial purposes, equal-sized onion rings are required (**Figure 3.4**). This recipe was created so that none of the onion would go to waste.

FIGURE 3.4 Frozen (left) and deep-fried (right) industrial onion rings.

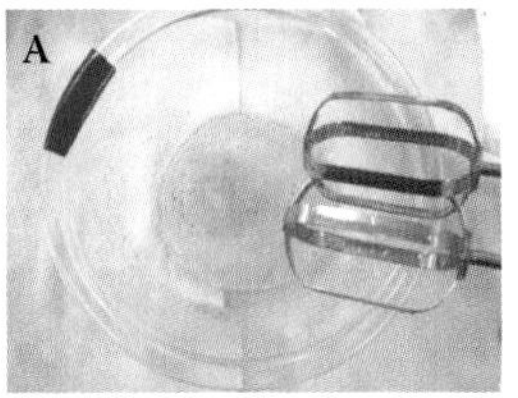

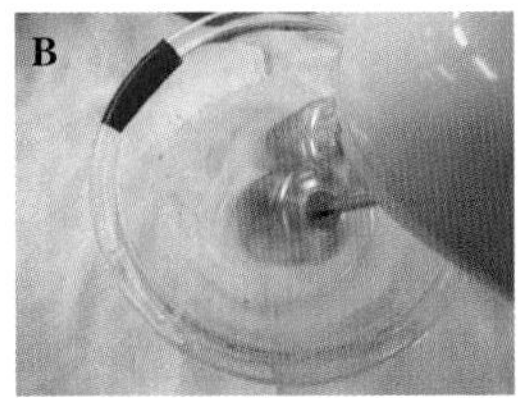

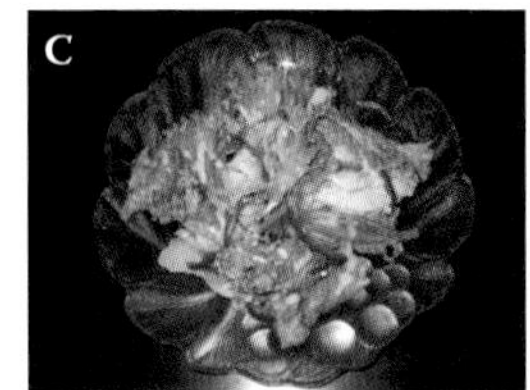

FIGURE 3.5 French dressing. (**A**) Mix sugar, salt, PGA, pepper, and mustard in a bowl with water and vinegar. (**B**) Add oil and blend. (**C**) Dress salad.

French Dressing (Figure 3.5)

Ingredients

(One large salad, serves 4–5)
25 g (2 Tbsp) vegetable oil†
30 mL (2 Tbsp) water
20 mL (1⅓ Tbsp) vinegar‡
5 g (1/2 Tbsp) sugar
2.5 g (1/2 tsp) salt
0.5 g propylene glycol alginate (PGA)
Pepper, mustard

†YOU CAN USE ANY TYPE OF OIL; OLIVE, CORN, SUNFLOWER, AND SO FORTH.

‡YOU CAN USE ANY TYPE OF VINEGAR; CIDER, BALSAMIC, GRAIN, RICE, WINE, AND SO FORTH.

Preparation

1. Mix sugar, salt, PGA, pepper and mustard in a bowl.
2. Add water and vinegar and mix vigorously with a hand mixer or a blender (**Figure 3.5A**).
3. Add oil gradually while stirring, and blend well until the solution emulsifies (**Figure 3.5B**) (**Hint 1**).
4. Pour French dressing over salad (**Figure 3.5C**).

pH = 2.69

FIGURE 3.6 Salad dressing-based product. Color depends on included ingredients.

Hint

1. All ingredients except the oil are mixed together first, and then oil is added gradually for homogeneous emulsification. Your salad dressing will be a success every time.

This recipe is based on the French "vinaigrette," resulting in a salad dressing with a color ranging from white (**Figure 3.6**, right) to brown, depending on vinegar used. French dressing made in the United States might contain tomato purée or tomato ketchup, making it more orange in color (**Figure 3.6,** left). Add tomato purée to this recipe and enjoy this variation on the classic vinaigrette.

Tips for the Amateur Cook and Professional Chef

- A combination of xanthan gum (Chapter 15) and sodium alginate produces flow properties that are intermediate between those of these two components.
- Sodium-alginate solution can be frozen and thawed without any change to its appearance or viscosity upon remelting.
- Addition of increasing amounts of alcohol to an aqueous alginate solution increases alginate solution viscosity and eventually causes precipitation of the alginate.
- PGA does not gel until the pH is below 3.0.
- Algin polymers undergo a cross-linking reaction with most polyvalent cations (except magnesium).
- If calcium is added too rapidly, the result is spot gelation and a discontinuous gel structure. To avoid this, use a sequestering agent and control the pH of the system.
- Enzymes such as protease, cellulase, amylase, and galactomannanase have no effect on the alginate molecule.
- Glycerol and glycols improve the flexibility of alginate films.

- Algin products can be used to improve the quality of frozen foods.
- Flavor release is improved in frozen fruit pies by replacing part of the pre-gelatinized starch with algin.

REFERENCES AND FURTHER READING

Andersen, I. L., Skipnes, O., Smidsrod, O., Ostgaard, K., and P. C. Hemmer. 1977. Some biological functions of matrix components in benthic algae in relation to their chemistry and the composition of seawater. *ACS Symp Ser.* 48:361–81.

Beriain, M. J., Gómez, I., Petri, E., Insausti, K., and M. V. Sarriés. 2011. The effects of olive oil emulsified alginate on the physico-chemical, sensory, microbial, and fatty acid profiles of low-salt, inulin-enriched sausages. *Meat Sci.* 88:189–97.

Boyle, J. L. 1959. The stabilization of ice cream and ice lollies. *Food Technol. Australia* 11:543.

Brownlee, I. A., Allen, A., Pearson, J. P., Dettmar, P. W., Havler, M. E., Autherton, R., and E. Onsoyen. 2005. Alginate as a source of dietary fiber. *Crit. Rev. Food Sci. Nutr.* 45:497–510.

Chavez, M. S., Luna, J. A., and R. L. Garrote. 1994. Cross-linking kinetics of thermally preset alginate gels. *J. Food Sci.* 59:1108–10.

Childs, W. H. 1957. Coated sausage. U.S. Patent No. 2,811,453.

Conti, E., Flaibani, A., Regan, O., and I. W. Sutherland. 1994. Alginate from *Pseudomonas fluorescens* and *P. putida*: Production and properties. *Microbiology* 140:1125–32.

Cottrell, I. W. and P. Kovacs. 1980. Alginates. In *Handbook of water-soluble gums and resins*, ed. R. L. Davidson, chap. 2. New York: McGraw-Hill.

Draget, K. I., Ostgaard, K., and O. Smidsrod. 1991. Homogeneous alginate gels: A technical approach. *Carbohydr. Polym.* 14:159–78.

Draget, K. I., Steinsvag, K., Onsoyen, E., and O. Smidsrod. 1998. Na and K-alginate; effect on Ca^{2+}-gelation. *Carbohydr. Polym.* 35:1–6.

Ensor, S. A., Sofos, J. N., and G. R. Schmidt. 1990. Optimization of algin/calcium binder in restructured beef. *J. Muscle Foods* 1:197–206.

Ernst, E. A., Ensor, S. A., Sofos, J. N., and G. R. Schmidt. 1989. Shelf life of algin/calcium restructured turkey products held under aerobic and anaerobic conditions. *J. Food Sci.* 54:1147–54.

Espevik, T. and G. Skjak-Braek. 1996. Application of alginate gels in biotechnology and biomedicine. *Carbohydr. Eur.* 14:19–25.

General Foods Corp. 1954. Confectionery product. British Patent No. 2,485,043.

Gibsen, K. F. 1957. Alginate composition for making milk puddings. U.S. Patent No. 2,808,337.

Gibsen, K. F. and L. B. Rothe. 1955. Algin, versatile food improver. *Food Eng.* 27:87–9.

Glicksman, M. 1962. Utilization of natural polysaccharide gums in the food industry. *Adv. Food Res.* 11:109–200.

Glicksman, M. 1962. Freezable gels. U.S. Patent No. 3,060,032.

Grant, G. T., Morris, E. R., Rees, D. A., Smith P. J. C., and D. Thom. 1973. Biological interactions between polysaccharides and divalent cations: The egg-box model. *FEBS Lett.* 32:195–8.

Grasdalen, H. 1983. Highfield 1H-NMR spectroscopy of alginate. Sequential structure and linkage conformations. *Carbohydr. Res.* 118:255–60.

Green, H. C. 1936. Fibrous alginic acid. U.S. Patent No. 2,036,934.

Harada, S. and M. Ikeda. 1995. Fried food and process for producing same. European Patent Application, EP 0 603,879 A2.

Hartwig, M. 1960. Method of reducing the swelling capacity of synthetic alginate skins. U.S. Patent No. 2,965,498.

Helgerud, O. and A. Olsen. 1955. A method for the preservation of food. British Patent No. 728,168.

Helgerud, O. and A. Olsen. 1958. Block freezing of foods (fish). U.S. Patent No. 2,763,557.

Henning, W. 1957. Canned herring. German Patent No. 1,004,470.

Hirst, E. L. and D. A. Rees. 1965. The structure of alginic acid. Part V. Isolation and unambiguous characterization of some hydrolysis products of the methylated polysaccharide. *J. Chem. Soc.* 1182–7.

Hunter, A. R. and J. K. Rocks. 1960. Cold milk puddings and method of producing the same. U.S. Patent No. 2,949,366.

Imeson, A. P. 1984. Recovery and utilization of proteins using alginates. In *Gums and stabilizers for the food industry 2*, ed. G. O. Phillips, D. J. Wedlock, and P. A. Williams, 189–201, Oxford: Pergamon Press.

Jenkins, D. J. A., Axelsen, M., Kendall, C. W. C., Augustin, L. S. A., Vuksan, V., and U. Smith. 2000. Dietary fiber, lente carbohydrates, and the insulin-resistant diseases. *Brit. J. Nutr.* 83:S157–63.

Kohler, R. and W. Dierichs. 1958. Stabilizers for frozen desserts. U.S. Patent No. 2,854,340.

Kohler, R. and W. Dierichs. 1959. Product and process for the production of aqueous gels. U.S. Patent No. 2,919,198.

Krefting, A. 1986. An improved method of treating seaweed to obtain valuable products therefrom. British Patent No. 11,538.

Krigsman, J. G. 1957. Alginic acid and the alginates applied to the food industry. *Food Technol. Australia* 9:183–5.

Kunz, C. E. and W. B. Robinson. 1962. Hydrophilic colloids in fruit pie fillings. *Food Technol.* 16:100–2.

Langmaack, L. 1961. Method of producing synthetic sausage casings. U.S. Patent No. 2,973,274.

Lebrun, L., Junter, G. A., Jouenne, T., and L. Mignot. 1994. Exopolysaccharide production by free and immobilized microbial cultures. *Enz. Microb. Technol.* 16:1048–54.

LeGloahec, V. C. E. and J. R. Herter. 1938. Treating seaweed. U.S. Patent No. 2,138,551.

Lelli, A. and P. Ferrero. 1995. Stabilizer for acidic milk drinks. European Patent Office 0639,335 (A1).

Liu, L., Kerry, J. F., and J. P. Kerry. 2007. Application and assessment of extruded edible casings manufactured from pectin and gelatin/sodium alginate blends for use with breakfast pork sausage. *Meat Sci.* 75:196–202.

Mancini, F. and T. H. McHugh. 2000. Fruit-alginate interactions in novel restructured products. *Nahrung* 44:152–7.

McDowell, R. H. 1960. Applications of alginates. *Rev. Pure Appl. Chem.* 10:1–15.

McDowell, R. H. and J. L. Boyle. 1960. Gelling of milk with alginate. British Patent No. 839,767.

McNeely, W. H. 1959. Algin. In *Industrial gums*, ed. R. L. Whistler, 55–82. New York: Academic Press.

Means, W. J. and G. R. Schmidt. 1986. Algin/calcium gel as a raw and cooked binder in structured beef steaks. *J. Food Sci.* 51:60–5.

Means, W. J. and G. R. Schmidt. 1986. Process for preparing algin/calcium gel structured meat products. U.S. Patent No. 4,603,054.

Means, W. J., Clarke, A. D., Sofos, J. N., and G. R. Schmidt. 1987. Binding, sensory and storage properties of algin, calcium-structured beef steaks. *J. Food Sci.* 52:252–62.

Mehyar, G. F., Han, J. H., Holley, R. A., Blank, G., and A. Hydamaka. 2007. Suitability of pea starch and calcium alginate as antimicrobial coatings on chicken skin. *Poultry Sci.* 86:386–93.

Merton, R. R. and R. H. McDowell. 1960. Powdered alginate jellies. U.S. Patent No. 2,930,701; British Patent No. 828,350.

Messina, B. T. and D. Pape. 1966. Ingredient cuts heat-process time. *Food Eng.* 8:48–51.

Miller, A. 1960. Composition and method for improving frozen confections. U.S. Patent No. 2,935,406.

Moncrieff, R. W. 1953. Stabilizing fruit drinks. *Food* 22:498–9.

Mountney, G. J. and A. R. Winter. 1961. The use of a calcium alginate film for coating cut-up poultry. *Poultry Sci.* 40:28.

Neiser, S., Draget, K. I., and O. Smidsrod. 1998. Interactions in bovine serum albumin-calcium alginate gel systems. *Food Hydrocolloids* 13:445–8.

Nishide, E., Mishima, A., Anzai, H., and N. Uchida. 1992. Properties of alginic acid from sulfated polysaccharides extracted from residual algae by the hot-water method. *Bull. College Agric. Vet. Med., Nihon University*, No. 49, 140–2.

Nussinovitch, A. 1997. *Hydrocolloid applications: Gum technology in the food and other industries*. London: Blackie Academic & Professional.

Nussinovitch, A. 2003. *Water-soluble polymer applications in foods*. Oxford, UK: Blackwell Publishing.

Nussinovitch, A. 2010. *Polymer Macro- and micro-gel beads: Fundamentals and applications*. New York: Springer.

Nussinovitch, A., Kopelman, I. J., and S. Mizrahi. 1990. Effect of hydrocolloids and minerals content on the mechanical properties of gels. *Food Hydrocolloids* 4:257–65.

Nussinovitch, A., Peleg, M., and M. D. Normand. 1989. A modified Maxwell and a non-exponential model for characterization of the stress relaxation of agar and alginate gels. *J. Food Sci.* 54:1013–6.

Nussinovitch, A. and M. Peleg. 1990. Strength-time relationships of agar and alginate gels. *J. Texture Studies* 21:51–60.

Nussinovitch, U. and A. Nussinovitch. 2011. Clinical aspects of alginates. In *Biodegradable polymers in clinical use and clinical development*, ed. A. J. Domb, N. Kumar, A. Ezra, 137–184. Hoboken, NJ: John Wiley & Sons.

Oates, C. G., Ledward, D. A., and J. R. Mitchell. 1987. *7th World Congr. Food Sci. Technol.*, Raffles City Singapore 28 Sep–2 Oct.

Ohling, R. A. G. 1959. Food preparation by cold gelation. Canadian Patent No. 574,261.

Olsen, A. 1955. Freezing fish in alginate jelly. *Food Manuf.* 30:267–70, 285.

Onsoyen, E. 1990. Marine hydrocolloids in biotechnological applications. In *Advances in fisheries technology and biotechnology for increased profitability*, paper from the 34th Atlantic Fisheries Technol. Conf. Seafood Biotechnol. Workshop, ed. M. N. Voight and J. R. Botta, 265–86. Lancaster, PA: Technomic Publishing Co. Inc.

Onsoyen, E. 1992. Alginates In *Thickening and gelling agents for food*, ed. I. Imeson, 1–23. Glasgow: Chapman & Hall.

Onsoyen, E. 1996. Commercial applications of alginates. *Carbohydr. Eur.* 14:26–31.

Pelkman, C. L., Navia, J. L., Miller, A. E., and R. J. Pohle. 2007. Novel dietary calcium-gelled alginate-pectin beverage reduced energy intake in nondieting overweight and obese women: Interactions with restraint status. *Am. J. Clin. Nutr.* 86:1595–602.

Peschardt, W.J. S. 1946. Manufacturing artificial edible cherries, soft sheets, and the like. U.S. Patent No. 2,403,547.

Poarch, A. E. and G. W. Tweig. 1957. Gel-forming compositions. U.S. Patent No. 2,809,893.

Pressman, R. 1957. Coatings for meat-wrapping sheets. U.S. Patent No. 2,811,454.
Rees, D. A. 1969. Structure, conformation and mechanism in formation of polysaccharide gels and networks. In *Advances in carbohydrate chemistry and biochemistry*, vol. 24. ed. M. L. Wolform and R. S. Tipson, 267–332. New York: Academic Press.
Rocks, J. K. 1960. Method and composition for preparing cold water desserts. U.S. Patent No. 2,925,343.
Saeki, H. and K. Inoue. 1997. Improved solubility of carp myofibrillar proteins in low ionic strength medium by glycosylation. *J. Agric. Food Chem.* 45:3419–22.
Sato, R., Sawabe, T., Kishimura, H., Hayashi, K., and H. Saeki. 2000. Preparation of neoglycoprotein from carp myofibrillar protein and alginate oligosaccharide: Improved solubility in low ionic strength medium. *J. Agric. Food Chem.* 48:17–21.
Schmidt, G. R. and W. J. Means. 1986. Process for preparing algin/calcium gel structured meat products. U.S. Patent No. 4,603,054.
Shand, P. J., Sofos, J. N., and G. R. Schmidt. 1993. Properties of algin calcium and salt phosphate structured beef rolls with added gums. *J. Food Sci.* 58:1224–30.
Shetty, C. S., Bhaskar, N., Bhandary, M. H., and B. S. Raghunath. 1996. Effect of film-forming gums in the preservation of salted and dried mackerel. *J. Sci. Food Agric.* 70:453–60.
Shiotani, H. and M. Hara. 1955. Ice cream stabilizer. Japanese Patent No. 3031 (*Chem. Abstr.* 51:13263e).
Slepchencnko, I. R., Knizhnik, E. B., and L. A. Piraeva. 1956. Production of calcium alginate films and their utilization in the freezing of meat. *Sbornik Stud. Rabot. Moskov. Teknol. Inst. Myasnoi i Molch. Prom.*, No. 4, 39 (*Chem. Abstr.* 53:13440f).
Smidsrod, O. 1972. Molecular basis for some physical properties of alginates in the gel state. *Faraday Disc. Soc.* 57:263–74.
Stanford, E. C. C. 1883. On algin: A new substance obtained from some of the commoner species of marine algae. *Chem. News* 47:254.
Stanford, E. C. C. 1884. On the economic applications of seaweeds. *J. Soc. Arts* 32:717.
Steiner, A. B. 1948. Algin gel-forming compositions. U.S. Patent No. 2,441,729.
Steiner, A. B. 1949. Alginate ice cream stabilizing composition. U.S. Patent No. 2,485,934.
Steiner, A. B. and W. H. McNeely. 1954. Algin in review. *Adv. Chem.* 11:68–82.
Stokke, B. T., Draget, K. I., Smidsrod, O., Yuguchi, Y., Urakawa, H., and K. Kajiwara. 2000. Small angle X-ray scattering and theological characterization of alginate gels. 1. Ca-alginate gels. *Macromolecules* 33:1853–63.
Strachan, C. C., Moyls, A. W., Atkinson, F. E., and D. Britton. 1960. Commercial canning of fruit pie fillings. *Can. Dept. Agr. Publ.* 1062.
Takeda, H., Iida, T., Okada, A., Ootsuka, H., Ohshita, T., Masutani, E., Katayama, S., and H. Saeki. 2007. Feasibility study on water solubilization of spawned out salmon meat by conjugation with alginate oligosaccharide. *Fisheries Sci.* 73:924–30.
Takeuchi, T., Murata, K., and I. Kusakabe. 1994. A method for depolymerization of alginate using the enzyme system of *Flavobacterium multivolum*. *J. Japanese Soc. Food Sci. Technol.* 41:505–11.
Tapia, M. S., Rojas-Grau, M. A., Carmona, A., Rodriguez, F. J., Soliva-Fortuny, R., and O. Martin-Belloso. 2008. Use of alginate- and gellan-based coatings for improving barrier, texture and nutritional properties of fresh-cut papaya. *Food Hydrocolloids* 22:1493–503.
Toft, K. 1982. Interactions between pectins and alginates. *Prog. Food Nutr. Sci.* 6:89–96.

Toft, K., Grasdalen, H., and O. Smidsrod. 1986. Synergistic gelation of alginates and pectins, in ACS Symposium Series No. 310 *Chemistry and function of pectins*, ed. M. L. Fishman and J. J. Jen. Washington: American Chemical Society.

Trout, G. R. 1989. Color and bind strength of restructured pork chops: Effect of calcium carbonate and sodium alginate concentration. *J. Food Sci.* 54:1466–70.

Trout, G. R., Chen, C. M., and S. Dale. 1990. Effect of calcium carbonate and sodium alginate on the textural characteristics, color and color stability of restructured pork chops. *J. Food Sci.* 55:38–42.

Weingand, R. 1957. Synthetic sausage casings. U.S. Patent No. 2,802,744.

Weingand, R. 1959. Process for producing synthetic sausage casing from alginates. U.S. Patent No. 2,897,547.

Whistler, R. L. and K. W. Kirby. 1959. Composition of alginic acid of *Macrocystis pyrifera*. *Hoppe-Seyler's Z. Physiol. Chem.* 314:46.

Wolff & Co. 1951. Sausage casings. British Patent No. 711,437.

Yilmazer, G., Carrillo, A. R., and J. L. Kokini. 1991. Effect of propylene glycol alginate and xanthan gum on stability of O/W emulsions. *J. Food Sci.* 56:513–7.

Yilmazer, G. and J. L. Kokini. 1992. Effect of salt on the stability of propylene glycol alginate/xanthan gum/polysorbate-60 stabilized oil in water emulsions. *J. Texture Studies* 23:195–213.

Yu, X. L., Li, X. B., Xu, X. L., and G. H. Zhou. 2008. Coating with sodium alginate and its effects on the functional properties and structure of frozen pork. *J. Muscle Foods* 19:333–51.

4

Carrageenan and Furcellaran

Introduction and Historical Background

Carrageenan is an intercellular matrix constituent in many species of red seaweed. The hydrocolloid forms bendable structures inside the alga, which assist in dealing with wave movement and strong currents. Minute amounts of carrageenan are employed to enhance viscosity or gel strength. Chinese evidence from ~600 B.C. shows various utilizations of red seaweed. One application for the cold-water red alga *Chondrus chrispos*, originating in Ireland, is to boil it in milk to obtain a condensed product for consumption. Cooked seaweed as a constituent of the dish "St. Patrick's soup" resulted from the Irish potato famine in the mid-19th century. The dried, bleached plants were introduced to the United States from Ireland, their appearance confirmed in Scituate, Massachusetts, in 1835. The first extraction of carrageenan from *C. chrispos,* in 1844, is credited to C. Schmidt. This was followed by a patentable extraction procedure using alcohol precipitation. The consequently accelerated commercial manufacture of these extracts brought about a profitable subsistence for numerous seaweed and gum sellers.

During World War II, the hunt for agar replacements was key to stimulating the commercialization of a new seaweed extract, furcellaran, with serendipity playing a crucial role in the expansion of this gum. It is said that a barber in Denmark was investigating a variety of seaweeds for use as a fluid permanent-wave product. He found that the seaweed *Furcellaria fastigiata* provides a slippery, tacky substance with fine viscosity-forming properties. This finding shaped the foundation of *Danish agar*, as the hydrocolloid was called. The seaweed *F. fastigiata* was growing in Danish, Norwegian, and Baltic waters. In 1960, the employees of the Estonian Laboratory of Marine Ichthyology determined that the seaweed *Furcellaria lumbricalis*, growing in the Baltic Sea, could be utilized to manufacture furcellaran. In 1964, the confectionery manufacturer Kalev initiated work on a technology for its extraction. In 1966, the foundation for producing furcellaran in Saaremaa, Kärla, was completed, and the preliminary output quantity of 40 tons was used by the confectionery industry. In 1989, Kalev's monopoly ended and Est-Agar became a small enterprise. In 1997, Est-Agar AS was created and all shares were bought by AS Saare Kalur. Today, furcellaran is extracted from the red seaweed *F. lumbricalis* and, to the best of our knowledge, Est-Agar AS is the world's only producer of this unique gum.

Structure

The structure of carrageenan was revealed step by step, beginning with a determination of the presence of D-galactose, then confirmation of the presence of ester-sulfate groups and the assumption that carrageenan is made up of two fractions. Key steps were taken to isolate and separate these two fractions (κ and λ): only κ-carrageenan was responsive to potassium-salt precipitation. Additional research and classification of these fractions continued. In 1967, the structure of a third carrageenan (ι) was revealed. Red seaweeds contain a family of gums composed of agar (**Chapter 2**), furcellaran, and the three kinds of carrageenan (κ, ι, and λ). Carrageenans are linear polysaccharides made of alternating β-1,3- and α-1,4-linked galactose residues. The basic repeating unit of the carrageenans is the disaccharide carrabiose. The 1,4-linked residues are usually, but not invariably, present as 3,6-anhydride. Differences in this basic structure can result from substitutions (either anionic or nonionic) on the hydroxyl groups of the sugar residues, and from the absence of a 3,6-ether linkage.

Carrageenans are exceedingly sulfated. Carrageenans extracted from *Gigartina* species occasionally contain pyruvate residues. Sulfated galactans from the family Grateloupiaceae may contain methoxyl groups, even though their alternating carrageenan structure is uncertain. Fractionation of the carrageenan from *C. chrispos* with potassium chloride results in the separation of κ- and λ-carrageenan: κ-carrageenan is precipitated by the potassium chloride, while the λ fraction remains in solution. In these fractions, half of the sugar units in κ-carrageenan were 3,6-anhydro-D-galactose, whereas λ-carrageenan contained little or none of this sugar. Furcellaran is an anionic, partially sulfated polysaccharide. It is composed of D-galactose, 3,6-anhydro-D-galactose, and D-galactose-4-sulfate. The structure of furcellaran is comparable to that of κ-carrageenan and has been described as a hybrid of κ/β-carrageenan complex. The crucial dissimilarity is that κ-carrageenan has one sulfate ester residue per two sugars, while furcellaran has one sulfate ester residue per three or four sugar residues.

Sources and Production

Carrageenan-yielding species occur in no less than seven families. The most important red seaweed sources for carrageenan are *C. crispos*, a small cold-water seaweed producing κ and λ types; the warm-water *Eucheuma* species producing κ and λ types, and the large cold-water *Gigartina* species from which κ- and λ-carrageenans are manufactured. The cold-water seaweeds are cropped yearly, whereas the warm-water seaweeds are cultivated every 3 months. Carrageenan-yielding algae are cultivated in the Philippines, Indonesia, Canada, the United States, France, Korea, Spain, Portugal, Morocco, Mexico, Chile, Denmark, and Brazil.

The value of the carrageenan depends upon the time and conditions under which the seaweed is cropped. Its fast drying to a suitable moisture content optimizes its capacity for preservation. Following transport, the dried seaweed is stocked in storehouses awaiting extraction. The seaweed that is destined for manufacture is examined and batches are chosen to attain the required product. Combination is essential to attaining consistent features, with a particular producer promoting ~200–300 blends. To manufacture the hydrocolloid, the carrageenan, which is present in a gel-like state in the alga at ambient temperature, must be heated in water to a temperature above its gel's melting point. Prior to extraction, the seaweed must be cleaned of sand. Alkaline conditions are required to lessen or eliminate acid-catalyzed depolymerization processes or modification of the galactan structure, resulting in enhanced gelling capability. In addition to different extraction techniques, each of these can have several modifications.

Three processes are most frequently employed: alcohol precipitation, freeze thawing, and gel press. In the alcohol process, the seaweed is washed to get rid of dirt and impurities. Extraction is then carried out with dilute alkali for several hours. Residues are separated out by centrifugation, followed by filtration through porous silica or activated charcoal to produce a 1–2% carrageenan solution. Evaporators are used to concentrate the carrageenan extracts 1.5- to 2-fold and isopropyl alcohol is utilized to get a coagulant, which is pressed dry by removing the highest potential quantity of excess fluid. The precipitated carrageenan is dried in a steam-heated drier, followed by particle-size reduction to the required state. The additional two processes (freeze thawing and gel press) are analogous to the first, apart from the means of recovering the carrageenan from solution. In the freeze-thaw method, following concentration by evaporation, the material is extruded through spinnerets into a cold solution of potassium chloride. The gelled threads are further dewatered by successive potassium-chloride washes then pressing. The gel goes through freezing, thawing, chopping, and rewashing in a potassium-chloride solution followed by air-drying. Since potassium chloride is employed, the procedure is specific for furcellaran and κ-carrageenan, both of which gel with potassium and display noticeable syneresis. The gel-press method skips the freeze-thaw cycle and makes use of pressure to dewater the gel.

The seaweed growing near the shores of Saaremaa is probably the very last resource of *F. lumbricalis* that is technologically functional. The amount of seaweed gathered in a given year is decided at the Estonian Marine Institute and on average, it is far below the maximum possible level so as not to harm the resource by harvesting more than the natural reproduction rate can support. In addition to the commercialized harvest, local people gather seaweed that storms have cast ashore, dry it in the sun, and sell it to the extracting factory.

In Est-Agar AS, furcellaran is extracted from raw *F. lumbricalis* and the liquid extract is purified by filtration. The liquid extract can be converted into flaky furcellaran by evaporation of water via drum drying. Appropriate release of the dried material from the dryer roll involves addition of a minute quantity of roll-stripping agents. The furcellaran can also be isolated from the liquid extract by precipitation with potassium chloride. This process gives a pure and concentrated product.

Accessible Types of Gum

Carrageenans are anionic polymers (they include half-ester sulfate groups). The free acid is unstable, and carrageenans are usually sold as a blend of sodium, potassium, and calcium salts. Marketable carrageenans are frequently sold in κ, ι, or λ form. In reality, each form contains varying quantities of the other two forms. For the purpose of gaining viscosity, λ-carrageenan is employed, while for gelling purposes, blended calcium and potassium salts of the κ and ι types are utilized.

Regulatory Aspects

In the United States, carrageenan is generally regarded as safe (GRAS) (21CFR 182.7255) and is endorsed as a food additive (21 CFR 172.620) by the Food and Drug Administration (FDA). In Europe, carrageenan is distinguished for its safety and value (having the designation E407). Given that carrageenan is a "natural product," in Japan it is not subject to food-additive regulations, aside from those of good manufacturing practice (GMP). In 1984, technologists from around the world validated the use of carrageenans in foods, with no need to specify acceptable daily intake (ADI). Nevertheless, the safety of carrageenan in the presence of acid has been questioned, owing to its depolymerization under those conditions. Furcellaran, the anionic partially sulfated polysaccharide, is classified together with carrageenan (E407).

Molecular Weight and Consistency

Carrageenan is commonly regarded as a high-molecular-weight linear polysaccharide. It is composed of repeating galactose units and 3,6-anhydrogalactose, both sulfated and nonsulfated, linked by alternating α-(1-3)- and β-(1-4)-glycosidic linkages. Marketable food-grade carrageenans usually have number-average molecular weights of ~200–400 x 10^3. Under 10^5 Daltons, the functionality of the carrageenans is for the most part lost. To market a comparatively standardized carrageenan, the raw materials need to be mixed together and adjusted by adding sugars to provide unvarying values of gel strength in water and milk, and invariable viscosities and stabilizing efficiencies in milk. After mixing, carrageenans are packaged and sold in polyethylene-lined fiber drums. Safety measures are taken to diminish caking of the powder as a result of moisture absorbance.

Solutions and Gels

Properties

κ-carrageenan dispersed in cold water swells to a great extent (except the sodium salt), but only barely dissolves. The degree of swelling depends on the process. The dissolution temperature is variable, depending on hydrocolloid concentration and on the cations carried with it, but the average dissolution temperature is ~49°C. Cold dispersions of ι-carrageenan swell and take on thixotropic properties. Heat is required for their dissolution. λ-carrageenan is entirely soluble in cold water, despite the cations associated with it. Organic solvents, for example, methanol, ethanol, acetone, and glycerin, slow the solubility of the carrageenans with adequate quantities of salts, which in turn can even prevent hydration and dissolution of κ- and λ-carrageenans. ι-carrageenan is unique in its ability to dissolve in hot salt solutions, but is less soluble in sugar solutions than κ- and λ-carrageenans. A highly viscous solution results from carrageenan dissolution. The viscosity is an outcome of its linear macromolecular structure and polyelectrolytic nature. Mutual repulsion among sulfate groups along the polymer chain causes the molecule to extend, and its hydrophilicity causes it to be enclosed by a sheath of immobilized water molecules. Both factors contribute to resistance to flow. Viscosity depends on concentration, temperature, the presence of other solutes, the molecular weight of the carrageenan and its type.

Preparation

Throughout the preparation of carrageenan solutions, measures are taken to reduce clumping: cold water and milk are employed for dispersion prior to heating; if equipment, for example a high-shear mixer, is utilized, the fine particles are poured gradually into the vortex; the carrageenan powder is mixed together with the sugar prior to its dispersal in water; sometimes fluids, for instance alcohols, are used to wet the carrageenan powder prior to its dispersal in water. Carrageenans are soluble in water at temperatures above 75°C. Regular mixing equipment can be utilized to deal with solutions of ~10% soluble carrageenan. Sodium salts of κ- and ι-carrageenan are soluble in cold water, whereas with calcium and potassium they display a variable degree of swelling and do not dissolve entirely. As previously noted, λ-carrageenan is completely soluble in cold water.

Furcellaran is soluble in pure water at temperatures higher than 75°C. The sodium salt of furcellaran is soluble in cold water. In milk, furcellaran is soluble at temperatures > 75°C, whereas in a sugar solution, it is soluble at temperatures > 90°C. Furcellaran is insoluble in salt solutions.

Viscosity

Commercial carrageenan solutions are accessible in viscosities ranging from ~5–800 mPa.s, as determined at 75°C and a concentration of 1.5% (w/w). If the determined viscosities are lower than 100 mPa.s, their flow properties are very close to Newtonian. Carrageenan solutions usually display pseudoplastic (shear-thinning) properties. In other words, with solutions it is essential to state the shear rate at which the viscosity measurements have been taken, and this rate should be suited to the precise function of that solution. The pseudoplastic performance of carrageenan can be portrayed by a power-law equation. As a general rule, viscosity amplifies almost exponentially with concentration. Such performance is characteristic of carrageenan and other linear charged polymers as an outcome of the interactions between such polymer chains. The presence of salts decreases the viscosity of carrageenan (ionic macromolecule) solutions due to a decline in electrostatic repulsion. Carrageenans, for example κ and ι types, may gel in the presence of potassium and calcium. On the whole, the higher the temperature, the lower the viscosity of the solution, and safety measures need to be taken to avoid thermal degradation of the polymers. At temperatures higher than the gelling point, both gelling and non-gelling solutions perform similarly. Nevertheless, in gelling carrageenans, if ions are employed as gel promoters, then at cooling there is an increase in the measured apparent viscosity.

Effect of Molecular Weight

The relationship between apparent viscosity and average molecular weight complies with the Mark–Houwink equation. Depolymerization can occur by acid-catalyzed hydrolysis, and this is related to 3,6-anhydride content. Usually, gelled hydrocolloids are more stable to acid than those in the sol state. The secondary and tertiary structures developed upon gelation may shield the glycosidic bonds from attack, enabling the use of carrageenans in low-pH systems, particularly if sufficient potassium chloride or additional potassium salt is present to preserve the gel state. Carrageenans are typically stable at around pH 9. Depolymerization depends on hydrogen-ion activity, and temperatures and rates can be approximated by simple mathematical relationships.

Gel Preparation and Mechanical Properties

Gel formation may be associated with crystallization or precipitation from solution. κ- and ι-carrageenans require heat for dissolution. Following cooling, and in the presence of positively charged ions (e.g., potassium or calcium), gelation occurs. Gel preparation involves dispersing the carrageenan plus salt (potassium chloride) in distilled water and heating to ~80°C, then adjoining the water evaporated throughout the heating process and little by little cooling to ambient temperature to induce gelation. Gel strength can be determined with a Universal Testing Machine. κ-carrageenan generates strong, rigid gels exhibiting some syneresis. It forms helices with K^+, which causes the helices to aggregate and the gel to contract, and to be easily broken. The manufactured gel is somewhat cloudy, but becomes clear with sugar. ι-carrageenan produces elastic gels. It forms helices with Ca^{2+} and limited aggregation contributes to elasticity, with no syneresis. The gel is clear and stable to freeze-thawing.

The association of molecular chains into double helices was proposed as a method of gel formation. Confirmation of double helices was provided by X-ray diffraction studies and by dimerization, which occurred throughout the cooling of solutions containing carrageenan segments that were too short to shape an extended network. At temperatures higher than the gelling point of the sol, the polymer exists in solution as random coils. The domain model provides a description of the gelation process: gelation is suggested to occur when domains aggregate to structure a three-dimensional (3D) network. κ- and ι-carrageenans require potassium or other gel-promoting cations to produce a 3D network. The presence of potassium, which has a preferred radius in the hydrated state as compared to bulkier cations, leads to decreased repulsion between the chains. An additional model based on a study using X-ray diffraction suggested an "egg-box" structure in which cations of preferred sizes lock neighboring chains into an aggregate. Nevertheless, this assumed model of carrageenan's primary structure is currently considered incorrect. The domain model proposes that cations are locked together in helical regions of adjacent domains to build up a network. This is looked upon as a revival of a previous idea, but on a tertiary, more than a secondary, level. Another model of carrageenan gelation assumes cation-induced aggregation of single helices. The presence of 6-sulfated 1,4-linked residues in the polymer chain of κ- or ι-carrageenan or furcellaran has been shown to detract from the strength of their gels. This is a consequence of kinks in the chain that slow down the structuring of double helices. These kinks can be eliminated by conversion of 6-sulfated residues into 3,6-anhydride by alkaline treatment, the melting temperatures of carrageenan gels being higher than their gelling temperatures. This is a function of charge density and is termed hysteresis. It has been studied by optical rotation and further elucidated by thermodynamics involving terms of a free energy surface with two minima.

The effect of gum concentration on gel yield stress and deformability modulus was evaluated for carrageenan, agar (**Chapter 2**), and alginate (**Chapter 3**) gels. Carrageenan was distinguished by maximal yield stress for a particular hydrocolloid concentration. Increasing the gum concentration beyond that point reduced the yield stress. Maximum yield stress was achieved for carrageenan gels at ~1.5% potassium chloride. Maximum strength was acquired at a similar potassium-chloride concentration for 1, 2, and 3% carrageenan gels. The force-deformation data can serve as an index for carrageenan gel strength and observed texture. The suggested method can even be used for carrageenan gels that do not break under the usual compression tests, for instance the unyielding gel manufactured from 2% κ-carrageenan, 2% locust bean gum (LBG) (**Chapter 8**), and 0.2% potassium chloride. Studies on the flexibility of carrageenan gels, and the association between their recoverable work and asymptotic relaxation modulus can be found elsewhere.

Furcellaran produces strong and brittle gels. Addition of sugar changes the gel texture from brittle to more elastic. The gel is thermoreversible. The gelling temperature increases with increasing gum concentration and content of Na^+ and sugar. The gel strength increases in parallel to an increase in pH and in concentrations of K^+ and Ca^+.

Reactivity with Proteins

Carrageenans are covalently bound to protein moieties in the intercellular matrix. These proteins can be removed by alkaline extraction of the polysaccharide from the algae. Electrostatic interactions between negatively charged carrageenans and positively charged ions on proteins (as in milk) can occur. Consequently, carrageenans can be regarded as stabilizers for proteinaceous systems including milk, protein precipitants, enzyme inhibitors, blood anticoagulants, and lipemia-clearing agents. The involved interactions can be specific or nonspecific. One of the characteristics of carrageenans' reactivity with milk is gelation. This takes place because κ-casein and carrageenan interact to form a complex, which aggregates into a 3D gel network. Specific interactions with κ-casein occur at pH levels higher than the isoelectric point of the protein. Nonspecific interactions can occur below this value. Carrageenans may interact with αs1 and β caseins. However, binding is much weaker than with κ-casein, and does not result in gel construction. κ-Casein isolated from a milk-protein system occurs as independent spherical aggregates (submicelles), which are about 20 nm in diameter. When carrageenan is present, they align into ~20-nm wide thread-like structures. In comparison, the carrageenan chains are ~2 nm wide. The κ-casein particles are connected to the hydrocolloid chain by electrostatic attraction, mainly with κ- and ι-carrageenans. Those carrageenans (but not λ-carrageenan) are able to suspend cocoa particles in chocolate milk, possibly because the formation of a gel network is required, in parallel to the fact that suspension ability holds only for high-molecular-weight carrageenans. κ-casein makes up 8–15% of the casein micelle. A fraction of the outer surface of the micelle is exposed and can interact with carrageenan. The ultrastructure of carrageenan–milk sols and gels has been reported. Detailed information on the interaction of sulfated polysaccharides with proteins can be acquired elsewhere. Synergistic effects of carrageenan are reported in **Chapter 16.**

Applications

Carrageenans are employed as binders and gelling, thickening, and stabilizing agents. Characteristic dairy applications for carrageenans are in milk gels, pasteurized and sterilized milk products, puddings and pie fillings, whipped products, cold-prepared milks, acidified milks, frozen desserts, and infant formulas. These hydrocolloids constitute one of the ingredients of ready-made preparations, or are part of a powder to be added to milk. Brittle gels or creamier products can be achieved, depending on the combination of κ- and ι-carrageenans and their blends with polyphosphates. In chocolate milks, 0.025–0.035% carrageenan is included prior to pasteurization and cooling, and the interactions create a delicate structure, which keeps the cocoa particles suspended; in addition, the drink is richer and has improved mouthfeel. In sherbets and ice creams, carrageenans are used as stabilizers to prevent the separation of fat or additional solids contained in the products. The presence of κ-carrageenan in ice-cream mixes and frozen ice cream improves the stabilizing effect under frozen storage conditions. The ability of κ-carrageenan to interact with caseins, leading to gelation, seems to be the main reason for its observed cryoprotection. The use of cellulose gum (**Chapter 5**) and carrageenan, either alone or in a mixture, is crucial for decreasing fat and calories in confections.

In the food industry, the uses of furcellaran fall into two main categories: milk-based products—milk puddings (flans), chocolate milks, custards, and various desserts; and water-based products—flan jelly, icing, piping jelly for cake decoration, wine cut jelly, marmalade and jam, fruit juices, confectionery, and meat products. Furcellaran can be used as a stabilizing, thickening, and gelling agent in the food, agricultural, cosmetics, and pharmaceutical industries. Furcellaran provides thermoreversible texture and is a medium to strong gelling polysaccharide for applications such as baked goods, confectionery, marmalade, jam, yoghurt, meat, and water- and milk-based desserts. Furcellaran mixed with other gums provides a unique soft and shiny texture, a medium–strong gel, creaminess, smoothness, homogeneity, a smooth cutting edge, outstanding mouthfeel, and superior flavor release. Suggested levels of furcellaran for use in different applications are 0.5–1.5% for baked goods, water-based desserts, and jams; 1.0–1.5% for inclusion in pudding, and 2.0–2.5% for marmalade candies.

Milk Applications

Carrageenans are employed as a dry component in powders for inclusion in milk and milk-manufactured goods. In milkshakes and instant-breakfast powders, carrageenan is used to suspend the constituents, making the beverage richer, with added body. A fine-mesh λ-carrageenan is employed for quick cold solubility. Creamy flans and custards are obtained by including κ-carrageenan. Mixtures of κ- and ι-carrageenan with tetrasodium pyrophosphate assist not only in inducing gelation but also in enhancing product firmness and creaminess. λ-carrageenan is preferred for cold-prepared mixes, and in blends with ι-carrageenan and tetrasodium pyrophosphate for syneresis prevention or decrease. The role of LBG (**Chapter 8**) and λ-carrageenan mixtures on the stability of whipped dairy creams in freezing-thawing processes was analyzed as a function of gum concentration. The gum's cryoprotective effect was evident through a notable reduction in changes in elastic properties of whipped creams during freezing and frozen storage, especially for λ-carrageenan concentrations greater than 0.085%, regardless of LBG level. This could suggest a special cryoprotective role for λ-carrageenan in dairy systems. In Dutch desserts, a blend of starch (**Chapter 14**) and carrageenan is employed to produce an enjoyable consistency following extended periods of storage. In addition, to develop strictly vegetarian desserts based on gelled systems with required physical structure and perceived texture, it is important to control the properties of the biopolymer mixtures and understand their behavior under different physicochemical conditions. Use of κ-carrageenan as an alternative to a starch–carrageenan blend results in a more consistent set. In chocolate-milk formulations consisting of about 1% cocoa and 6% sugar, vanillin is added as a flavor modifier, and carrageenan reacts with κ-casein following pasteurization and throughout cooling to create a stable suspension, producing a tastier beverage. The addition of carrageenans at a level of 0.025–0.040% is sufficient to suspend calcium phosphate in chocolate milks. The gum inclusion restrains whey-off without imparting any palpable gel structure or flavor to the milk.

Being related to carrageenan, furcellaran has good stabilizing properties in chocolate milk at concentrations of 0.05% or less. In ice creams and sherbets, which can be classified to a certain extent as frozen foams, LBG, guar gum (**Chapter 8**) and carboxymethyl cellulose (CMC) (**Chapter 5**) are employed as key stabilizers, that is, to avoid the separation or irregular distribution of fat and other solids and to prevent or reduce the formation of ice and lactose crystals. A stabilizer composed of 10–30% (w/w) guar gum, 5–20% (w/w) carrageenan, and/or xanthan gum (**Chapter 15**) and 60–80% (w/w) of an emulsifier chosen among fatty acid mono- or diglycerides, lactic or citric acid esters of fatty acid mono- or diglycerides, sorbitan mono- or tristearates, and lecithins, has been claimed to allow the manufacture of a pourable, aerated dairy dessert with a shelf life of 3 weeks without sagging or separating out.

When κ-carrageenan is employed as a secondary stabilizer, it can prevent whey-off throughout storage—before the mixture has been frozen, and afterward when it goes through a freeze-thaw cycle. Whole milk has an improved appearance and oral texture if low levels of carrageenan are included as emulsified fat stabilizers. Carrageenans (as little as 0.005% κ-carrageenan) have also been employed in sterilized and evaporated milks to avoid fat and protein disjointing. In products that are diluted with water before eating, for example, infant formulas, κ-carrageenan, modified to some extent with alkali, is used to prevent separation of fat and protein. In cottage-cheese manufacture, carrageenan and LBG are utilized to prevent separation and produce a curd with "cling." In the manufacture of hard cheeses, carrageenan produces better yield by affecting the precipitation of whey proteins. Whey proteins are coagulated by food-grade phosphates or κ-carrageenan and the aggregates are incorporated into casein coagulates for cheese manufacture. In imitation cheeses, ~2.5% κ-carrageenan is utilized as a binder (gelling agent) of water, hydrogenated fat, and sodium caseinate. It also serves to improve sliceability and shredability. For customary cheeses, a dry mixture of carrageenans and microcrystalline cellulose (**Chapter 5**) is dispersed in a portion of the milk prior to pasteurization. Use of calcium chloride and an amplified quantity of rennet is recommended, as is doubling the quantity of the starter culture. Annatto and lipolytic enzymes can be added to improve color and flavor. Inclusion of carrageenan or gellan gum (**Chapter 10**) at 250 and 500 mg/kg increased yields (fresh weight) of cheddar cheese by 4 to 17% as compared to traditional methods without gum. Gellan gum gave higher yields than carrageenan. Cheese treated with carrageenan included higher levels of nitrogen than that with or without added gellan gum. Carrageenan caused decreases of 10 and 22–26% in total solids and nitrogen, in that order, relative to controls. New frozen yoghurt has been claimed to have the technological characteristics of ice cream and the qualities of yoghurt, including freshness, viable lactic acid bacteria, and a distinctive flavor. This product was prepared from fresh, semiskim, pasteurized, and homogenized milk, to which selected starter cultures, sugar, guar gum, thickening agents, carrageenan, and mono- and diglycerides of fatty acids had been added. An alternative gelling/stabilizing system for better mousse texture was developed, based on pectin (**Chapter 13**), starch, and carrageenan.

Furcellaran is soluble in hot milk and upon cooling, the solution sets into a smooth and glassy product. The largest use of furcellaran is for blanc manger pudding, which is produced by mixing powders of furcellaran, sugar, flavor, color and, if desired, other ingredients in cold milk or cream and stirring while heating. No boiling is required to produce this dish. It is important to note that the American consumer prefers milk products that are based on cooked starch. In milk pudding, furcellaran can be utilized in combination with starch. If used without starch, the product tends to synerese while a small amount of added starch (3–5 g starch per pint) will eliminate this problem.

Pudding desserts have also been prepared with κ-carrageenan, skim milk powder, native maize starch, sucrose, and water. A mixture design was used to study the effects of varying concentrations of carrageenan, skim milk powder, and starch, while the amounts of water and sucrose were kept constant. Mixtures were heated for 20 min at 90°C. In the defined concentration range, the exclusion effect of starch had an important influence on the dessert's properties. This effect was more pronounced than the effect of milk protein concentration.

Water Applications

Characteristic water applications for carrageenans are water gelation, fat stabilization, as thickening agents and for suspension and bodying purposes. ι-carrageenan is employed to imitate gelatin gels (**Chapter 9**), but without their high melting temperature. The consistency of carrageenan-based gels can be improved by adding agar (**Chapter 2**) and LBG (**Chapter 8**) to the formulations. Techniques to clarify LBG have been developed to preserve the transparent appearance of carrageenan gels. To overcome the insolubility in cold water, sodium salts of κ- and ι-carrageenans, which can dissolve in the cold, are employed. As an alternative to the traditional fruit jellies based on sugar and pectin recipes, mixtures of κ- and ι-carrageenan and noncaloric sweetener are employed to build a low-calorie product. To reduce acid hydrolysis, the acid is added as late in the process as possible. Sorbet with a smooth texture can be achieved using a mix of carrageenan, LBG, and guar gum (**Chapter 8**) or pectin (**Chapter 13**). The carrageenan-to-sorbet ratio should be optimized to reduce the potential for flavor masking (at an excessively high ratio) or the manufacture of a grainy product (excessively low amount of additives). In a blend with guar gum, κ-carrageenan is utilized to produce pet foods. Seaweed flour is used to replace extracted carrageenans in such preparations. κ- or ι- carrageenans offer consistency, gloss, and advanced adhesiveness to both sauces and relishes. In reduced-calorie salad dressings, carrageenans are utilized to impart the oral sensation of a high-oil system; for high-protein salad dressings further protective colloids are needed. In a mixture with xanthan gum (**Chapter 15**), carrageenan stabilizes oil-in-water emulsions, or decreased-calorie spreads. Additional water applications include pie fillings. When added with sodium chloride, sweetness is diminished in κ-carrageenan systems, probably due to the content of endogenous cations (Ca^{2+}, K^+, and Na^+) controlling Na^+ mobility. The sweetest systems, including lactose and/or xanthan gum, undergo maximum flavor enhancement by NaCl. A quite unique application of carrageenan in mashed potatoes includes the addition of blends of κ-carrageenan and xanthan gum. When blends (each biopolymer at 1.5 g/kg) were included in the formulation, the product exhibited very acceptable sensory quality. Carrageenan provided the appropriate texture, while xanthan gum imparted creaminess and good mouthfeel to the product.

Meat and Fish Applications

Carrageenans (κ and ι) and their blends with LBG (**Chapter 8**) were utilized in gefilte fish (a dish made from a mixture of poached, ground deboned fish, such as carp, whitefish and/or pike) to reach the desired taste and consistency. A mixture of κ-carrageenan, LBG, and potassium chloride is employed to coat fish in order to prevent freezer burn and mechanical disintegration during manufacturing. Mixing fish gelatin (**Chapter 9**) and marine polysaccharides leads to products with improved gel strength, and gelling and melting temperatures. Complexes of fish gelatin and κ-carrageenan at 60°C are probably stabilized by electrostatic interactions. κ-carrageenan is employed in sausages to reduce their fat content without affecting their flavor. Addition of carrageenan to turkey meat sausages causes a decrease in emulsion stability, and an increase in water-holding capacity, hardness, and cohesiveness of the formulated sausages. Carrageenan's presence has no significant effect on sausage taste; however, it improves sausage appearance and texture. Blends of high-gelling whey protein concentrates and carrageenan, in combination with dry additions of tapioca starch (**Chapter 14**), improved the final texture of low-fat sausages. Addition of those fat-replacing ingredients as a preformed gel resulted in a more homogeneous, juicier product, similar to the full-fat commercial control. This was achieved through efficient water entrapment and the gel's ability to produce a less dense texture in the low-fat product. In the manufacture of ham and throughout pumping and tumbling, κ-carrageenan acts together with protein (**Chapter 16**) to attach free water, thereby preserving moisture and soluble solids content. The pumping solution is composed of brine, wherein gums are dispersed following salt dissolution. Gum dissolution takes place throughout cooking. Treatments utilizing κ-carrageenan and starch in hams showed that the incorporation of carrageenan increases yield, decreases purge, and results in a sensory perception of reduced juiciness. Increasing starch increased the perception of juiciness. There was no synergistic effect on moisture retention by a combination of starch and carrageenan. Poultry processing suffers from several difficulties, for example protein denaturation, textural alterations, moisture loss, fat oxidation, breakdown of meat texture, and so forth. These problems can be lessened by including phosphates, salts, starches, proteins, and carrageenan in the brine solution, which is then incorporated into the meat by injection or tumbling. Low-fat ground-beef patties had high-quality taste, good consistency, and high satisfaction scores relative to control patties (ground beef and carrageenan alone), when a blend of κ- and ι-carrageenans having different viscosities and gelling characteristics were used jointly with other water binders taken from: xanthan gum/LBG, pea flour, alginate (**Chapter 3**) or modified starch. Thus, nonmeat components aided in manufacturing a better low-fat beef product. The utilization of carrageenan, starch, and milk-soy proteins in low-fat, high-added-water bologna has also been reported to improve its overall acceptability.

Recipes with Carrageenan and Furcellaran

Milk Pudding (Figure 4.1)

Ingredients

(Serves 5)

1.5–2.0 g (1 tsp) κ-carrageenan powder†

300 mL (1⅕ cups) water

30 g (3⅓ Tbsp) white sugar

200 mL (⅘ cup) milk

†FOR ι-CARRAGEENAN OR FURCELLARAN, USE 4.0–5.0 G OR 2.0–4.0 G, RESPECTIVELY.

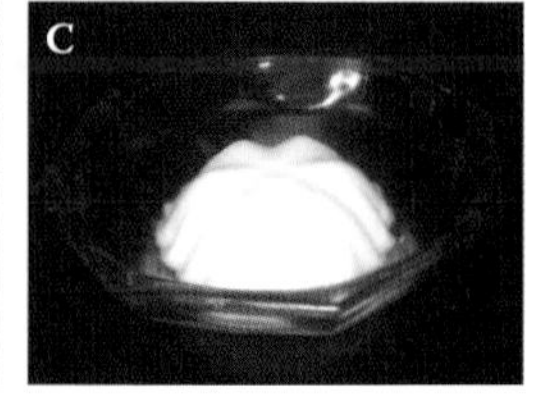

FIGURE 4.1 Milk pudding. (**A**) Heat the carrageenan and sugar in water until fully dissolved. (**B**) Add lukewarm milk to the carrageenan solution. (**C**) Cool until set and serve.

Preparation

1. Mix carrageenan and white sugar in a pan, and add water.
2. Heat the carrageenan dispersion at 80–90°C until fully dissolved (**Figure 4.1A**), then remove from heat.
3. Add lukewarm milk (30–40°C) to the carrageenan solution (**Figure 4.1B**) (**Hint 1**).
4. Pour the solution into moistened molds (**Hint 2**), and cool at less than 20°C until it sets (**Figure 4.1C**).

pH = 6.65–6.85

Hints

1. If you pour cold milk into the carrageenan solution, the jelly will set only partially.
2. Moistening the molds makes it easier to remove the jelly once it has set.

This recipe has a smooth texture; you can change the texture by changing the amounts of κ-carrageenan, ι-carrageenan, or furcellaran.

Flan (Figure 4.2)

Ingredients

(Serves 5)
1 g (½ tsp) κ-carrageenan powder†
50 g (a little under ⅓ cup) white sugar
445 mL (1¾ cups) milk
Yellow food coloring, vanilla flavoring (optional)

†FOR ι-CARRAGEENAN OR FURCELLARAN, USE 3.0 G OR 1.5 G, RESPECTIVELY.

Preparation

1. Mix carrageenan and white sugar in a pan (**Figure 4.2A**), and add milk to it gradually with stirring (**Figure 4.2B**) (**Hint 1**).
2. Heat the carrageenan dispersion to 80–90°C until fully dissolved, and remove from heat.
3. Add yellow food coloring and vanilla flavoring to the milk solution if you like (**Figure 4.2C**).
4. Pour the solution into moistened molds, and cool them at less than 20°C until set (**Figure 4.3D**).

pH = 6.63

FIGURE 4.2 Flan. (**A**) Mix carrageenan and white sugar in a pan. (**B**) Add milk gradually. (**C**) Add coloring and/or flavoring. (**D**) Pour the solution into moistened molds, cool, and allow to set.

Hint

1. Add milk little by little to avoid lumps.

This recipe can be used for an egg-free dessert jelly: "flan" is the same as caramel custard (**Chapter 7**). Using carrageenan enables making flan without eggs.

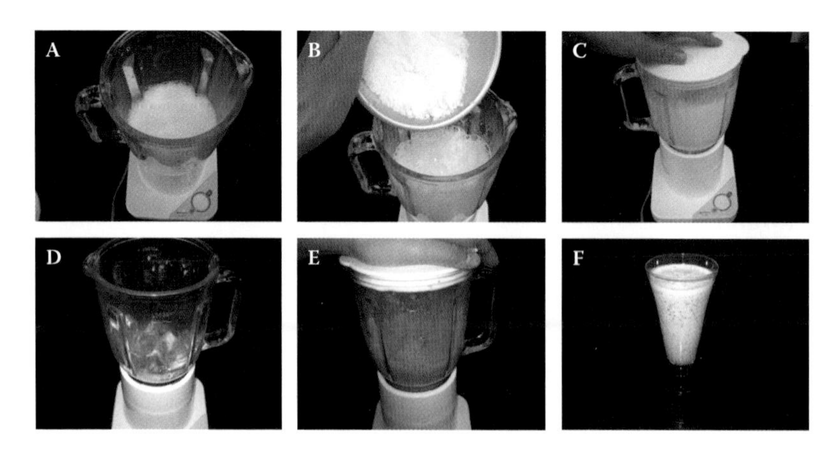

FIGURE 4.3 Chocolate drink. (**A**) Pour milk in blender. (**B**) Add mixed powder to milk. (**C**) Blend. (**D**) Add a mixture of chocolate, espresso, and ice to the blender. (**E**) Blend the mixture. (**F**) Pour into a glass.

Chocolate Drink (Figure 4.3)

Ingredients

(Serves 5)
0.3 g ($^3/_{20}$ tsp) κ-carrageenan powder
500 mL (2 cups) milk
90 g (10 Tbsp) white sugar
55 mL (⅔ cups) skim milk
750 g ice
75 g bar chocolate
5 shots (150 g) espresso (cold)

Preparation

1. Mix carrageenan, skim milk and white sugar in a bowl.
2. Pour milk into a blender (**Figure 4.3A**), add the powder mixture (**1**) (**Figure 4.3B**) and blend (**Figure 4.3C**). Chill in refrigerator for 1 day.
3. Put one-fifth of bar chocolate, 1 shot espresso, and 100 g ice into a blender (**Figure 4.3D**), and blend until chocolate and ice are grain-sized (**Figure 4.3E**) (**Hint 1**). Add one-fifth of the chilled mixture (**2**), blend, and serve in a glass (**Figure 4.3F**).

pH = 6.57

Hint

1. Be careful not to blend too much to enjoy the texture of crushed chocolate and ice.

Ham (Figure 4.4)

Ingredients

(Serves ~20)
1 kg pork
400 mL (1⅗ cups) water
4 g (2 tsp) κ-carrageenan powder
20 g (4 tsp) salt
10 g (a little over 1 tsp) white sugar
Onion powder, garlic powder†
Smoking chips

†YOU CAN CHOOSE YOUR FAVORITE SPICES.

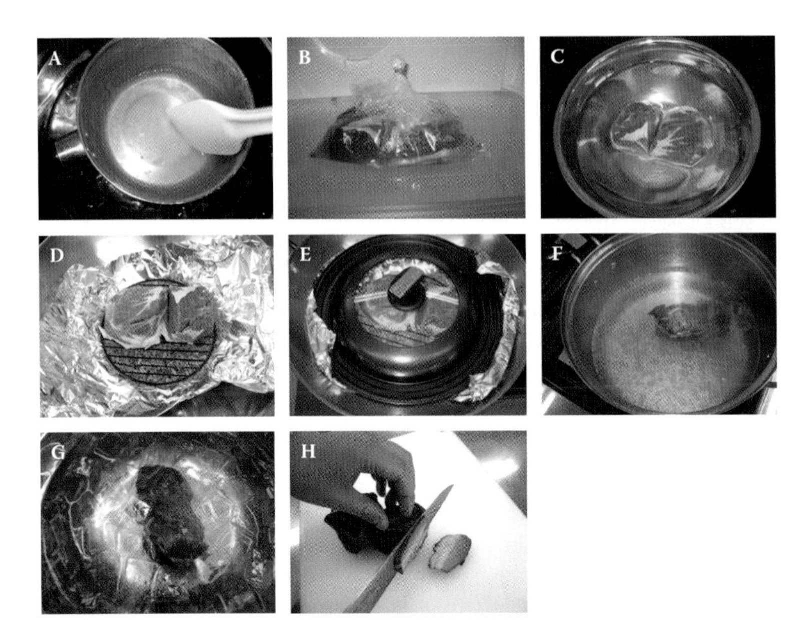

FIGURE 4.4 Ham. (**A**) Add onion and garlic powder to a prepared mixture of carrageenan, salt, and sugar; stir over heat and allow to cool. (**B**) Cure the meat in the prepared brine solution in the refrigerator for 1 or 2 days. (**C**) Soak the meat in water. (**D**) and (**E**) Smoke the meat. (**F**) Boil the ham. (**G**) Cool it. (**H**) Dry and slice the prepared meat.

Preparation

1. Mix carrageenan, salt, and white sugar in a pan, add water, heat for 5–10 min. with stirring, and remove from heat.
2. Add onion powder and garlic powder, and cool to 5°C (**Figure 4.4A**).
3. Rinse pork and blot dry with a paper towel; place in a plastic bag, and pour in the brine (**2**); expel the air, close bag with rubber bands, and refrigerate for 1 or 2 days (**Figure 4.4B**).
4. Put water in a bowl, and soak the meat (**3**) in it for 3 h (**Figure 4.4C**) (**Hint 1**).
5. Smoke the meat for 1 h (**Figure 4.4D,E**) (**Hint 2**).
6. Boil the ham (**5**) at 70°C for 2 h (**Figure 4.4F**), and cool it immediately in an ice-water bath (**Figure 4.4G**). Let rest for 30 min.
7. Dry with a paper towel, and slice (**Figure 4.4H**).

pH of the brine = 7.61

Hints

1. You can tie the meat with string or stuff it into a casing tube if you want to shape the ham. The casing tube preserves the meat for longer periods.

2. This process is optional, to achieve a smoked flavor. Details of the smoking method are provided in **Chapter 6**.

Adding carrageenan improves the texture of the ham because the carrageenan and protein interact, increasing the ham's elasticity, and decreasing syneresis.

Tips for the Amateur Cook and Professional Chef

- A more elastic (less brittle) carrageenan gel can be produced by incorporating LBG into the recipe.
- Agar, carrageenan, LBG, and potassium chloride yield a superior gel in terms of cohesiveness, strength, and eating qualities.
- κ-carrageenan added to ι-carrageenan can modify the elasticity of the produced gel, in a manner similar to LBG.
- In cheese spreads, carrageenan combined with LBG prevents fat or whey separation.
- Added at 0.05–0.30%, carrageenan improves the resistance of spaghetti to breakdown during cooking.
- Fatty meat and fish products can be protected from oxidative rancidity by using a 0.5% carrageenan solution as a protective coating.
- 0.3–0.4% carrageenan added to whole and sliced strawberries before packing in sugar syrup and freezing results in defrosted fruits with improved appearance and gloss, better shape, and firmer texture.
- Syneresis can be eliminated from furcellaran milk pudding preparations by adding 3–5 g starch per pint.
- Furcellaran at about 0.05% is used to stabilize fruit pulp in pulp-containing juices and soft drinks.

REFERENCES AND FURTHER READING

Arbuckle, W. S. 1972. *Ice cream*, 2nd ed. Westport, CT: Avi Publishing Co.

Arnott, S., Scott, W. E., Rees, D. A., and C. G. A. McNab. 1974. ι-Carrageenan molecular structure and packing of polysaccharide double helices in oriented fibres of divalent cation salts. *J. Mol. Biol.* 90:253–67.

Ayadi, M. A., Kechaou, A., Makni, I., and H. Attia. 2009. Influence of carrageenan addition on turkey meat sausages properties. *J. Food Eng.* 93:278–83.

Baker, G. L. 1949. Edible gelling compositions containing Irish moss extract, locust bean gum, and an edible salt. U.S. Patent No. 2,466,146.

Baker, G. L. 1954. Gelling compositions. U.S. Patent No. 2,669,519.

Barisas, L., Rosett, T. R., Gao, Y., Schmidt, S. J., and B. P. Klein. 1995. Enhanced sweetness in sweetener–NaCl-gum systems. *J. Food Sci.* 60:523–7.

Bayley, S. T. 1955. X-ray and infrared studies of κ-carrageenan. *Biochim. Biophys. Acta* 17:194–205.

Bjerre-Petersen, E., Christensen, J., and P. Hemmingsen. 1973. Furcellaran. In *Industrial gums*, 2nd ed., ed. R. L. Whistler and J. N. BeMiller, 123–36. New York: Academic Press.

Bourgade, G. 1871. Improvement in treating marine plants to obtain gelatin. U.S. Patent No. 112,535.

Bryce, T. A., Clark, A. H., Rees, D. A., and D. S. Reid. 1982. Concentration dependence of the order-disorder transition of carrageenans. Further confirmatory evidence for the double helix in solution. *Eur. J. Biochem.* 122:63–9.

Bryce, T. A., McKinnon, A., Morris, E. R., Rees, D. A., and D. Thom. 1974. Chain conformations in the sol-gel transitions, and their characterisation by spectroscopic methods. *J. Chem. Soc. Faraday Disc.* 57:221–9.

Bullens, C., Krawczyk, G., and L. Geithman. 1995. Cheese products with reduced fat content involving use of carrageenans and microcrystalline cellulose. *Latte* 20:177–80.

Bullock, K. B., Huffman, D. L., Egbert, W. R., Bradford, D. D., Mikel, W. B., and W. R. Jones. 1995. Nonmeat ingredients for low-fat ground beef patties. *J. Muscle Foods* 6:37–46.

Camacho, M. M., Navarrete, N. M., and A. Chiralt. 2001. Stability of whipped dairy creams containing locust bean gum/λ-carrageenan mixtures during freezing-thawing processes. *Food Res. Int.* 34:887–94.

Christiansen, E. 1959. Danish agar. In *Industrial gums*, ed. R. Whistler, 51–4. New York: Academic Press.

Dexter, D. R., Sofos, J. N., and G. R. Schmidt. 1993. Quality characteristics of turkey bologna formulated with carrageenan, starch, milk and soy protein. *J. Muscle Foods* 4:207–23.

Dolan, T. C. S. and D. A. Rees 1965. The carrageenans. Part II. The positions of the glycosidic linkages and sulphate esters in λ-carrageenan. *J. Chem. Soc.* 3534–9.

Dybing, S. T. and D. E. Smith. 1998. The ability of phosphates or κ-carrageenan to coagulate whey proteins and the possible uses of such coagula in cheese manufacture. *J. Dairy Sci.* 81:309–17.

Fernandez, C., Canet, W., and M. D. Alvarez. 2009. Quality of mashed potatoes: Effect of adding blends of kappa-carrageenan and xanthan gum. *Eur. Food Res. Technol.* 229:205–22.

Glicksman, M. 1966. Frozen gels of Eucheuma. U.S. Patent No. 325,062.

Glicksman, M. 1969. *Gum technology in the food industry*. New York: Academic Press.

Glicksman, M. 1983. Red seaweed extracts (agar, carrageenans, furcellaran). In *Food hydrocolloids*, vol. 2., ed. M. Glicksman, 73–113. Boca Raton: CRC Press.

Glicksman, M., Farkas, E., and R. E. Klose. 1970. Cold-water-soluble Eucheuma gel mixtures. U.S. Patent No. 3,502,483.

Grindrod, J. and T. A. Nickerson. 1968. Effect of various gums on skim milk and purified milk proteins. *J. Dairy Sci.* 51:834–41.

Guiseley, K. B., Stanley, N. F., and P. A. Whitehous. 1980. Carrageenan. In *Handbook of water-soluble gums and resins*, ed. R. L. Davidson, 5–1 to 5–30. New York: McGraw-Hill.

Haas, P. 1921. The nature and composition of Irish moss mucilage. *Pharm J.* 106:485.

Hansen, P. M. T. 1982. Hydrocolloid-protein interactions: Relationship to stabilization of fluid milk products. A review. In *Gums and stabilizers for the food industry*, ed. G. O. Phillips, D. J. Wedlock, and P. A. Williams, 127–38. Oxford: Pergamon Press.

Haug, I. J., Draget, K. I., and O. Smidsrød. 2004. Physical behavior of fish gelatin-k-carrageenan mixtures. *Carbohydr. Polym.* 56:11–9.

Hood, L. F. and J. E. Allen. 1977. Ultrastructure of carrageenan-milk sols and gels. *J. Food Sci.* 42:1062–5.

Izzo, M., Stahl, C., and M. Tuazon. 1995. Using cellulose gel and carrageenan to lower fat and calories in confections. *Food Technol.* 49:45–6, 48–9.

Jensen, T. W. 1995. Alternative gelling system for mousse. *Scand. Dairy Information* 9:22–3.

Kampmann, R. 1995. Carrageenans in dressings. *Internationale Zeitschrift-fur-Lebensmittel-Technik, Marketing, Verpackung und Analytik* 46:44, 46–7.

Kanombirira, S. and K. Kailasapathy. 1995. Effects of interactions of carrageenan and gellan gum on yields, textural and sensory attributes of cheddar cheese. *Milchwissenschaft* 50:452–8.
Keeney, P. G., and M. Kroger. 1974. Frozen dairy products. In *Fundamentals of dairy chemistry*, 2nd ed., ed. B. H. Webb, A. H. Johnson, and J. A. Alford, 873–913. Westport, CT: Avi Publishing Co.
Laos, K., Brownsey, G. J., and S. G. Ring. 2007. Interactions between furcellaran and the globular proteins bovine serum albumin and β-lactoglobulin. *Carbohydr. Polym.* 67:116–23.
Lawson, C. J. and D. A. Rees. 1970. An enzyme for the metabolic control of polysaccharide conformation and function. *Nature* 227:390–3.
Lewis, J. G., Staley, N. F., and G. G. Guist. 1988. Commercial production and applications of algal hydrocolloids. In *Algae and human affairs*, ed. C. A. Lembi, and J. R. Waaland, 205–36. Cambridge: Cambridge University Press.
Lin, C. F. 1977. Interaction of sulfated polysaccharides with proteins. In *Food colloids*, ed. H. D. Graham, 320–46. Westport, CT: Avi. Publishing Co.
Lin, C. F. and P. M. J. Hansen. 1970. Stabilisation of casein micelles by carrageenan. *Macromolecules* 3:269–74.
Litex Co. 1960. *Danish Agar*. Copenhagen: Litex Co.
Lyons, P. H., Kerry, J. F., Morrissey, P. A., and D. J. Buckley 1999. The influence of added whey protein/carrageenan gels and tapioca starch on the textural properties of low fat pork sausages. *Meat Sci.* 51:43–52.
McCandless, E. L. and M. R. Gretz. 1984. Biochemical and immunochemical analysis of carrageenans of the Gigartinaceae and Phyllophoraceae. *Hydrobiologia* 116/117:175–8.
Mueller, G. P. and R. A. Rees. 1968. Current structural views of red seaweed polysaccharides. In *Proc. Drugs from the sea,* ed. H. D. Freudenthal, 241–55. Washington DC: Marine Technology Society.
Nadison, J. 1995. The interaction of carrageenan and starch in cream desserts. *Scand. Dairy Information* 9:24–5.
Nielsen, B. J. 1976. Function and evaluation of emulsifiers in ice cream and whipped emulsions. *Gordian* 76:200–25.
Nishinari, K., Watase, M., Miyoshi, E., Takaya, T., and D. Oakenfull. 1995. Effects of sugar on the gel-sol transition of agarose and κ-carrageenan. *Food Technol. Chicago* 49:90–6.
Nunes, M. C., Raymundo, A., and I. Sousa. 2006. Gelled vegetable desserts containing pea protein, κ-carrageenan and starch. *Eur. Food Res. Technol.* 222:622–8.
Nussinovitch, A. 1997. *Hydrocolloid applications: Gum technology in the food and other industries*. London: Blackie Academic & Professional.
Nussinovitch, A. 2003. *Water-soluble polymer applications in foods.* Oxford: Blackwell Publishing.
Nussinovitch, A. 2010. *Polymer macro- and micro-gel beads: Fundamentals and applications*. New York: Springer.
Nussinovitch, A., Kaletunc, G., Normand, M. D., and M. Peleg. 1990. Recoverable work vs. asymptotic relaxation modulus in agar, carrageenan and gellan gels. *J. Texture Studies* 21:427–38.
Nussinovitch, A., Kopelman, I. J., and S. Mizrahi. 1990. Effect of hydrocolloid and minerals content on the mechanical properties of gels. *Food Hydrocolloids* 4:257–65.
Nussinovitch, A., Kopelman, I. J., and S. Mizrahi. 1990. Evaluation of force deformation data as indices to hydrocolloid gel strength and perceived texture. *Int. J. Food Sci. Technol.* 25:692–8.

Painter, T. J. 1966. The location of the sulphate half-ester groups in furcellaran and κ-carrageenan. In *Proc. 5th Int. Seaweed Symp.*, ed. E. G. Young and J. L. McLachlan, 305–13. London: Pergamon Press.
Paoletti, S., Smidsrod, O., and H. Grasdalen. 1984. Thermodynamic stability of the ordered conformation of carrageenan polyelectrolytes. *Biopolymers* 23:1771–94.
Parolis, H. 1981. The polysaccharides of *Phyllymenia hieroglyphica* and *Pachymenia hymantophra*. *Carbohydr. Res.* 93:261–7.
Payens, T. A. J. 1972. Light scattering of protein reactivity of polysaccharides especially of carrageenans. *J. Dairy Sci.* 55:141–50.
Prabhu, G. A. and J. G. Sebranek. 1997. Quality characteristics of ham formulated with modified corn starch and kappa-carrageenan. *J. Food Sci.* 62:198–202.
Rees, D. A. 1972. Mechanism of gelation in polysaccharides systems. In *Gelation and gelling agents, British Food Manufacturing Industries Research Association, Symp. Proc.* No. 13, London, 7–12.
Rotbart, M. 1984. Carrageenan isolation, characterization and upgrading of functional properties. MSc thesis, Technion-Israel Institute of Technology, Israel.
Rotbart, M., Neeman, I., Nussinovitch, A., Kopelman, I. J., and U. Cogan. 1988. The extraction of carrageenan and its effect on the gel texture. *Int. J. Food Sci. Technol.* 23:591–9.
Schmidt, C. 1844. Uber phlanzenschleim und bassorin. *Annal. Chem. Pharm.* 51:29–62.
Smith, D. B. and W. H. Cook. 1953. Fractionation of carrageenan. *Arch. Biochem. Biophys.* 45:232–3.
Snoeren, T. H. M. 1976. κ-Carrageenan. *A study on its physiochemical properties, sol-gel transition and interaction with milk proteins*. PhD Thesis, Nederlands Instituut voor Zuivelonderzoek, Ede, The Netherlands.
Snoeren, T. H. M., Both, P., and D. G. Schmidt. 1976. An electron-microscope study of carrageenan and its interaction with κ-casein. *Neth. Milk Dairy J.* 30:132–41.
Snoeren, T. H. M., Payenes, T. A. J., Jeunink, J., and P. Both. 1975. Electrostatic interaction between κ-carrageenan and κ-casein. *Milchwissenschaft* 30:393–6.
Soukoulis, C., Chandrinos, I., and C. Tzia. 2008. Study of the functionality of selected hydrocolloids and their blends with κ-carrageenan on storage quality of vanilla ice cream. *LWT - Food Sci. Technol.* 41:1816–27.
Spano, A. 1994. A new product for the ice cream manufacturer. *Latte* 19:396–402.
Stainsby, G. 1980. Proteinaceous gelling systems and their complexes with polysaccharides. *Food Chem.* 6:3–14.
Stanley, N. F. 1987. Production, properties and use of carrageenans. In *Production and utilization of products from commercial seaweeds*, ed. D. J. McHugh, 97–147. Rome: FAOUN.
Stanley, N. F. 1990. Carrageenans. In *Food gels*, ed. P. Harris, 79–119. London and NY: Elsevier Applied Science.
Thomas, W. R. 1992. Carrageenan. In *Thickening and gelling agents for food*, ed. A. Imson, 25–39. Glasgow: Blackie Academic & Professional.
Tilly, G. 1995. Stabilizer composition enabling the production of a pourable aerated dairy dessert. European Patent Application EP 0 649 599 A1.
Towle, G. A. 1973. In *Industrial gums*, ed. R. L. Whistler and J. N. BeMiller, 83–114. New York: Academic Press.
Verbeken, D., Thas, O., and K. Dewettinck. 2004. Textural properties of gelled dairy desserts containing κ-carrageenan and starch. *Food Hydrocolloids* 18:817–23.

5

Cellulose Derivatives

Introduction

The most plentiful organic substance in nature is cellulose, making up approximately one-third of the world's vegetative matter. An estimated 10^{11} tons of cellulose is synthesized annually. Cellulose, hemicellulose, and lignins are the chief components of most land plants, providing structural support. Cellulose can be converted into hundreds of manufactured goods, which influence all phases of our everyday lives. Cellulose content in vegetative tissues varies: for example, wood includes ~40–50% and cotton 85–97%. Cellulose is a linear polymer of D-glucose monomers joined by β-D-1,4 linkages and arranged in repeating units of cellobiose, each composed of two anhydroglucoses. Every anhydroglucose encloses three hydroxyl groups, which are accessible for reaction. Cellulose has an extended molecular-chain length and the hydrogen-bonding ability of the three hydroxyl groups is particularly high. Spatially, the polymer is arranged in long thread-like molecules. These molecules align to shape fibers, many regions of which are highly ordered and have a crystalline structure as a result of lateral association by hydrogen bonding. These crystalline regions confer rigidity and strength.

The degree of polymerization of cellulose depends on its source: values can range from 1,000–15,000 for native cellulose to 200–3,200 for commercial-grade purified cotton lint and wood pulp; different values can be found elsewhere. The polymer has a maximum three degrees of substitution (i.e., the average number of substituted groups per anhydroglucose unit: DS). Products with a broad variety of functional properties can be created by controlling the DS and its type. Partial substitution is preferred and numerous derivatives are possible; nevertheless, only a few of them are of interest to the manufacturing and food industries. Cellulose is not soluble in water, owing to its extensive intra- and intermolecular hydrogen-bonded crystalline domains. The manufacture of water-soluble cellulose derivatives, in contrast to that of polymers based on petrochemical resources, begins with a preformed polymer backbone of either wood or cotton cellulose, instead of a monomer. To make it suitable for food utilization, cellulose can be transformed into a soluble compound via its derivatization and disruption of the hydrogen bonds.

Manufacture

Cellulose ethers are manufactured by preparing alkali cellulose and then subjecting it to alkylation or alkoxylation. Specific solvents are required to solubilize cellulose. The commercial derivatization reactions carried out heterogeneously at temperatures of 50–140°C under nitrogen atmosphere can result in high molecular weights. Access to all hydroxyls is not equal, and the allocation of substituents is therefore inconsistent. Swelling, crystalline-region disruption, reaction uniformity, and catalysis of alkoxylation reactions may result from using soda to form alkali cellulose. Acetone and inert diluents (isopropanol) are employed to disperse the cellulose, and to restrain the reaction kinetics, making it easy to recover the product and provide heat transfer. Two kinds of reactions are used to manufacture cellulose ethers: Williamson etherification (reaction of alkali cellulose with organic halide) and alkoxylation, relating to reaction with an epoxide. Information on these reactions can be found elsewhere. In the manufacture of methylcellulose (MC), the use of dimethyl sulfate as an alternative to methyl chloride results in the creation of various by-products.

Cellulose derivatives can be produced by reacting alkali cellulose with either methyl chloride to form MC, propylene oxide to form hydroxypropylcellulose (HPC), or sodium chloroacetate to form sodium carboxymethylcellulose (CMC). Blended derivatives, for instance, methylhydroxypropylcellulose (MHPC), can be formed by combining two or more of these reagents. Of the numerous possible derivatives, which have been studied and produced, CMC, MC, MHPC, and HPC are exploited in food production, as are other modified forms of cellulose which have been found to have functional hydrocolloidal properties and significance in more than a few food applications. CMC is the principal cellulose-derived hydrocolloid. It is an anionic polymer, which is vital for viscosity-forming applications and is able to react with charged molecules within specific pH ranges. HPC, a nonionic cellulose ether, is soluble in water below 40°C and in polar organic solvents. Microcrystalline cellulose (MCC) is an acid-hydrolyzed, pure α-cellulose material, which has viscosity-forming and water-absorptive properties and is utilized in frozen and dairy-food products.

Properties of Methylcellulose (MC) and Methylhydroxypropylcellulose (MHPC)

Commercial MC products have an average DS of 1.4 to 2.0. A minimum DS of ~1.4 is required for solubility in water, and at a DS of 2.0–2.2, solubility in organic systems is achieved. Cellulose products come in a range of particle sizes with a white to off-white color. The purity of the cellulose products depends on their utilization, either technical or, at higher purity, in foods. For food purposes, less than 1% sulfated ash and residual heavy metals, as specified in the European Pharmacopoeia, are permitted, while a maximum ash content of about 2.5% can be found in cellulose products destined for other uses. MC is metabolically inert and has a neutral taste and odor. Commercial MHPC products have an average methyl DS of 1.0–2.3. MC and MHPC are soluble in cold water but insoluble in hot water. When solutions of these compounds are heated, gel structures can form, at gelation temperatures ranging from 50 to 90°C. Solutions of MC can be prepared by dispersing the powder in hot water (80–90°C), adding cold water (0-5°C) or ice to the final volume, and agitating until a smooth texture is achieved. MHPC products might require cooling to 20–25°C or less. Stirring is recommended for enhanced dissolution. MC and MHPC solutions in cold water are smooth, clear, and pseudoplastic.

Solution viscosity declines with increasing temperature to the thermal gel point, after which viscosity increases sharply until the flocculation temperature is reached. The temperatures for 0.5% solutions are 50–75°C for MC and 60–90°C for MHPC. Flocculation results from the weakening of hydrogen bonding between polymer and water molecules and the strengthening of interactions between polymer chains. Gels are shaped as a result of phase separation and are prone to shear thinning. If the temperature is lowered, the original solution is restored. The thermal gel point is manipulated by the kind and degree of substitution. The flocculation temperature decreases with increasing salt concentration and increases with increasing alcohol concentration. The viscosity of MC and MHPC is steady over a very wide range of pHs (2–13). Low salt concentrations have a small effect on viscosity, whereas higher levels (7% sodium chloride, 15% potassium chloride, or 4% sodium bicarbonate) cause salting out of the polymer solution. At temperatures >140°C, MC and MHPC powders darken, and at temperatures >220°C, they decompose. In the dry state, they are resistant to microorganisms, but if solutions are to be stored, preservatives such as sodium benzoate or potassium sorbate are advised.

Dilute (0.1%) MC or MHPC decreases the surface tension of water by ~40% at 20°C. MC surface activity and water-retention properties are useful in emulsion-type sauces, as well as in whipped toppings and creams.

Hydroxypropylcellulose (HPC)

Cellulose reacted with aqueous sodium hydroxide followed by propylene oxide goes through an alkoxylation reaction to yield HPC. Etherification is catalyzed by the alkali. Propylene oxide reacts with water to form poly(propylene glycol) by-products, so the quantity of water is minimized to improve yields. The reaction is carried out at 70–100°C for 5–20 h in stirred autoclaves in the presence of organic diluents, neutralized, and the HPC is washed with hot water (70–90°C), then dried and ground to supply an off-white, tasteless, granular powder. Food-grade HPC is no less than 99.5% pure, with a sulfated ash content of under 0.2%. It is a physiologically inert polymer, which is edible, thermoplastic, and nonionic. It is soluble in water below 40°C and in many organic solvents, such as methanol, ethanol, and propylene glycol. Aqueous, lump-free solutions can be formed by dry-blending the HPC with additional powders prior to dispersion and blending with glycerin or hot water to create a slurry, which is agitated and then added to cold water for dissolution.

HPC aqueous solutions are pseudoplastic, demonstrating slight or no thixotropy and comparatively high stability throughout shear degradation. Viscosity decreases with increasing temperature, by ~50% for each 15°C rise. HPC precipitates from solution at temperatures of 40–45°C. No gelation takes place during the dissolution-to-precipitation transition, in contrast to MC and MHPC. Precipitation is reversible and the polymer can be redissolved at 40°C. The presence of salts or organic substances in the solution decreases the precipitation temperatures, due to competition for water in the system. The pH of a 1% HPC solution can range from 5.0 to 8.8. HPC is nonionic, and consequently its viscosity remains stable at pH 2 to 11. To dampen the decrease in viscosity due to degradation, solutions are stored at a pH of 6.0–8.0. The surface tension of a 0.1% HPC solution is 440 μN/cm, as compared to 741 μN/cm for water, and HPC can consequently function as an emulsifying and whipping additive in creams and whipped toppings. HPC solutions blend well with nearly all natural and synthetic hydrocolloids. A synergistic effect on viscosity is detected upon blending with CMC. Since HPC is highly substituted with hydroxypropyl, it is resistant to microbial attack. Nevertheless, preservatives should be added to solutions stored for prolonged periods. HPC's thermoplasticity renders it processable by all fabrication methods. It has outstanding film-forming properties, for example: elasticity and lack of tackiness, and it shows good-quality heat sealability and acts as a barrier to oil and fat, all of which are useful for many food, chemical, and pharmaceutical applications.

Microcrystalline Cellulose (MCC)

MCC only came into its own in the food industry in the late 1960s. The chief uses of this material are for pharmaceutical tableting and food stabilization. MCC is a purified native cellulose and not a chemical derivative. During its production, the dissolving α-cellulose pulp is treated with a dilute mineral acid and the cellulose microfibril is unhinged. Hydrolysis is carried out until polymerization levels off. The paracrystalline regions, which have a disordered molecular structure, are weakened as the acid selectively etches this area from around the compactly arranged crystalline cellulose. Shear then releases the cellulose aggregates, which are the basic raw materials for commercial MCC products. Two chief types of MCC are manufactured: the first is a spray-dried, aggregated, porous, plastic and sponge-like MCC powder, and the second is a water-dispersible colloidal cellulose. The first type's main functions lie in its ability to serve as a binder or disintegrant for tablets, as a flow aid for cheeses, as a carrier for flavors, and as a fiber. The colloidal MCC's most important functions are as an emulsion stabilizer, a thixotropic thickener, a controller of moisture, a foam stabilizer, an ice-cream controller, a suspension agent, and a provider of cling. The spray-dried MCC is sold in various size ranges. These ranges (average particle sizes of 18 to 90 μm) have an effect on the flow properties of the powder used in tableting, in addition to its oil- and water-absorptive capacities and bulk densities. As a result of its nondigestibility, powdered MCC can be employed in low-calorie foods or as a source of food fiber. The powder is white, odorless, tasteless, free-flowing, and has a heavy metal content of less than 10 ppm, that is, less than 8 mg per 5 g water-soluble substances, 0.1% residue upon ignition, and a pH range for the different types of 5.0 to 7.0.

Colloidal MCC is a food-grade submicron particle, which can be utilized as a stabilizer. It consists of a bundle of short-chain cellulose units, with a molecular weight of a few million. Different types of commercial colloidal MCC are manufactured and their main use depends on their properties. These different products can be dispersed by homogenization, vigorous agitation, spoon stirring, and so forth, again depending on their properties. MCC has no swellable or hydration-like starch or other gum but is reliant on shear and accessible water. Heat has a small effect on its dispersion properties while acids, salts, and supplementary ingredients play a major role. Dispersions of colloidal MCC are flocculated by minute quantities of electrolytes, cationic polymers, and surfactants. MCC dispersions blend well with nearly all nonionic and anionic polymers. CMC, HPMC, and xanthan gum (**Chapter 15**) are the finest protective colloids for MCC dispersions. Given that dispersion (peptization) is very important, chief factors influencing this practice rely on the type of MCC, the hardness of the water, the amount of shear available, and the order of component addition. MCC has to be one of the first constituents added to the food system: it cannot be coated with fat because the agglomerate will not swell and discharge the microcrystals. MCC requires substantial quantities of water and cannot compete with other gums: the smaller the number of electrolytes in the system, the more effective the colloidal material and the higher the percentage of it that is released.

MCC gels are highly thixotropic and have a finite yield value at low concentrations. The construction of a network by solid-particle linkage accounts for the manufactured yield value and elasticity. If shear is applied, the gel shears and thins. Resting allows the gel to reform a network. The inclusion of CMC decreases the thixotropic nature of the gels and results in decreased yield values. Gum addition modifies the rheological behavior of MCC gels. Temperature has a minimal effect on the viscosity of MCC dispersions. The colloidal MCC functions after being peptized (dispersed), such that individual microcrystals are discharged into the aqueous phase of the food. Stabilization occurs when an adequate number of microcrystals have been dispersed (they do not hydrate) into this aqueous phase. Dispersion of the insoluble 0.2-μm cellulose particles is assisted by CMC, which is employed throughout processing to prevent hydrogen bonding and to achieve better taste. The CMC's fast hydration assists in linking the crystallites together after rehydration to shape a thixotropic gel network, which is resistant to heat, shear, and pH as low as 3.5. In emulsions, the microcrystals are positioned at the interface just around the fat globules, conferring emulsion stabilization. The result of microcrystal structuring is practical for foods such as whipped toppings, ice creams, and extruded foods. Briefly, colloidal MCC functions in foods as an opacifier, suspension aid, ice-crystal controller, emulsion stabilizer, thixotropic thickener, foam stabilizer, heat stabilizer, and nonnutritive gelling agent. In addition to its emulsion-stabilization effect, colloidal MCC is useful in preserving product texture, particularly of acidic foods, throughout high-temperature processing. An additional benefit of MCC is its discoloration at these temperatures at low pH, and the minute effect temperature has on its viscosity. Furthermore, the colloidal dispersions are pourable suspensions or smooth thixotropic gels; consequently, they can create different textures. Starch–MCC blends at ratios of 3–4:1 reduce, by 20–25%, the amount of starch necessary to achieve the same performance. This system masks flavor less, improves flow control, creates greater resistance to breakdown and shearing, and improves heat stability as compared to starch alone.

Carboxymethylcellulose (CMC)

General Information

Sodium CMC is a water-soluble, anionic, linear polymer, generally known as CMC. It was developed in Germany during World War I as a possible alternative for gelatin. In the '30s, CMC was used to eliminate the redeposition of soil on fabric during washing and rinsing. CMC was adsorbed to the fabric by hydrogen bonding and owing to its anionic character repelled the dirt, which was also negatively charged. CMC also improved the effectiveness of synthetic detergents, thereby promoting renewed attention to its production after the war. It was first manufactured commercially by Kalle and Co. in Wiesbaden-Biebrich in the late 1930s. Hercules developed a commercial process for its manufacture in 1943 and by 1946, it was in full-scale commercial production. Additional production resulted from the gum's approval by the FDA as a food additive. The development of specialized types of CMC followed quickly as an outcome of its film-forming ability. In industries where highly purified types of CMC are preferred (i.e., foods, pharmaceuticals, and cosmetics), they are referred to as *cellulose gums*. Requirements for the characteristics and purity of these gums can be found in the Food Chemicals Codex, as well as FDA and FAO publications. CMC is sold as a white to buff-colored, tasteless, odorless, loose powder. The gum is employed for more applications than any other branded water-soluble polymer. CMC is used in detergents, drilling fluids, paper, mining, textiles, foods, coatings, and cosmetics around the world. The viscosity-forming ability of CMC, its moisture-binding capacity, dissolution abilities, and texturization capabilities have advanced its broad use in foods such as extrudates, emulsions, and frozen desserts, among many others.

Chemical Nature and Manufacture

Cellulose is a linear polymer of β-anhydroglucose units, each containing three hydroxyl groups. CMC is produced by treating cellulose with aqueous sodium hydroxide, followed by reaction with monochloroacetic acid or sodium monochloroacetate in compliance with the Williamson etherification reaction. Esters of monochloroacetic acid have also been used, and substitute processes described. If only one of the three hydroxyl groups is carboxymethylated, the DS is 1.0. Technical-grade CMC has a purity of 94–99%, and that employed in foods is at least 99.5% pure. Additional physical properties include 8% moisture content when packed, a browning temperature of 227°C, a charring temperature of 252°C, and a bulk density of 0.75 g/mL. Biological oxygen demand (BOD) following 5 days of incubation is 11,000 ppm for high-viscosity CMC with DS 0.8, and 17,300 ppm for low-viscosity CMC with DS 0.8. A 2% CMC solution has a specific gravity of 1.0068 at 25°C and a refractive index of 1.3355. The typical pH of a 2% solution is 7 and the surface tension of a 1% solution at 25°C is 71 dyn/cm, with a density of 1.59 g/mL and a refractive index of 1.515. A DS of 0.4–1.4 is customary for commercial CMC, and can be adjusted to higher values for specialized products. If the DS is below 0.4, the resultant CMC is not water-soluble. Food-grade CMC (2%) has a DS of 0.65–0.95 and a minimum viscosity of 25 mPa.s. Viscosities can be managed by oxidative degradation of the crude product with hydrogen peroxide to get low-viscosity CMC; the higher the DS, the higher the solubility of the polymer.

Chemical and Physical Properties

Solution Properties

CMC is soluble in hot and cold water, making it fairly multipurpose. It can be soluble in water-miscible organic solvents as well, for instance, water ethanol. Low-viscosity types are more tolerant to increasing ethanol concentrations than their higher viscosity counterparts, enduring up to 50% ethanol or 40% acetone. This property is significant in alcoholic beverages and instant bar blends where viscosity and clarity are required. Sodium CMC is a carboxylic acid salt. Its dilute solution has a neutral pH and virtually all of its carboxylic acid groups are in the sodium salt form, with very few in the free-acid form. At pH 3.0 or lower, CMC reverts to an insoluble free-acid form. The pK_a (acid dissociation constant, i.e., a quantitative measure of the strength of an acid in solution) of sodium CMC varies from 4.2 to 4.4 and differs to some extent with DS. Upon dissolution, CMC goes through dispersion and hydration steps. Without good dispersion, clumps are formed with a swollen external skin made up in part of hydrated gum. To prevent this, the gum should be added at a sufficiently slow rate for every particle to separate. Additional clump-prevention techniques include dry blending in sugar or other water-miscible nonsolvent such as glycerin, sorbitol, or propylene glycol, or dispersion of the gum in oil followed by addition of the mixture, including an emulsifier, to an aqueous system. Dissolution can be achieved by supplying the gum through a smooth-walled funnel into a water-jet eductor where it is dispersed by turbulent water flow, which generates a suction effect. Such an operation allows 80–90% instant wetting and hydration of the gum. The addition of minute quantities of dioctyl sodium sulfosuccinate to the gum prior to its dissolution in water decreases its propensity to clump. CMC dissolves when the hydroxyl groups are substituted by CMC groups; in the cellulose the chains disconnect, and the unreacted hydroxyl groups become accessible to association with water. Inclusive solubility and subsequent solution properties are preserved as the gum chains are held apart by Coulomb repulsion, which permits the entrance of water and full interaction with the chain hydroxyls. Salts tend to diminish gum hydration and its corresponding viscosity.

Viscosity

Almost certainly, the most practical property of CMC is its capacity to impart viscosity. CMC displays the typical rheological properties of linear polymers. Nearly all solutions of CMC are pseudoplastic, that is, the higher the shear rate, the larger the decrease in measured viscosity. Most products having DS values below ~1.0 are also thixotropic, possibly resulting from the numerous aggregates in the aqueous dispersion. In other words, viscosity depends on shear rate and time. Thixotropy is also a function of uniformity of substitution. Consistency of substitution also increases tolerance to acidic systems and dissolved ions. The presence of salts in solution represses the disaggregation of CMC and, as a result, has an effect on its viscosity. Higher viscosities can be achieved by dissolving CMC in a glycerin–water mixture. An additional effective approach to achieving high viscosities is to merge CMC with nonionic cellulose derivatives. Such combinations in a 1% solution offer about twice the viscosity expected from the individual ingredients. CMC flow features vary from thixotropic to smooth. The smooth-flowing type is preferred in food systems such as syrups or frostings where a smooth texture is needed; thixotropic CMCs are preferred in sauces and purées, which are grainier. A finely ground CMC is optimal where viscosity in liquid and hydration need to be rapidly achieved. Coarse pulverization serves when the dispersion is to be fabricated from cellulose derivatives. The pH of a 1% CMC solution is characteristically in the range of 7.0–8.5. Solution viscosity is more or less uninfluenced at pH 5.0–9.0, whereas at pH < 3.0, viscosity may increase and precipitation of the free-acid form of CMC may occur, and at pH > 10 there is a minor reduction in viscosity. Consequently, cellulose gums should not be utilized in highly acidic food systems. CMC solutions enclosing 5% acetic acid or 1.0% citric or lactic acids can be stored for months at room temperature with no marked change in viscosity.

CMC solutions can be heat treated, for example, at 80°C for 30 min. or 100°C for 1 min. to destroy potentially polymer-degrading bacteria. This treatment has essentially no influence on the CMC. Intended for extended periods of storage, the inclusion of preservatives is advised and the pH should be kept at between 7 and 9, while being kept away from elevated temperatures, oxygen, and sunlight. The viscosity of CMC solutions decreases with increasing temperature, but this decrease is reversible if the heating is not prolonged; otherwise, the damage is lasting. If salts are present in solution, CMC disaggregation is repressed, and viscosity can be manipulated dependent on whether the metal is mono-, di- or trivalent. Monovalent ions (except for Ag^{+}) have little consequence, divalent ions (Ca^{2+}, Mg^{2+}) lower viscosity, and trivalent ions (Al^{3+}, Cr^{3+}, Fe^{3+}) can render the CMC insoluble or cause gelation via their complexing with carboxylic groups. The order of addition of salts to the CMC solution is significant in terms of the resultant viscosity. Following dissolution of the gum, salts have little effect on viscosity. If the salt is dissolved prior to addition to the CMC, it restrains disaggregation and lowers viscosity.

CMC's broad range of applications is an outcome of its compatibility with many other food ingredients. In some cases, mixing CMC with additional gums results in increased viscosity. CMC can act together with proteins, since it is a protein-reactive material. CMC, in contrast to MC, MHPC, and HPC, is not highly surface active. Due to its substantial ability to attach water, it is employed in gels, pie fillings, and other foods to reduce syneresis or to enhance water-binding ability. CMC gelation can be favored by careful selection of ion concentration, regulating the pH, and controlling ion release via the use of suitable chelating agents. Suitable ions serve as cross-linking agents for polymer chains via salt formation with neighboring anionic groups. There are many different potential gel compositions. High DS cellulose gum gelation occurs with trivalent cations. Probable gelation of adequately low DS cellulose gums is possible when they are subjected to high-shear conditions. The lowest DS type displays the firmest gel strength, with a smooth, greasy texture.

Stability and Physical Data

Cellulose-gum solutions remain stable, with no decrease in viscosity, in the absence of bacterial contamination. Heat treatment, for instance, 1 minute at 100°C, is required to prevent such contamination. Continuing storage requires the addition of a suitable preservative, for example, sodium benzoate, sorbic acid, or sodium propionate. In foods that are designed to enclose cellulases, the enzyme must be inactivated to prevent a severe decrease in viscosity. Given that the attack on cellulose gum is assumed to take place in solution, its storage as a dry powder where moisture is kept at a minimum ensures no undesirable changes. Subsequent to its manufacture using solvents, the product is in an aseptic state. Because microbial attack of the gum results from microbial growth, simple sanitary measures can guarantee its stability. Other stability factors, which influence cellulose degradation are ultraviolet (UV) radiation, and the presence of molecular oxygen and heavy metals which provide catalysts for oxidative degradation. Consequently, chelators, for example sodium citrate or hexametaphosphate, are employed to stabilize cellulose gum in foods which entrain air and contain heavy metals. The cellulose gum is classified into three particle-size designations: regular, coarse, and fine. On the one hand, cellulose gum is a hygroscopic material, and appropriate packaging is required for producers and consumers. On the other hand, in foods that require moisture retention to preserve eating quality (e.g., cakes and frostings), the higher DS grade gums are the best water binders.

Food Applications

Cellulose derivatives are employed in numerous edible items. The use of sodium CMC for food purposes is on the rise, particularly in developed countries where the requirement for expediency has grown since the beginning of the 1950s. Many purposes, that is, binding, viscosity formation, stabilization, or moisture retention, can be achieved by the addition of cellulose derivatives in foodstuffs at levels of 0.1–0.5%, but commonly not more than 1%. Texturization is advantageous in desserts and low-calorie food items; viscosity forming in bakery batters, pet foods, gravies, low-calorie foods, soups, sauces, snack foods, beverages, and desserts, to name a few. The maximal permissible inclusion of CMC in ice cream is 0.5%. Stabilizers other than CMC are locust bean gum (LBG) and guar gum (**Chapter 8**), gelatin (**Chapter 9**), alginate (**Chapter 3**), and carrageenan (**Chapter 4**). These increase viscosity by rapid hydration and delay the formation or enlargement of ice crystals. If no or merely minute quantities of 0.05% CMC are added to ice cream, iciness is more prevalent than in manufactured foods with 0.2% CMC. Viscosity is enhanced and melting rate decreases in ice cream blends with increasing CMC concentration. The inclusion of cellulose gum in ice cream also protects it from thermal shock and enhances its mouthfeel. Marketable blend stabilizers based on blends of guar gum, LBG, and CMC generate lower apparent viscosities than the pure polysaccharides. A high-viscosity CMC–LBG blend is recommended for consistency and flavor properties of ice cream. To improve the smoothness and chewiness of ice milk, a sucrose–corn syrup mix can be utilized together with a higher concentration of stabilizer than in ice creams. For instance, CMC, guar gum, and LBG are used in combination with carrageenan, the latter to avoid the separation of whey protein (**Supplemental G**). A blend of CMC with additional stabilizers is also proposed for stabilizing soft ices. CMC can also be used to stabilize the consistency of sherbet.

To retain pulp in fruit juice-based beverage suspensions throughout storage, CMC is included within their composition. The concentration that will produce optimal stability is dependent on the soluble solids content. The overall amount required is small because the viscosity of the product is high from the start. There are many beverages with suspended water-insoluble solid particles, such as cocoa, powdered green tea, and calcium carbonate, in which the suspension stability and resuspension property of the solid particles in the beverage are relevant to the development of commercial products. Cocoa and CMC particles cohere tightly and the aggregated particles further interact weakly with the milk component, leading to stabilization of the entire cocoa beverage system. CMC also decreases or prevents the formation of oily rings in bottlenecks. CMC is routinely added to fruit-product combinations following their hydration. If additional hydrocolloids, for instance, xanthan gum (**Chapter 15**) and propylene glycol alginate (PGA) (**Chapter 3**) are employed, then low-viscosity CMC is preferable, whereas if CMC is the only component, it should be of high viscosity.

CMC is recommended in dry blends for acidic milk drinks and for the production of sparkling drinks. CMC is also utilized to provide body to low-calorie drinks. In the manufacture of instant hot beverages, HPC exploited as a suspending agent mixed with a warm liquid (e.g., alcohol) facilitates consistent release over extended periods of use. Cellulosic derivatives are also used to thicken alcohols employed for flavoring or for cocktails.

Cellulosic derivatives are used in baked goods to overcome difficulties connected with flour quality and to improve the quality of the prepared product. CMC can be employed in cake mixes as a batter binder. CMC and other cellulose ethers rapidly enhance viscosity and decrease air inclusion, thereby preventing undesirable decreases in viscosity throughout food preparation. The addition of CMC to cakes improves the homogeneity of the product, particularly if raisins, pieces of crystallized fruit, or chocolate chips are mixed into the preliminary mixture. The moisture content retained by the cakes subsequent to such an addition is increased, resulting in a less stale product. CMC is utilized to adjust paste consistency if portions of raisins, grains, or supplementary components are brought together into the product mixture. In addition, water retention is enhanced. In contrast, MC derivatives tend to lose moisture during baking and shaping of the gel used for the manufacture of gluten-free breads. A lesser amount of gluten increases dough-mixing time, decreases loaf volume, and alters the consistency to an undesirable roughness. The inclusion of up to 0.5% MHPC with low-gluten flour will improve these parameters. If other types of flour, for instance rice or potato, are used for gluten-free breads, then a mixture of CMC and MHPC in the flour will provide characteristics comparable to wheat breads. High-fiber breads can be manufactured by including a blend of flour, bran, CMC, and guar gum. CMC and MC derivatives are employed in additional baked goods, for instance, low-calorie cakes, biscuits, and ice-cream cones. A dust-free, dried anticaking agent based on cellulose has also been described. CMC is used in fruit pies and pastry fillings to improve water retention, and in doughnuts to decrease oil absorption during frying and to improve consistency. CMC is an anionic hydrocolloid, which reacts with proteins at their isoelectric pH to shape a complex. Formation is influenced by pH, molecular weight, CMC concentration, and salt content. The soluble complex (resulting from the reaction between CMC and casein or other milk proteins) is stable to heat treatment and storage. Consequently, preparations of acidified, pasteurized dairy products or dairy drinks become feasible when the gum (0.2–0.5%) is dissolved directly in the acidic milk product, or in the milk prior to the addition of acid or fruit. At neutral pH, milk proteins (**Supplement G**) can react with CMC, causing whey separation in low-viscosity mixes such as ice creams and ice milks. Separation can be avoided by reacting carrageenan with the caseinates (**Supplement G**) instead. At a pH of ~3.2, CMC is utilized to precipitate whey proteins (**Supplement G**).

CMC molecules appear to interact through attractive electrostatic forces with whey proteins in neutral and slightly acidic pH environments, leading to the formation of soluble protein–polysaccharide hybrids. At lower pH, limited or extensive whey protein–CMC interactions might take place, leading to modulation of the stability of the whey protein concentrate emulsions during ageing or subsequent heat or freeze treatment. Milk desserts favor a combination of starches (**Chapter 14**) and carrageenans (**Chapter 4**) to yield textural improvements. Combinations of MHPC and CMC can be utilized as a bulking agent. Whipped cream based on milk and vegetable proteins (**Supplement K**) can be improved by the addition of HPC.

MCC gel is an exceptional foam stabilizer. Its inclusion enables reducing fat by 3–4% in the item for consumption. At cellulose gel concentrations > 1% and after the addition of sucrose, thixotropic gels are formed. Thus gelled sucrose systems show altered rheological properties following MCC supplementation. Special foams for foam-mat drying were prepared from clarified apple juice by adding various concentrations of two different types of foaming agents—a protein (egg white) and a hydrocolloid—and whipping for different times; the most solid foams included 0.2% MC and 2 to 3% egg white. The degree of foam solidity could be predicted from its air volume fraction and average bubble size, independent of its formulation and method of preparation. MCC at a concentration of ~0.4% assists in maintaining the unique consistency of frozen desserts. Additional applications involve the thickening of food systems while preserving the desired mouthfeel, improving clingability of starch sauces, a decrease in calories and increase in fiber of particular foods, whitening effects, improvement of suspensions of chocolate solids in sterilized drinks, stabilization of emulsions, and its use in different formed extruded foods. Cellulose gels may also be employed in the manufacture of *surimi* (**Chapter 6**). The gel strength of commercial *surimi* was increased by the addition of phosphate blends. Phosphate blends also allow nonisothermal flow of *surimi*, enabling the use of extrusion machinery and processing methods to manufacture final products on site.

Coated extruded and structured products can potentially make use of cellulose derivatives, as binders of miniature portions of meat, fish, and potatoes for instance, in the manufacture of meat sticks and many potato-based products. MC-based coating plasticized with sorbitol might serve as an alternative for healthier potato chips. The most effectual formulation was 10 g/L MC with 7.5 g/L sorbitol, which decreased oil absorption by 30%. Possible reactivity of MC or CMC with proteins and cohesion ability are vital in such products. The coating batter of such products can also include cellulosic derivatives. This enhances batter adhesion to the product, stabilizes it in subsequent freeze-thaw cycles, and lessens oil incorporation during deep-fat frying. MC and MHPC films are tough and clear, soluble in water and insoluble in organic liquids, fats, and oils. This and thermogelation are beneficial in the frying of extruded food items. Predusting of a product prior to cooking or reheating can yield crispier products following microwave cooking. In coating nonstructured products, CMC can be added at a maximum of 4% to the batter. The predust should contain at least 20% MHPC, having 27–31% methoxyl groups and 6–12% hydroxypropyl groups. Because cellulose derivatives decrease oil absorption, they are also employed as components or formers of edible barriers in composite foods such as ice-cream cones and ice cream, tomato paste and pizza bases. In sausage casings, CMC improves peelability in high-speed peeling machines. Additional important uses are the thickening of emulsified and nonemulsified sauces to prevent separation, and to preserve the texture and appearance of sauces or soups that are heated prior to eating, making use of the thermogelation ability of cellulose derivatives. In other words, the thermogelation ability of MC and HPMC makes them helpful as viscosity formers in sauces that require heating, as the hydrocolloids prevent the sauces from becoming fluid during the process, thereby preserving the product's expected consistency and appearance. Products that are consumed hot should be prepared cool to avoid problems with MC or MHPC hydration at concentrations of 0.25–1.0%.

Additional applications include coating food products such as meat and poultry, where water-absorption properties are required, and as a water binder in semi-humid products. Antimicrobial packaging systems (films) based on olive leaf extract and MC were found to have an antimicrobial effect against *Staphylococcus aureus* on Kasar cheese and may have applications in a variety of food products owing to their comparable cost, efficient antimicrobial activity, few regulatory concerns, and environmental friendliness. Cellulose derivatives can additionally be used in the low-calorie products (bakery, meat, and dairy) most desired by consumers.

Recipes with Cellulose Derivatives

Parsley Spaghetti (Figure 5.1)

Ingredients

(Serves 5)
50 g parsley
400 mL (1⅗ cups) water
15 g (3⅓ Tbsp) MC powder
5 g (1 tsp) salt
24 g (2 Tbsp) olive oil
Hot water, syringe
Hot soup (optional)

Preparation

1. Chop parsley in a food processor, add water, and process (**Figure 5.1A**). Pour mixture into a bowl.
2. Whisk MC into parsley dispersion (**Figure 5.1B**) until it thickens, and then mix with a hand mixer (**Figure 5.1C**) (**Hint 1**).
3. Whisk or mix in salt and olive oil (**Figure 5.1D**), and then refrigerate overnight.
4. Boil water in a pot. Draw refrigerated mixture into a syringe (**Figure 5.1E**), and inject it into the boiling water (**Figure 5.1F**).
5. If desired, prepare hot soup and add to parsley spaghetti (**Figure 5.1G**) (**Hint 2**).

pH = 6.02

Hints

1. Use caution when mixing the liquid dispersion with a hand mixer to avoid splattering.
2. You can also inject the dispersion directly into boiling soup.

FIGURE 5.1 Parsley spaghetti. (**A**) Grind parsley, add water and process. (**B**) Whisk MC into the parsley dispersion. (**C**) Blend with a hand mixer. (**D**) Add salt and olive oil, and stir. (**E**) Draw mixture into a syringe. (**F**) Squeeze out the mixture into boiling water. (**G**) Serving suggestion: parsley spaghetti in soup.

This is a modern gastronomy recipe. Other green leaves, or peas, can be used instead of the parsley. Use hot liquids such as soup to prepare the spaghetti because gelation occurs at higher temperatures using MC.

Takoyaki (Octopus Dumpling) (Figure 5.2)

Ingredients

(Serves 5)

5 g (a little over 1 Tbsp) MC powder

70 g (a little over ⅔ cup) weak flour

65 g (a little under ⅔ cup) strong flour

10 g (a little over 1 Tbsp) white sugar

10 g (a little over 2 Tbsp) dried bonito powder

700 mL (2⅘ cups) Japanese broth†

2 eggs

100 g boiled octopus

40 g (a little under ½ cup) fried batter (**Figure 5.2A**, bottom left)

50 g green onion

FIGURE 5.2 *Takoyaki* (octopus dumpling). (**A**) Bottom left: fried batter; bottom right: finely chopped pickled ginger; top: green onion rounds. (**B**) Mix MC, flours, white sugar, and dried bonito powder in a bowl. (**C**) Pour egg dispersion into the powder mixture and stir. (**D**) Cut octopus into bite-size pieces. (**E**) Prepare frying equipment and hot plate. (**F**) Pour batter. (**G**) Layer on octopus pieces, green onion, fried batter, and pickled ginger. (**H**) Tuck into sphere form and fry. (**I**) Serve.

50 g pickled ginger (**Figure 5.2A**, bottom right)

Vegetable oil, *takoyaki* sauce, mayonnaise, dried bonito flakes, green laver

†**WATER CAN SUBSTITUTED FOR BROTH.**

Preparation

1. Mix MC, both flours, white sugar, and dried bonito powder in a bowl (**Figure 5.2B**).
2. In a separate bowl, beat eggs and mix with Japanese broth.
3. Pour the egg dispersion (**2**) into the powder mixture (**1**) while stirring (**Figure 5.2C**).
4. Cut the octopus into bite-size pieces (**Figure 5.2D**), cut green onion into rounds (**Figure 5.2A**, top), and finely chop pickled ginger (**Figure 5.2A**, bottom right).
5. Heat iron plate for *takoyaki* (**Figure 5.2E**), and grease with vegetable oil by spreading with a paper towel (**Hint 1**). Pour the batter (**3**) (**Figure 5.2F**), and layer on octopus pieces, green onion (**4**), fried batter, and pickled ginger (**Figure 5.2G**).

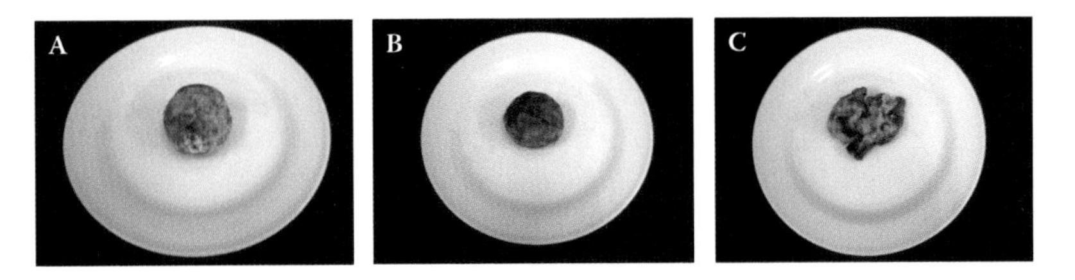

FIGURE 5.3 Adding MC to *takoyaki* makes for a softer, more liquid texture, and better freeze-thaw resistance. (**A**) Frozen *takoyaki*. (**B**) Thawed *takoyaki*. (**C**) Inside of thawed *takoyaki*.

6. Fry the batter (**3**) tucking or rolling it into a spherical form with a bodkin or fork until fully grilled (**Figure 5.2H**) (**Hint 2**).
7. Serve *takoyaki* with *takoyaki* sauce and mayonnaise, garnished with dried bonito flakes and green laver (**Figure 5.2I**).

pH of batter = 6.43

Hints

1. Heat vegetable oil well, and spread it all over the plate to prevent sticking.
2. When the batter browns on the bottom and sides, roll it over. Repeat the process until fully grilled into perfect *takoyaki* balls.

This recipe is famous in the Osaka region of Japan, where *takoyaki* shops abound. It is served as a snack food.

Adding MC to this recipe leads to a softer, more liquid texture, with better resistance to freezing and thawing (**Figure 5.3**).

Soy Burger Patties (Figure 5.4)

Ingredients

(Serves 5)
5 g (a little over 1 Tbsp) MHPC powder (**Hint 1**)
100 g soy protein flakes
250 mL (1 cup) water
75 g shortening
10 g (1⅔ Tbsp) cornstarch
10 g (a little over 1 Tbsp) potato starch
100 g onion
5 g (a little under 1 tsp) salt
Black pepper, nutmeg, garlic powder†
Ketchup (optional)

†OR USE YOUR FAVORITE SPICES.

FIGURE 5.4 Soy burger patties. (**A**) Mix MHPC, soy protein flakes and water. (**B**) Stir-fry cut onion and cool. (**C**) Add cornstarch, potato starch, cut onion, salt, and spices to the MHPC mixture and combine. (**D**) Form patties. (**E**) Bake or fry. (**F**) Garnish before serving. (**G**) Soy burgers.

Preparation

1. Mix MHPC and soy protein flakes in a bowl, add water with stirring (**Figure 5.4A**), and refrigerate overnight.
2. Chop onion finely and stir-fry in shortening, then cool (**Figure 5.4B**) (**Hint 2**).
3. Mix the MHPC mixture (**1**), cornstarch, potato starch, cut onion, salt, and spices in a bowl (**Figure 5.4C**), and knead well by hand.
4. Form the mixture into patties (**Figure 5.4D**), and bake at 200°C or fry in vegetable oil (**Figure 5.4E**).
5. Garnish with ketchup if desired (**Figure 5.4F**).

$$pH = 6.18$$

Hints

1. MC powder will give the same result.
2. Raw onion can be used for a fresh and crispier texture.

This recipe is good for vegetarian or macrobiotic diets. It is also good for soy burgers (**Figure 5.4G**).

Pineapple Ice Cream (Figure 5.5)

Ingredients

(Serves 5)
500 mL (2 cups) milk
1 g (¼ tsp) CMC powder
6 g (1 tsp) glycerol monostearate (GMS)
12 g (2 tsp) cornstarch
45 g (3¾ Tbsp) granulated sugar
200 g (⅘ cup) heavy cream
200 g (6 slices) canned pineapple
60 g (4 Tbsp) canned pineapple syrup
Pineapple flavoring, yellow food coloring (optional)

Preparation

1. Mix CMC, GMS, and cornstarch in a pan, add 100 mL milk, and heat with stirring (**Figure 5.5A**) until GMS dissolves and the mixture boils (more than 90°C).
2. Remove the CMC mixture from the heat, and add granulated sugar and the rest of the milk to it, then cool to 5°C (**Figure 5.5B**).
3. Cut canned pineapple into small pieces.
4. Add heavy cream, pineapple syrup, the cut pineapple, pineapple flavoring, and yellow food coloring to the cooled mixture (**2**) (**Figure 5.5C**) and mix.

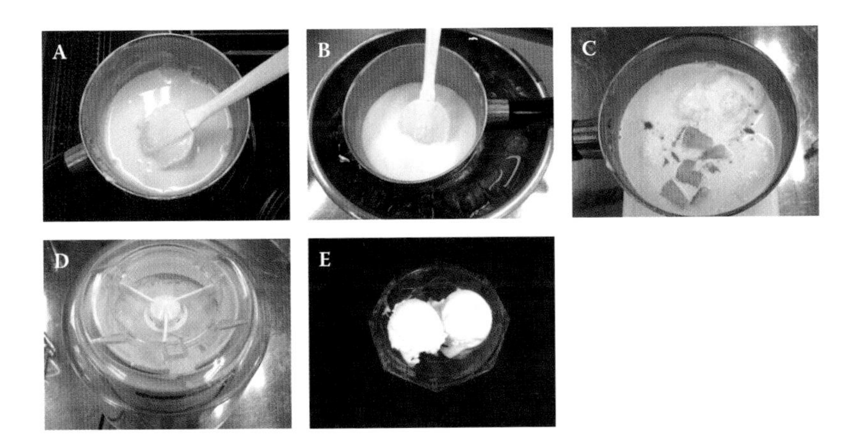

FIGURE 5.5 Pineapple ice cream. (**A**) Heat milk with CMC, GMS, and cornstarch while stirring. (**B**) Add granulated sugar and the remaining milk and allow to cool. (**C**) Add cut pineapple, cream, pineapple syrup, flavoring, and food coloring to the cooled mixture. (**D**) Pour ice-cream mixture into an ice-cream maker and follow manufacturer's instructions (**E**) Serve.

5. Pour the ice-cream mix (**4**) into an ice-cream maker (**Figure 5.5D**) (**Hint 1**), prepare according to the manufacturer's instructions and serve (**Figure 5.5E**).

pH = 5.83

Hint

1. If you do not have an ice-cream maker, put the ice-cream mix in a stainless steel container and freeze for 1 hour until the edge of ice cream begins to set. Whisk the half frozen ice cream, and return to freezer. Repeat this process several times.

This is an egg-free ice cream. You can use fresh pineapple, or other fruits such as mango, coconut, pear, orange, strawberry, and so forth.

Sugarpaste (Figure 5.6)

Ingredients

(Makes ~10 sugarpaste crafts)
40 g egg white
40 g (a little under 2 Tbsp) corn syrup
625 g (6¼ cups) icing sugar (powdered sugar)
2 g (1⅓ tsp) CMC powder
Shortening, food coloring

FIGURE 5.6 Sugarpaste. (**A**) Mix egg white, corn syrup, and sifted sugar. (**B**) Knead in CMC powder. (**C**) Soft, smooth and pliable paste. (**D**) Knead in food coloring. (**E**) Roll out paste. (**F**) Cut out desired shapes.

Preparation

1. Put egg white and corn syrup in a bowl, sift icing sugar in (**Figure 5.6A**) and stir.
2. Knead in CMC powder (**Figure 5.6B**) until the paste becomes soft, smooth, and pliable.
3. Knead in shortening if the paste is dry and cracked (**Figure 5.6C**) (**Hint 1**), add food coloring if desired (**Figure 5.6D**) (**Hint 2**).
4. Cover a cutting board with plastic wrap and roll out the paste (**Figure 5.6E**); cut out the desired shapes (**Figure 5.6F**).

pH = 6.81

Hints

1. Add more icing sugar or CMC powder if the paste is soft and sticky.
2. To store the sugarpaste, place in a plastic bag and then in an airtight container, and store at room temperature or in the refrigerator. Knead it before use. This paste can be frozen for up to 3 months.

This recipe is used to create cake decorations. Another idea is to use it for sugarpaste crafts.

Low-Fat Whipped Cream (Figure 5.7)

(Makes 200 g)

Ingredients

0.6 g (⅖ tsp) HPC powder
60 mL (4 Tbsp) water
140 g (a little over ½ cup) heavy cream

Preparation

1. Mix HPC and water with a hand mixer for 15 min (**Figure 5.7A**) (**Hint 1**).
2. Whip heavy cream with a hand mixer (**Figure 5.7B**) and add the HPC solution with continued mixing (**Figures 5.7C and 5.7D**).

pH = 6.33

Hint

1. Mix as vigorously as possible, and stir for a longer time for a creamier HPC dispersion.

This recipe provides a method for reducing calories in whipped heavy cream. HPC can be used as a substitute for fat.

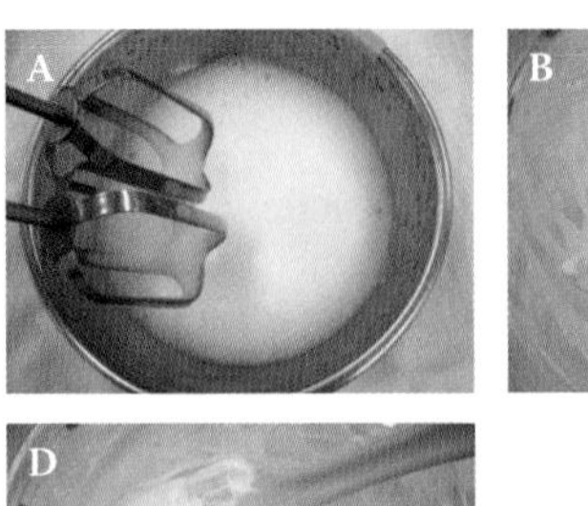

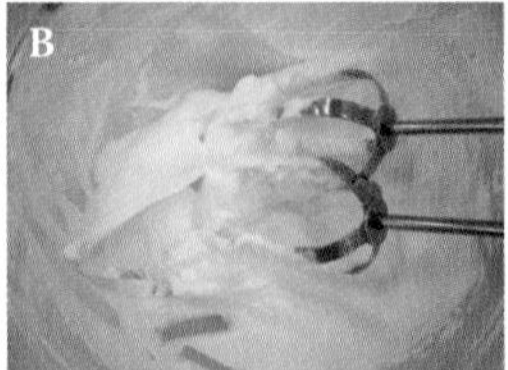

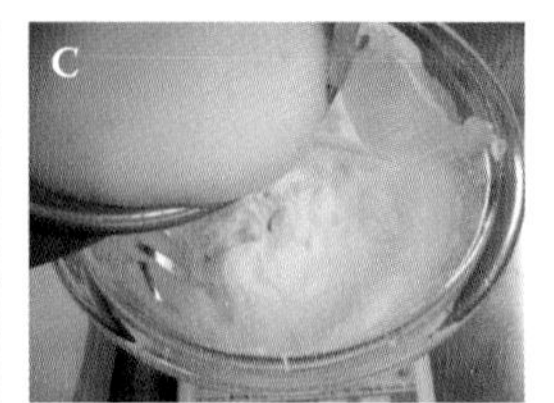

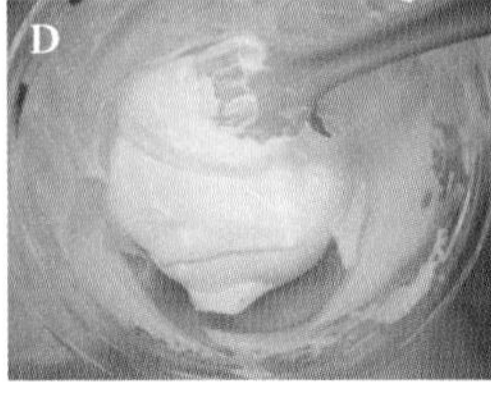

FIGURE 5.7 Low-fat whipped cream. (**A**) Mix together HPC and water. (**B**) Whip heavy cream. (**C**) Add HPC solution to the whipped cream. (**D**) Mix to combine HPC solution.

Tips for the Amateur Cook and Professional Chef

- CMC concentration (%) based on final total weight should be 0.16–0.20 for ice cream, sherbets, dry ice-cream mixes, ice milk, and soft frozen desserts, whereas for fruit purées or chocolate syrups, 0.75–1.0 is required.
- The clarity and sheen of cherry, blueberry, and raspberry fruit pie fillings are improved by addition of high-viscosity HPMC at a level of 0.5% of the total weight of the filling.
- Lower-viscosity HPC grades are typically used in whipped toppings at 0.2–0.3% of the topping.
- HPC is soluble in ethanol. It gives clear solutions in aqueous ethanol at concentrations of less than 50%.
- In salad dressings, CMC is typically used at a concentration of ~1.0%.
- CMC levels for use in soft drinks are typically 0.025–0.5% of the drink. For buttermilk drinks at pHs 4.5–4.6, 0.3–0.4% CMC is required.

REFERENCES AND FURTHER READING

Abdel-Baky, A. A., El-Fak, A. M., Abo El-Ela., and A. A. Farag. 1981. Fortification of Domiati cheese milk with whey proteins/carboxymethylcellulose complex. *Dairy Ind. Int.* 46:29, 31.

Anon. 1985. Methylcellulose in low-gluten bread. *Food Feed Chem.* 17:576.

Aqualon Co. 1987. *Klucel HPC, physical and chemical properties*. Wilmington, DE: Aqualon Co.

Aqualon Co. 1988. *Cellulose gum, physical and chemical properties*. Wilmington, DE: Aqualon Co.

Aqualon Co. 1989. *Culminal MC, MHEC, MHPC, physical and chemical properties*. Wilmington, DE: Aqualon Co.

Ayana, B. and K. N. Turhan. 2009. Use of antimicrobial methylcellulose films to control *Staphylococcus aureus* during storage of Kasar cheese. *Packaging Technol. Sci.* 22:461–9.
Baird, G. S. and J. K. Speicher. 1962. Carboxymethylcellulose. In *Water-soluble resins,* ed. R. L. Davidson and M. Sittig, 69–88. New York: Reinhold Publishing Corp.
Bayfield, E. G. 1962. Improving white layer cake quality by adding CMC. *Bakers Digest* 36:50–2, 54.
Bernal, V. M. and D. W. Stanley. 1989. Technical note: Methylcellulose as a binder for reformed beef. *Int. J. Food Sci. Technol.* 24:461–4.
Brown, W. and D. Henley. 1964. The configuration of the polyelectrolyte sodium carboxymethylcellulose in aqueous sodium chloride solutions. *Macromol. Chem. Phys.* 79:68–88.
Butler, R. W. 1962. Free acid cellulose ether film. U.S. Patent No. 3,064,313.
Butler, R. W. and G. I. Keim. 1965. Insolubilizing CMC with cationic epichlorohydrin-modified polyamide resin. U.S. Patent No. 3,224,986.
Butler, R. W. and E. D. Klug. 1980. Hydroxypropylcellulose. In *Handbook of water-soluble gums and resins*, ed. R. L. Davidson, chap. 13. New York: McGraw-Hill.
Chappell, R. A. 1995. Low dust powdered cellulose. U.S. Patent No. 5,391,382.
Cottrell, J. I. L., Pass, G., and G. O. Phillips. 1979. Assessment of polysaccharides as ice cream stabilisers. *J. Sci. Food Agric.* 30:1085–8.
D'Amico, L. R., Waring, S. E., and J. M. Lenchin. 1989. Composition for preparing freeze-thaw microwaveable pre-fried foodstuffs. U.S. Patent No. 4,842,874.
DeButts, E. H., Hudy, J. A., and J. H. Elliott. 1957. Rheology of sodium carboxymethylcellulose solutions. *Ind. Eng. Chem.* 49:94–8.
Deleon, J. R. and M. G. Boak. 1984. Method for preventing separation in fruit juice-containing products. U.S. Patent No. 4,433,000.
Desmarais, A. J. 1973. Hydroxyalkyl derivatives of cellulose. In *Industrial Gums,* 2nd ed., ed. R. L. Whistler, 649–73. New York: Academic Press.
Donges, R. 1990. Nonionic cellulose ethers. *Brit. Polym. J.* 23:315–26.
Dow Chemical Co. 1974. *Handbook on methocel cellulose ether products.* Midland, MI: Dow Chemical Co.
Foda, Y. H., Mahmoud, R. H., Gamal, N. F., and S. Y. Kerrolles. 1987. Special bread for body-weight control. *Ann. Agric. Sci.* 32:397–407.
Francis, P. S. 1961. Solution properties of water-soluble polymers. *J. Appl. Polym. Sci.* 5:261–70.
Ganz, A. J. 1966. Cellulose gum—A texture modifier. *Manuf. Confect.* 46:23–33.
Ganz, A. J. 1973. Some effects of gums derived from cellulose on the texture of foods. *Cereal Sci. Today* 18:398.
Ganz, A. J. 1974. How cellulose gum reacts with protein. *Food Eng.* June:67–69.
Ganz, A. J. 1977. Cellulose hydrocolloids. In *Food Colloids*, ed. H. Graham, 382–417. Westport, CT: AVI Press.
Glicksman, M. 1969. *Gum technology in the food industry.* New York: Academic Press.
Glicksman, M. 1986. In *Food hydrocolloids*, vol. III, 3–121. Boca Raton, FL: CRC Press Inc.
Glicksman, M., Frost, J. R., Silverman, J. E., and E. Hegedus. 1985. High quality, reduced-calorie cake. U.S. Patent No. 4,503,083.
Goff, H. D. and V. G. Davidson. 1994. Controlling the viscosity of ice cream mixes at pasteurization temperatures. *Modern Dairy* 73:12, 14.
Greminger, G. K. Jr. and K. L. Krumel. 1980. Alkyl and hydroxyalkylcellulose. In *Handbook of water-soluble gums and resins,* ed. R. L. Davidson, 3-1 to 3-25. New York: McGraw-Hill.

Hansen, P. M. T., Hildalgo, J., and I. A. Gould. 1971. Reclamation of whey protein with carboxymethylcellulose. *J. Dairy Sci.* 54:830–4.

Hercules Inc. 1978. *Cellulose gum-chemical and physical properties*. Wilmington, DE: Hercules Inc.

Heuser, E. 1944. *The chemistry of cellulose*, 379–91. New York: John Wiley.

Higgins, T. E. and D. P. D. Madsen. 1986. Cellulosic casing with coating comprising cellulose ether, oil and water-insoluble alkylene oxide adduct of fatty acids. U.S. Patent No. 4,596,727.

Hood, H. P. 1981. Whipped sour cream in aerosol can. *Food Eng*. 53:62.

Huber, C. S. and D. M. Rowley. 1988. Soft serve frozen yoghurt mixes. U.S. Patent No. 4,737,374.

Huber, C. S., Rowley, D. M., and J. W. Griffiths. 1989. Soft frozen water ices. U.S. Patent No. 4,826,656.

Ingram, P. and H. G. Jerrard. 1962. Measurement of relaxation times of macromolecules by the Kerr Effect. *Nature* 196:57–8.

Isogai, A. and R. H. Atalla. 1991. Amorphous cellulose stable in aqueous media: Regeneration from SO_2-amine solvent system. *J. Polym. Sci. A*, 29:113–9.

Keeney, P. G. 1982. Development of frozen emulsions. *Food Technol.* November:65–70.

Keller, J. 1984. *Sodium carboxymethylcellulose*. Special Report, NY State Agricultural Experimental Station, No. 53:9–19.

Kester, J. J. and O. Fennema. 1989. An edible film of lipids and cellulose ethers barrier properties to moisture vapor transmission and structural evaluation. *J. Food Sci*. 54:1383–9.

Klug, E. D. 1965. Sodium carboxymethylcellulose. In *Encyclopedia of polymer science and technology*, vol. III, 520–39. New York: Interscience Publishers Inc.

Klug, E. D. 1966. Mixed cellulose ethers. U.S. Patent No. 3,278,521.

Klug, E. D. 1967. Alkyl hydroxyalkyl cellulose ethers containing sulfoalkyl substituents U.S. Patent No. 3,357,971.

Koupantsis, T. and V. Kiosseoglou. 2009. Whey protein–carboxymethylcellulose interaction in solution and in oil-in-water emulsion systems. Effect on emulsion stability. *Food Hydrocolloids* 23:1156–63.

Krassig, H. A. 1985. Structure of cellulose and its relation to properties of cellulose fibers. In *Cellulose and its derivatives: Chemistry, biochemistry and applications*, ed. J. F. Kennedy, G. O. Phillips, D. J. Wedlock, and P. A Williams, 3–25. New York: Halsted Press.

Marteli, M. R., Carvalho, R. A., Sobral P. J. A., and J. S. Santos. 2008. Reduction of oil uptake in deep fat fried chicken nuggets using edible coatings based on cassava starch and methylcellose. *Ital. J. Food Sci.* 20:111–7.

McCormick, C. L. and P. A. Callais. 1987. Derivatives of cellulose in LiCl and N,N-dimethylacetamide solutions. *Polymer* 28:2317–22.

Moore, L. J. and C. F. Shoemaker. 1981. Sensory textural properties of stabilized ice cream. *J. Food Sci.* 46:399–402, 409.

Nielsen, G. R. and G. M. Pigott. 1994. Gel strength increased in low-grade heat-set surimi with blended phosphates. *J. Food Sci*. 59:246–250.

Nussinovitch, A. 1997. *Hydrocolloid applications: Gum technology in the food and other industries*. London: Blackie Academic & Professional.

Nussinovitch, A. 2003. *Water soluble polymer applications in foods.* Oxford, UK: Blackwell Publishing.

Nussinovitch, A. 2010. *Polymer macro- and micro-gel beads: Fundamentals and applications*. New York: Springer.

Raharitsifa, N., Genovese, D. B., and C. Ratti. 2006. Characterization of apple juice foams for foam-mat drying prepared with egg white protein and methylcellulose. *J. Food Sci.* 71:E142–51.

Samuels, R. M. 1974. Quantitative structural analysis of mechanical behavior in polycrystalline polymers. *Appl. Polym. Symp.* 24:37–43.

Sanderson, G. R. 1981. Polysaccharides in foods. *Food Technol.* July:50–7.

Sanz, T., Fernandez, M. A., Salvador, A., Munoz, J., and S. M. Fiszman. 2005. Thermogelation properties of methylcellulose (MC) and their effect on a batter formula. *Food Hydrocolloids* 19:141–7.

Shenkenberg, D. R., Chang, J. C., and L. F. Edmondson. 1971. Developed milk orange juice. *Food Eng.* April:97–98, 101.

Sirett, R. R., Eskritt, J. D., and E. J. Derlatka. 1981. Dry beverage mix composition and process. U.S. Patent No. 4,264,638.

Stelzer, G. I. and E. D. Klug. 1980. Carboxymethylcellulose. In *Handbook of water-soluble gums and resins*, ed. R. L. Davidson, 4-1 to 4-28. New York: McGraw-Hill.

Suderman, D. R., Wiker, J., and F. E. Cunningham. 1981. Factors affecting adhesion of coating to poultry skin. *J. Food Sci.* 46:1010–1.

Tavera-Quiroz, M. J., Urriza, M., Pinottia, A., and N. Bertola. 2012. Plasticized methylcellulose coating for reducing oil uptake in potato chips. *Sci. Food Agric.* 92:1346–53.

Thomas, W. R. 1986. Microcrystalline cellulose. In *Food Hydrocolloids*, vol. III, ed. M. Glicksman, 9–43. Boca Raton, FL: CRC Press Inc.

Vincent, A. and S. Harrison. 1987. Stabilizing dressings and sauces. *Food Trade Rev.* October:527–8, 531.

Ward, K. Jr. and P. A. Seib. 1970. Cellulose, lichenan and chitin. In *The carbohydrates—Chemistry and biochemistry*, 2nd ed., vol. 2A, ed. W. Pigman, D. Horton, and A. Herp, 413–45. New York: Academic Press.

Whistler, R. L. 1973. *Industrial gums*, 2nd ed. New York: Academic Press.

Whistler, R. L. and J. R. Zysk. 1978. Carbohydrates. In *Encyclopedia of chemical technology*, 3rd ed., vol. 4, ed. M. Grayson and E. Eckroth, 535–55. New York: John Wiley & Sons.

Yaginuma, Y. and T. Kijima. 2006. Effects of microcrystalline cellulose on suspension stability of cocoa beverage. *J. Dispersion Sci. Technol.* 27:941–8.

Zecher, D. and R. Van Coillie. 1992. Cellulose derivatives. In *Thickening and gelling agents for food*, ed. A. Immson, 40–65. Glasgow: Blackie Academic & Professional.

Zejian, L., Baodong, Z., and C. Lijiao. 1994. Test on the rheological behavior of beverage stabilizer and beverage quality control. *Trans. Chinese Soc. Agric. Eng.* 10:110–3.

6

Curdlan

Historical Background

During studies on the production of succinoglucan by *Alcaligenes faecalis* var. *myxogenes* strain 10C3, a mutant was discovered which could not create this acidic extracellular polymer. Instead, it produced a neutral polysaccharide, which was insoluble in neutral solution and which, when heated in aqueous suspension to 100°C for a few minutes, formed a firm and resilient gel. The polysaccharide was named *curdlan* by its discoverer, Prof. Tokuya Harada, in 1964. The name was coined from the word *curdle,* describing its gelling behavior at high temperatures.

Production

Curdlan is produced via fermentation. The method makes use of the above-mentioned mutant bacterial strain, which can be isolated from the soil. A typical process for the manufacture of curdlan includes fermentation of the raw materials, formation of a solution with NaOH and stirring, precipitation by alcohol, removal of the strain, clarification, sieving, crushing, and air-blast drying to obtain the final product. Commercial curdlan might include cellular debris, proteins, and nucleic and other organic acids. The hydrocolloid curdlan is currently produced by Kirin Kyowa Foods Co., Ltd., Tokyo, Japan, at more than 100 tons per annum.

Chemical Structure

Curdlan is a microbial polysaccharide produced by *A. faecalis* var. *myxogenes,* made up of D-glucosyl residues connected by β-1,3-linkages. In other words, the polysaccharide is characterized by repeating glucose subunits joined by a β-linkage between the first and third carbons of the glucose ring. The structure of curdlan is similar to that of agarose (**Chapter 2**), carrageenan (**Chapter 4**), and gellan gum (**Chapter 10**). Nevertheless, in contrast to κ- and ι-carrageenans and gellan gum, curdlan is a neutral hydrocolloid devoid of acidic components. The average degree of polymerization of curdlan is ~450. The degree of polymerization, or DP, is usually defined as the number of monomeric units in a macromolecule, polymer or oligomer molecule. For a homopolymer, there is only one type of monomeric unit. Its average molecular weight is ~44,000 to 77,000. Electron microscopy studies of the gel structures of the polymer of various molecular weights suggest that the gel is composed of long, 100–200-Å wide microfibrils, and indicates an apparent relationship between the length of the microfibrils and gel strength.

Regulatory Status and Toxicity

Curdlan was launched in Japanese markets in 1989 to improve the texture and water-holding capacity of processed foods and to produce novel foods. Curdlan is used as a food additive (emulsifier and stabilizer), for which a petition has been filed and a regulation issued for a permitted substance as an optional ingredient in standardized foods. Curdlan has been approved by the U.S. Food and Drug Administration (FDA) as a direct food additive.

Functional Properties

Solution Properties and Conformations

Curdlan is a tasteless, odorless white powder, which swells upon addition of water. The most important structure is a long linear chain; however, curdlan forms additional complex tertiary structures as a result of both intra- and intermolecular hydrogen bonding.

Curdlan dissolves in an alkaline aqueous solution but not in water at ambient temperature. The insolubility in water results from hydrogen-bonded crystalline domains similar to those in cellulose. 1,3–β-D-glucans with DP < 25 have been reported to be soluble in water, as has curdlan in water at elevated temperatures. An aqueous suspension of curdlan becomes clear upon heating at temperatures >55°C. In aqueous NaOH solution, curdlan's conformation changes from a triple helix to random coil depending on the concentration of NaOH. Transition occurs at a concentration of 0.19–0.24 M NaOH. A significant enhancement in viscosity was identified in the range of 0.05–0.1 M NaOH in line with the solvation of triple helices; subsequently, there is a viscosity reduction above 0.25 M NaOH as the triple molecules are dissociated into single chains of lower molecular weight.

Aqueous Suspension Properties

Curdlan can only be suspended in water, due to its insolubility. A well-dispersed suspension can be achieved by using physical means such as homogenization. Such suspensions are stable and no separation is observed when left standing. A paste-like suspension is formed. Curdlan preparations at 38°C demonstrate Newtonian behavior at 0.1% and plastic behavior at concentrations >0.2%.

Gelation

Although curdlan is insoluble in water, its aqueous suspension is capable of forming two types of heat-induced gel, depending on the temperature: a "low-set gel" is formed when the aqueous suspension is heated to between 55 and 60°C and then cooled to under 40°C. This gel is thermoreversible, and junction zones are formed by hydrogen bonds, analogous to agar–agar (**Chapter 2**) or gelatin (**Chapter 10**); a "high-set gel" is formed by heating the aqueous suspension above 80°C. The gel is thermoirreversible, and the junction zones in the gel are formed by hydrogen bonds and as well as by hydrophobic interactions. The high-set gel is stable against high-temperature treatment, for instance retorting, and remains tasteless, odorless, and colorless. The incipient temperature for thermoirreversible gelation is lowered by an increase in curdlan concentration. Furthermore, curdlan forms thermoirreversible gels in food systems, even when processed at lower than 80°C, due to increases in its components.

Gel formation of curdlan at 100–130°C and performance of frozen curdlan gels were studied by determining gel strength and syneresis. Even when aqueous suspensions of curdlan were heated to 100°C or higher, they formed a gel, the strength of which increased with temperature. Curdlan gel was also stable against freezing and thawing. The syneresis of a 4% curdlan gel after freezing and thawing decreased from 20.6% to about 2.1% by adding 5% waxy corn starch (**Chapter 14**) and to 8.9% by addition of 20% sucrose. Encouraged by these findings, new applications for foods have been developed, especially those subject to retorting and/or freezing.

Curdlan crystallizes directly into two or more polymorphic forms. Its native form, which is frequently imperfectly crystalline, readily yields a highly crystalline, hydrated polymorph upon annealing in the presence of water, and upon additional annealing under vacuum, changes to an anhydrous crystalline structure. The latter was revealed by X-ray diffraction analysis to be a triplex (three-stranded) structure. The structure of the triplex is stabilized by extensive interstrand and interhelix hydrogen bonds.

Commercial Food Applications

Curdlan can be part of unique food products and its food applications are numerous. A few examples are noodles, *kamaboko* (boiled fish paste), sausages and hams, a variety of processed cooked foods, processed rice cakes, cakes, ice cream and jellies. Curdlan can be used in edible films and dietetic foods and in numerous fabricated foods such as noodle-shaped *tofu*, processed *tofu* (frozen, retorted, or freeze-dried), within frozen thin-layered gel foods, in frozen konjac-like gel foods, and in heat-resistant cheese foods. Curdlan is used for texture modification, as a binding agent, for shape retention, and as a gelling agent that is stable against heating and freeze-thawing. In addition, it functions in film formation and as a low-energy ingredient. The level of use is often as high as 0.2 to 2.0%. Nevertheless in processed rice cakes, jellies, fabricated foods, and especially edible films and dietetic foods, the level is much higher, up to 10%, and in dietetic foods even higher than that.

Curdlan is a useful food additive not only as a texture improver in sausage but as a fat mimetic as well. A nonfat sausage using a curdlan-based fat mimetic, including two ingredients together with curdlan, was manufactured and evaluated for its quality. Viscoelastic properties of the nonfat sausage with the curdlan-based fat mimetic system were very close to those of the control (20% fat), particularly when reheated. The characteristics of curdlan gel in meat products were investigated in a model system, by means of differential scanning calorimetry (DSC) and by measuring dynamic viscoelasticity. The following conclusions were drawn: curdlan forms a nearly complete gel in a meat gel prepared at 75°C, which is a practical heating temperature. The gel within the meat reveals strong thermoirreversibility; the results of measuring the viscoelasticity of curdlan gels suggest that the rheological properties of a 75°C gel become similar to those of a thermoirreversible 90°C gel with increasing ionic strength, due to inhibition of hydrogen bond formation in the cross-linking of junction zones. Curdlan in meat products strongly entraps any free or released water within its gel structure during the heating process.

Honey and jam are sticky products that are difficult to handle when used for sandwiches, bread, and so forth. Shaping, and including these adhesive products within thin-layered curdlan gels, simplifies their handling and makes them more convenient. Using curdlan, thin-layered gels flavored with honey, strawberries, and so forth, were manufactured. These products can be used as a filling for bread, sandwiches, and the like. Following the same methodology, thin-layered gels flavored with mayonnaise, pizza sauce, tomato ketchup, strained vegetables such as spinach, carrots, seasoned minced meat and fish, such as chicken and tuna, can also be easily prepared. Curdlan (0% to 1% w/w) can decrease oil uptake and moisture loss in doughnuts during deep-frying. This result can probably be attributed to the thermal gelling property of curdlan, and the heat-induced gel during frying probably functions as an oil and moisture barrier. Cellulose derivatives (**Chapter 5**) were less effective than curdlan in this respect.

Curdlan can form a gel even when large quantities of oils and fats are included. Curdlan used in noodle dough decreases the leakage of soluble ingredients and thus, in addition to softening the noodles, it results in clearer soup broths. Noodle-shaped soy milk gel is prepared with curdlan. In addition, it can be retorted at 110°C for 20 min., and thus this product is packaged in retort pouches with water. The noodles can be served cold or, if desired, with soup seasoned with bonito stock and soy sauce since the shape is not changed by heating. Tofu that contains curdlan retains its smooth texture after freezing and thawing. This function is used in *Koori* tofu in traditional Japanese dishes (**Figure 6.1A,B** shows freeze-dried and simmered *tofu* respectively). *Surimi* (**Figure 6.2A,B**) products that include curdlan show improved elasticity. Other improvements in the texture of frozen sweet products such as ice creams and cakes have also been reported.

FIGURE 6.1 *Koori* tofu (Freeze-dried *tofu*). (**A**) *Koori* tofu. (**B**) Simmered *Koori* tofu.

FIGURE 6.2 *Surimi*. (**A**) Package of commercial *surimi*. (**B**) Commercial *surimi*.

Recipes with Curdlan

Udon (Japanese Noodles) (Figure 6.3)

Ingredients

(Serves 4)

100 g (a little under ⅘ cup) strong flour

300 g (a little under 2⅕ cups) weak flour

FIGURE 6.3 *Udon* (Japanese noodles). (**A**) Knead dough. (**B**) Put dough in plastic bag and knead it. (**C**) and (**D**) Roll out dough to 2 mm thickness. (**E**) Cut piled dough into 3–6 mm wide strips. (**F**) Sprinkle flour on noodles. (**G**) Boil noodles. (**H**) and (**I**) Serving suggestions.

180 mL (a little over ⅔ cup) hot water (60°C)

12 g (⅔ Tbsp) salt

2 g (⅘ tsp) curdlan

Weak flour

Preparation

1. Mix strong flour, weak flour, and curdlan in a bowl.
2. Heat water (60°C) and salt well, and add to the flour mixture (**1**).
3. Knead the dough (**2**) until it holds together (**Figure 6.3A**), put it in a plastic bag, and knead by stepping on it repeatedly for about 15 minutes (**Figure 6.3B**).
4. Cover kneaded dough (**3**) with a damp dish towel or plastic film, and let it sit for 15 minutes at 25°C.
5. Knead the dough (**4**) for 10 minutes until it holds together, at 25°C.
6. Sprinkle a rolling pin and cutting board with weak flour; roll out the dough (**5**) to 1–2 mm thickness (**Figure 6.3C,D**), cut to 20 cm width, and pile up the dough slices.
7. Sprinkle weak flour on a knife, and cut the piled dough (**6**) into 3–6 mm width (**Figure 6.3E**) (**Hint 1**), then sprinkle weak flour on noodles (**Figure 6.3F**).
8. Boil water (about 10 times the volume of the noodles) in a large pot, add noodles, and boil for 10 minutes (**Figure 6.3G**).
9. Rinse noodles with cold water so that they will not stick.
10. Serve with Japanese soup (cold or hot), and garnish with green onion, kamaboko, and so on (**Figure 6.3H,I**).

pH of noodles = 5.13

pH of boiled noodles = 6.83

Hint

1. It is easy to cut noodles uniformly on a light wooden cutting board (**Figure 6.3E**).

Japanese people like noodles that do not break easily. If you want noodles to have elasticity, you should increase the ratio of strong flour to weak flour and knead well. Furthermore, adding curdlan makes *udon* elastic, improves resistance to freezing, and gives noodles a better texture.

Kamaboko (Fish Cake) (Figure 6.4)

Ingredients

(Makes 3 cakes)

350 g white fish fillets† (**Hint 1**)

7 g (a little over 1 tsp) salt (2% of fish)†

FIGURE 6.4 *Kamaboko* (fish cake). (**A**) Remove the skin. (**B**) Blot fish dry and grind with curdlan. (**C**) Add salt and other ingredients while grinding continuously. (**D**) Spread the ground fish mass out on a board. (**E**) Heap fish meat. (**F**) Prepare fish cake. (**G**) Heat fish cake in a steamer. (**H**) Turn off heat and leave fish cake in steamer.

2 g (⅘ tsp) curdlan
10 g (a little over 1 tsp) potato starch
7 g (a little over 2 tsp) white sugar
20 g egg white
1 g (¼ tsp) sodium glutamate

†**YOU CAN ALSO USE COMMERCIAL FISH PASTE (*SURIMI*) (FIGURE 6.2A,B) AS A SUBSTITUTE FOR THE WHITE FISH FILLETS AND SALT.**

Preparation

1. Wash the white fish fillets.
2. Remove skin (**Figure 6.4A**), soak fish meat in ice water, replacing the ice water a few times (**Hint 2**).
3. Blot fish meat dry with dish towel or paper towel.
4. Grind fish meat and curdlan in a food processer (**Figure 6.4B**) or use an earthenware mortar and pestle.

5. Add salt to the ground fish in a few steps, and continue to grind until it reaches the consistency of meat for dumplings (**Hint 3**).
6. Add potato starch, white sugar, egg white, and sodium glutamate to the ground fish meat, and continue to grind (**Figure 6.4C**).
7. Spread the ground fish meat on a cutting board with a knife (**Figure 6.4D**), heap to form a fish cake (5 cm × 15 cm × 5 mm) (**Figure 6.4E**), and shape the fish cake with a wetted knife (**Figure 6.4F**).
8. Cover airtight with plastic wrap, and refrigerate for 20 minutes.
9. Heat the fish cake in a steamer (80–90°C) for 30 minutes (**Figure 6.4G**), turn off the heat, and leave in steamer for another 10 minutes (**Figure 6.4H**).
10. Remove the fish cake from the steamer, and cool with running water.

Hints

1. Many kinds of fish are used to make *kamaboko*, such as sea bream, tuna, horse mackerel, mackerel, sardine, squid, codfish, shark, and so forth. One of the most popular fish used for this purpose is Alaskan pollock.
2. Soaking in ice water removes fishy smell and water-soluble protein. This process is important for making smooth *kamaboko*.
3. Lowering the temperature to around 0°C is important for making *surimi*. The salt-soluble proteins actin and myosin are dissolved by adding salt and become actomyosin, which is of high molecular weight and makes the *surimi* stickier.

Kamaboko is a type of cured *surimi* product. *Surimi* can be eaten steamed, grilled, fried and deep-fried. These processed *surimi* products are widely used in East and Southeast Asia, where *surimi* products are usually eaten with noodle soups.

Sausages (Figure 6.5)

Ingredients

(Makes 30 sausages)
700 g lean pork
300 g pork fat
18 g (1 Tbsp) salt
4 g (a little over 1 tsp) white sugar
100 g crushed ice
5 g (2 tsp) curdlan
White pepper, garlic powder, nutmeg, sage, and so on
Salted sausage sheep casing

FIGURE 6.5 Sausages. (**A**) Put lean pork and pork fat with salt and sugar in a bowl and cover with plastic film. (**B**) Wash and soak the natural casing. (**C**) Grind lean pork. (**D**) Knead ground lean pork, and ice. (**E**) Add pork fat, ice and spices to the meat and mix. (**F**) Prepare feeding option. (**G**) Stuff the casing. (**H**) Continue the stuffing. (**I**) Tie off the filled casing. (**J**) Dry the surface of the stuffed product. (**K**) Smoke the sausages. (**L**) Boil the sausages and serve.

Preparation

1. Cut lean pork and pork fat into 2-cm cubes.
2. Put the cut lean pork, 12 g salt and 3 g white sugar into a bowl, and put the cut pork fat, 6 g salt and 1 g white sugar into another bowl. Cover each mixture with plastic film (**Figure 6.5A**), and store in the refrigerator for 1–2 nights.
3. Wash a salted sausage sheep casing a few times, and soak it in water for 1 hour (**Figure 6.5B**).
4. Grind stored pork fat and lean pork mixtures separately (**Figure 6.5C**) (**Hint 1**).
5. Mix curdlan, white pepper, garlic powder, sage, and so on.
6. Knead the ground lean pork (**4**) and 70 g crushed ice well to a sticky consistency (**Figure 6.5D**).
7. Add the ground pork fat (**4**), the curdlan and spice mixture (**5**), and 30 g crushed ice to the lean pork and mix gently (**Figure 6.5E**).
8. Fit the soaked casing (**3**) onto a stuffer or a piping bag nozzle (**Figure 6.5F**), fill the stuffer or piping bag with the pork mixture (**7**) (**Figure 6.5G**), pipe out a little of the pork mixture, tie the end of the casing, and continue to pipe out pork mixture without air (**Figure 6.5H**).
9. Twist the casing with the pork mixture every 8 cm, and tie off the end of the casing (**Figure 6.5I**).
10. Dry the surface of the stuffed sausages for 1 hour (**Figure 6.5J**).

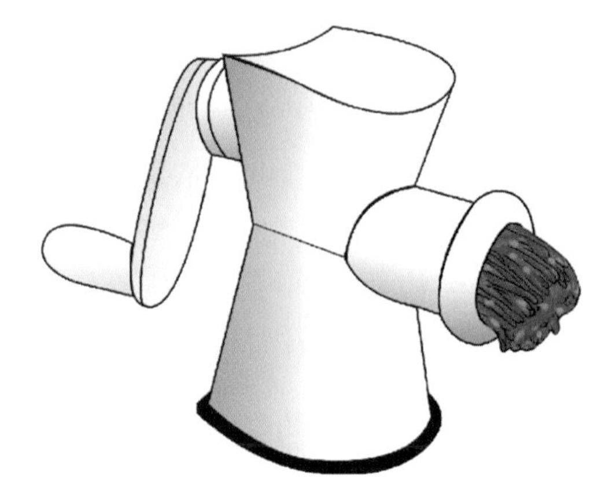

FIGURE 6.6 A mincer for meat grinding.

11. Smoke the sausages for 15 min–1 hour (**Figure 6.5K**) (**Hint 2**).
12. Boil the smoked sausage (**11**) at 75–80°C for about 30 minutes, until the temperature at the center of the sausage reaches 68–72°C (**Figure 6.5L**).

pH of raw pork mixture = 5.31

Hints

1. A mincer is usually used to grind the meat (**Figure 6.6**), although a food processor can be used in this recipe (**Figure 6.5C**).
2. There are several types of fish or meat smoking kits to choose from, depending on the quantity of fish or meat to be smoked, or how often smoked fish or meat is prepared. If you smoke fish or meat often, you can use a home smoker (**Figure 6.7B**), or you can make a home smoker using cardboard (**Figure 6.7A**). For those who do not smoke meat or fish often, it can be prepared in a deep frying pan such as a Chinese wok, with aluminum foil, wire netting, and smoking chips (**Figure 6.5K**), layer the frying pan

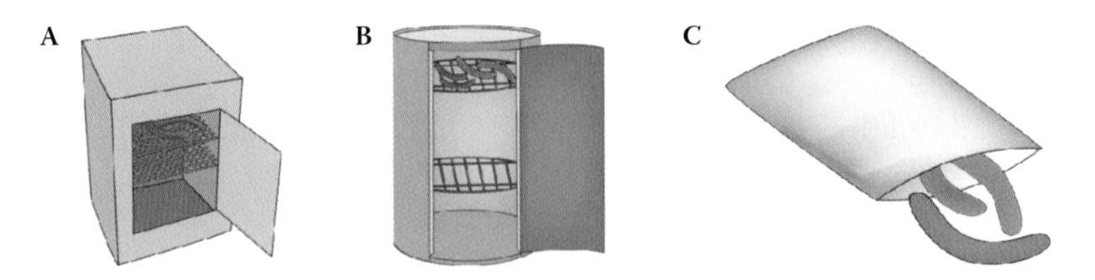

FIGURE 6.7 Smoking appliances. (**A**) Handmade cardboard home smoker. (**B**) Home smoker. (**C**) Smoking bag.

with aluminum foil, put the smoking chips on it, heat for 10 minutes until smoking, and then place fish or meat on wire netting over the frying pan. You can also use a smoking bag with a charcoal briquette (**Figure 6.7C**).

Curdlan is used in this recipe to improve water-holding capacity. In addition, curdlan can be used as a fat mimetic, or fat content can be reduced by adding curdlan and water.

There are many kinds of sausage in the world, and its origin is not certain. Sausage was created so that the whole pig could be consumed, including blood and internal organs, and so that it could be stored from winter to spring.

Doughnuts (Figure 6.8)

Ingredients

(Makes 12)

110 g ($\frac{4}{5}$ cup) strong flour

110 g ($\frac{4}{5}$ cup) weak flour

1.5 g ($\frac{3}{5}$ tsp) curdlan

12 g (1 Tbsp) baking powder

1 g ($\frac{1}{6}$ tsp) salt

60 mL (4 Tbsp) milk

40 g unsalted butter, softened

60 g (a little over $\frac{1}{4}$ cup) granulated sugar

1 egg, beaten

Vanilla flavoring, strong flour, vegetable oil for deep-frying, powdered sugar

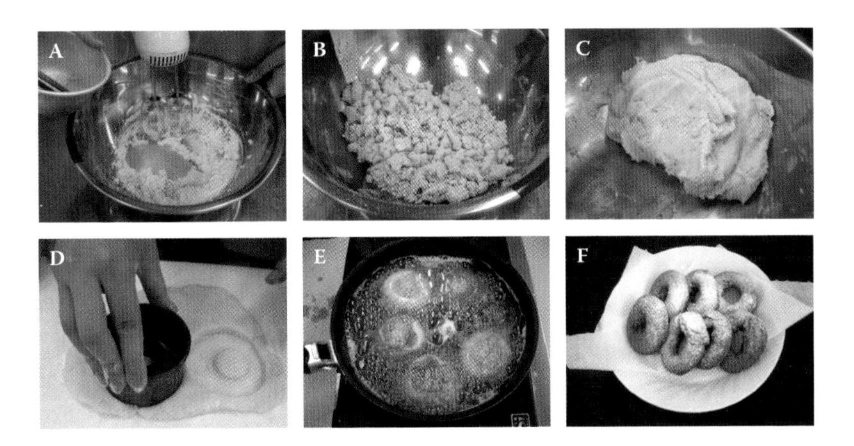

FIGURE 6.8 Doughnuts. (**A**) Combine butter, sugar and egg while stirring. (**B**) Knead dough until crumbly. (**C**) Combined dough. (**D**) Make rounds from rolled out dough. (**E**) Deep-fry the rounds. (**F**) Sprinkle with powdered sugar.

Preparation

1. Mix strong flour, weak flour, curdlan, baking powder, and salt in a bowl, and sieve.
2. Put the softened butter in a bowl, blend, add granulated sugar, and continue to mix until the mixture becomes whitish, then add the beaten egg with stirring in a few steps (**Figure 6.8A**).
3. Mix dry ingredients (**1**) and butter mixture (**2**) with a spatula.
4. When the dough becomes crumbly (**Figure 6.8B**), mix in milk and vanilla flavoring until just combined (**Figure 6.8C**).
5. Sprinkle strong flour on a rolling pin and cutting board, and roll out the dough to 10 mm thickness; cut double rounds with a cutter (**Figure 6.8D**).
6. Heat vegetable oil to 180°C, dust doughnuts with strong flour, and then shake off excess (**Hint 1**), and deep-fry them for 90 seconds each side (**Figure 6.8E**).
7. Drain on paper towel and cool.
8. Sprinkle powdered sugar on doughnuts (**Figure 6.8F**).

pH of dough = 5.30

Hint

1. Dusting with too much flour will cause oil to bubble and spatter.

Curdlan is added in this recipe to decrease oil uptake during deep-frying. Doughnuts are very popular in many countries. Originally deep-fried, today's doughnuts can also be baked or cooked (jelly doughnuts).

Kinugoshi Tofu (Soybean Curd) (Figure 6.9)

Ingredients

(Makes 500 g)

500 g (a little over 2 cups) soybean milk†

2 g (⅘ tsp) curdlan

80 mL (⅓ cup) water

5 g (1 tsp) bittern‡

†A METHOD FOR MAKING *TOFU* FROM SOYBEAN IS GIVEN IN SUPPLEMENTAL K. SOYBEAN MILK HAS TO CONTAIN MORE THAN 10% SOYBEAN SOLIDS.

‡THE BITTERN USED IN THIS RECIPE CONTAINS ABOUT 0.4% MAGNESIUM CHLORIDE SOLUTION. THE CONCENTRATION OF THE BITTERN DEPENDS ON THE TYPE USED.

FIGURE 6.9 *Tofu* (soybean curd). (**A**) Stir together water, bittern, and curdlan. (**B**) Pour soybean milk and curdlan into a mold and remove bubbles. (**C**) Steam the mold. (**D**) Remove *tofu* from mold and immerse in water.

Preparation

1. Put lukewarm water (40–50°C) and bittern in a bowl and add curdlan with stirring for 1–2 minutes (**Figure 6.9A**).
2. Heat soybean milk to 40°C (**Hint 1**).
3. Pour the hot soybean milk (**2**) into the curdlan dispersion (**1**) and mix gently.
4. Pour the heated soybean milk and curdlan dispersion (**3**) into a mold, remove bubbles (**Figure 6.9B**), and steam for 15–20 minutes (at 80–90°C) (**Figure 6.9C**).
5. Remove the mold from the steamer, remove *tofu* from the mold, and put it in water for 30 minutes (**Figure 6.9D**).

pH of *kinugoshi tofu* = 6.02

pH of unheated *tofu* = 6.02

pH of soybean milk = 6.60

pH of bittern = 7.29

Hint

1. If the temperature of the soybean milk is more than 40°C, it will begin to set as soon as it is mixed with the bittern.

This *tofu* is called *kinugoshi tofu* in Japan, which is directly translated to "silk-like *tofu*." This *tofu* is softer than *momen tofu* (**Supplemental K**). *Momen* means "cotton."

Multilayered Jelly (Figure 6.10)

Ingredients

(Serves 1)

A { 30 mL (2 Tbsp) peach juice†
1.5 g (⅗ tsp) curdlan
20 g (a little over 2 Tbsp) white sugar }

50 mL (⅕ cup) hot water (more than 80°C)

A { 30 g (2 Tbsp) grape drink†
1.5 g (⅗ tsp) curdlan
20 g (a little over 2 Tbsp) white sugar }

50 mL (⅕ cup) hot water (more than 80°C)

B { 30 mL (2 Tbsp) orange juice†
1.5 g (⅗ tsp) curdlan
20 g (a little over 2 Tbsp) white sugar }

50 mL (⅕ cup) hot water (more than 80°C)

†YOU CAN USE YOUR FAVORITE FRUIT JUICE OR DRINK.

Preparation

1. First, make peach jelly (**ingredients A**). Heat peach juice to 40°C.
2. Pour hot peach juice into a bowl, add curdlan with stirring (**Figure 6.10A**), and blend for 2–3 minutes keeping at 40°C (**Hint 1**).

FIGURE 6.10 Multilayered jelly. (**A**) Add curdlan to heated juice and stir. (**B**) Add hot water to the curdlan dispersion, while stirring. (**C**) Pour the dispersion into a receptacle and let cool and form a gel. (**D**) Prepare another layer of the jelly and pour it over the first layer of gel. (**E**) Repeat with an additional fruit dispersion, pour over second layer of the product, and let cool. (**F**) Steam the jelly.

3. Add 50 g hot water and white sugar to the peach juice–curdlan dispersion, stir well for 1 minute keeping at 60°C (**Figure 6.10B**) (**Hint 2**), and remove air bubbles with a spoon.
4. Pour dispersion into a bottle or can, and cool to less than 20°C to set the jelly (**Figure 6.10C**).
5. Next, make grape jelly (**ingredients B**). Follow steps (**1**) to (**3**), cool the curdlan–grape dispersion down to 40°C, pour it on the peach jelly (**4**), and cool to less than 20°C (**Figure 6.10D**).
6. Finally, make orange jelly (**ingredients C**) in the same way, pour curdlan–orange dispersion on the grape jelly, and cool to less than 20°C (**Figure 6.10E**).
7. Put a lid on the bottle or can, and heat or steam at 85°C for 30 minutes to stick the jelly layers together (**Figure 6.10F**).

pH of peach jelly = 4.08

pH of grape jelly = 3.47

pH of orange jelly = 3.88

Hints

1. Curdlan powder should be swollen well at 40°C to set jelly.
2. Curdlan should be dispersed well at 60°C so as not to melt jelly when the curdlan dispersion is heated.

The method of preparing canned or bottled jelly using curdlan has been patented.

Tips for the Amateur Cook and Professional Chef

- Gel strength increases with curdlan concentration.
- Gel strength for a moiety formed at 55°C stays almost constant at 60–80°C. It increases with further heating to 100°C.
- Gel strength formed by heating for a few minutes at high temperature (i.e., 90°C) is much greater than that obtained for a gel heated for a few hours at lower temperature (i.e., at 70°C).
- Gel strength does not change between pH values of 3 and 10.
- A high-set gel is much stronger and more resilient, and subject to less syneresis than a low-set gel.
- Curdlan gel texture is somewhere between the hard, brittle texture of agar and the soft, elastic texture of gelatin.
- Curdlan-based gels have superior candying effect and are therefore recommended for making imitation fruits for cake decorations.
- In fruit gels and yoghurt, curdlan levels of 0.5 to 5.0% gum were used in aqueous compositions which formed gels upon heating to 55 to 80°C, followed by cooling.

REFERENCES AND FURTHER READING

Anonymous. 1996. 21 CFR 172—Food additives permitted for direct addition to food for human consumption: Curdlan. *Fed. Regist.* 61:65941–2.

AOAC. 1990. *Official methods of analysis*, 15th ed. Washington, DC: Association of Official Analytical Chemists.

Chuah, C. T., Sarko, A., Deslandest, Y., and R. H. Marchessault. 1983. Triple-helical crystalline structure of curdlan and paramylon hydrates? *Macromolecules* 16:1375–82.

Clamen, A. and B. L. Drasinger. 1974. Fermentation process for the simultaneous production of protein and biopolymers. U.S. Patent 3,856,626.

FMC Corp. 1979. Spreadable heat-resistant jelly. U.S. Patent 3,947,604.

Funami, T., Yada, H., and Y. Nakao. 1998. Curdlan properties for application in fat mimetics for meat products. *J. Food Sci.* 63:283–7.

Funami, T., Yada, H., and Y. Nakao. 1998. Thermal and rheological properties of curdlan gel in minced pork gel. *Food Hydrocolloids* 12:55–64.

Funami, T., Yotsuzuka, F., Yada, H., and Y. Nakao. 1998. Thermoirreversible characteristics of curdlan gel in a model reduced fat pork sausage. *J. Food Sci.* 63:575–9.

Funami, T., Funami, M., Tawada, T., and Y. Nakao. 1999. Decreasing oil uptake of doughnuts during deep-fat frying using curdlan. *J. Food Sci.* 64:883–8.

Funami, T., Funami, M., Yada, H., and Y. Nakao. 1999. Gelation mechanism of curdlan by dynamic viscoelasticity measurements. *J. Food Sci.* 64:129–32.

Harada, T. 1977. Production, properties and application of curdlan. In *Exocellular microbial polysaccharides. ACS Symp. Ser.* 45, ed. P. A. Sandford and A. Laskin, 265–83. Washington: American Chemical Society.

Harada, T. 1979. Curdlan: a gel-forming β-1,3-glucon. In *Polysaccharides in food*, ed. J. M. V. Blanshard and J. R. Mitchell, 283–300. London: Butterworths.

Harada, T., Masada, M., Fujimori, K., and I. Maeda. 1966. Production of a firm, resilient gel-forming polysaccharide by a mutant of *Alcaligenes faecalis* var. *myxogenes* 10C3. *Agric. Biol. Chem.* 30:196–8.

Harada, T., Terasaki, M., and A. Harada. 1993. Curdlan. In *Industrial gums*, 3rd ed., ed. R. L. Whistler and J. N. BeMiller, 427–45. San Diego: Academic Press.

Ikeda, T., Moritaka, S., Sugiura, S., and T. Umeki. 1976. Layered jelly desserts. U.S. Patent 3,969,536.

Kasai, N. and T. Harada. 1980. Ultrastructure of curdlan. In *Fiber diffraction methods. ACS Symp. Ser.* 141, ed. A. D. French and K. H. Gardner, 363–84. Washington: American Chemical Society.

Kimura, H., Kusakabe, K., Tokuda, K., Miyawaki, M., and H. Nakatani. 1975. Edible jelly-like products containing polysaccharide having β-1, 3 pyranose. U.S. Patent 3,980,027.

Kimura, H., Kusakabe, K., Tokuda, K., Miyawaki, M., and H. Nakatani. 1975. Shaped polysaccharide articles and a method for producing them. U.S. Patent 3,899,480.

Kimura, H., Sato, S., Nakagawa, T., Matsukura, A., Suzuki, T., Asai, M., Kanamaru, T., Shibata, M., and S. Yamatodani. 1973. New thermo-gellable polysaccharide. U.S. Patent 3,754,925.

Kimura, H., Sato, S., Nakagawa, T., Nakatani, H., Matsukura, A., Suzuki, T., Mitsuko, A., Kanamuru, T., Shibata, M., and S. Yamatodani. 1974. Thermogellable polysaccharide. U.S. Patent 3,822,250.

Konno, A., Kimura, H., and T. Nakagawa. 1978. Gel formation of curdlan. *Nippon Nougei Kagaku Kaishi* 52:247–50.

Konno, A., Okuyama, K., Koreeda, A., Harada, A., Kanazawa, Y., and T. Harada. 1994. Molecular association and dissociation in formation of curdlan gels. In *Food hydrocolloids, structures, properties, and functions,* ed. K. Nishinari and E. Doi, 113–8. New York: Plenum Press.

McIntosh, M., Stone, B. A., and V. A. Stanisich. 2005. Curdlan and other bacterial (1→3)-β-D-glucans. *Appl. Microbiol. Biotechnol.* 68:163–73.

Misaki, M., Tsujimoto, Y., Nakagawa, T., Sukenari, J., and S. Moritaka. 1974. Method for preparing jellified foods. U.S. Patent 3,857,975.

Miwa, M., Nakao, Y., and K. Nara. 1994. Food application of curdlan. In *Food Hydrocolloids, Structures, Properties, and Functions.* ed. K. Nishinari and E. Doi, 119–24. New York: Plenum Press.

Nakao, Y. 1991. Characteristic of curdlan and its application to foods. *Jap. Food Sci.* 30:35–45.

Nakao, Y., Konno, A., Taguchi, T., Tawada, T., Kasai, H., Toda, J., and M. Terasaki. 1991. Curdlan: Properties and application to foods. *J. Food Sci.* 56:769–72,776.

Nussinovitch, A. 1997. *Hydrocolloid applications: Gum technology in the food and other industries.* London: Blackie Academic & Professional.

Nussinovitch, A. 2003. *Water soluble polymer applications in foods.* Oxford, UK: Blackwell Publishing.

Nussinovitch, A. 2010. *Polymer macro- and micro-gel beads: Fundamentals and applications.* New York: Springer.

Nussinovitch, A. and M. Peleg. 1991. Model for calculating the compressive deformability of double layered curdlan gels. *Biotechnol. Progr.* 7:272–4.

Takeda Chemical Ind. K. K. 1975. Jelly foods made from heat gelling polysaccharides by contracting at least two separate dispersion and heating. West German Patent 2,514,797.

Takeda Chemical Ind. K. K. 1975. Jelly foods made from heat gelling polysaccharides contacting at least two separate dispersions and heating. Netherlands Patent 7,504,264.

Takeda Yakuhin Kogyo, K. K. 1976. Gelled foodstuff production using polysaccharide consisting mainly of β-1,3 glucose units as gelling agents. British patent 1,431,354.

7

Egg Proteins

Historical Background

Eggs have been consumed since the dawn of man, with ostrich and chicken eggs being the most common. Eggs are fairly easy to obtain and serve as a superb protein resource. Eggs can be used in a large variety of recipes, from simply boiled, fried, or stuffed to serving as ingredients in complex pastries, custards, or meringue. Early on in man's history, female game birds (jungle fowl) were probably perceived as a source of both meat and eggs. Man discovered that by taking the eggs from the nest, they could persuade the female jungle fowl to lay extra eggs and to keep on laying eggs for an extended season. Jungle fowl were domesticated in India by 3200 B.C.E. Evidence from China and Egypt demonstrates that fowl were laying eggs for human consumption in around 1400 B.C.E. Archaeological confirmation of egg consumption dates back to the Neolithic age. The Romans found egg-laying hens in England, Gaul, and among the Germans. The first domesticated fowl reached North America with the second voyage of Columbus in 1493.

The Structure of the Egg

Egg is one of the most nourishing and multipurpose of foods. The egg contains a complete diet for the embryo and a principal source of food for the chick's first few days of life. In freshly laid eggs, the shell is entirely filled. An air cell forms inside the egg with moisture loss and contraction of the contents throughout cooling (**Figure 7.1**). A premium egg has only a small air cell. The yolk is centrally positioned in the albumen and is enclosed by the colorless vitelline membrane. The germinal disc, that is, the place where fertilization took place, is attached to the yolk. On opposite ends of the yolk are two whitish cord-like entities known as chalazae (**Figure 7.1**). These hold the yolk in the core of the albumen. A large portion of the albumen is thick. The albumen is surrounded by two shell membranes and the shell itself. The shell includes thousands of pores that allow the egg to "breathe."

The Composition of the Egg

The white portion of the hen egg is composed of ~88% water, 10.6% protein, 0.8% carbohydrates, and 0.6% minerals; the fresh yolk is 51.1% water, 16% protein, 33.6% fat, 0.6% carbohydrates and 1.7% ash.

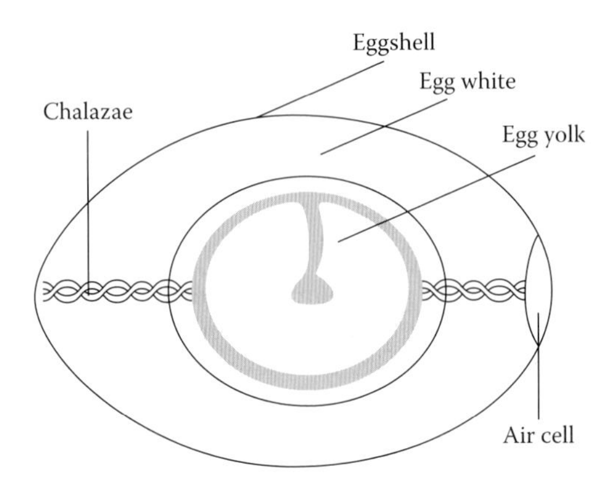

FIGURE 7.1 Egg structure.

Egg Yolk

The yolk amounts to 36% of the egg's total weight, containing ~50 to 52% dry matter, depending on the age of the laying hen and length of storage. Lipids are the main ingredient of the dry matter and the ratio of lipid to protein is ~2:1. Yolk lipids are composed of ~62% triglycerides, 33% phospholipids, and less than 5% cholesterol. Less than 1% of the yolk lipids are carotenoids, which provide its color. Proteins are either free or found in lipoprotein assemblies. Lipids and proteins interact to form low- and high-density lipoproteins, the main ingredients of the yolk. The yolk can be separated into clear yellow fluid (plasma) and granules. The granules consist of circular complexes with diameters of 0.3 to 2.0 μm. The dry matter content of the granules is ~44%, made up of ~64% proteins, 31% lipids, and 5% ash. Plasma comprises 78% of yolk dry matter and is composed of 85% low-density lipoproteins (LDLs) and 15% livetins. LDLs are spherical particles surrounded by a monofilm of phospholipids and protein. The phospholipids play an important part in the stability of the LDL structure due to hydrophobic associations. The cholesterol within the phospholipid film contributes to an increase in its rigidity.

Egg White

The egg white comprises ~60% of the total egg weight. During storage, carbon dioxide release induces a pH increase from 7.5 (when laid) to 9.5 after a few days. This is presumed to be the cause for egg white liquefaction due to dissociation of the ovomucin–lysozyme complex. Proteins represent more than 90% of the dry matter of egg white. Egg white proteins are mostly globular, and acidic or neutral, apart from lysozyme and avidin, which are highly alkaline. More than 50% of the total protein is ovalbumin, a globular protein with half of its amino acids hydrophobic and one-third of them electrically charged. Ovotransferrin comprises 13% of the total proteins. It is the most heat-sensitive egg white protein, although complexation of iron or aluminum significantly augments its heat stability. Ovomucoid is to a great extent glycosylated protein. It comprises up to 25% of the carbohydrates and is heat resistant at acidic pH.

Ovomucin is also a high molecular weight, highly glycosylated protein. It can react electrostatically with other egg white proteins.

Essential Nutrients and Value of Eggs

Eggs are particularly important as a source of protein. One egg has ~6 to 7 g of protein. The emulsified fat in the yolk is easily digested. The ratio of unsaturated to saturated fats is about 2 to 1, which is believed to be beneficial. Oleic acid is the major unsaturated fat, and it has no effect on blood cholesterol. Eggs contain vitamin A, and the B vitamins—that is, thiamine, riboflavin, and niacin, and vitamin D. Each of these vitamins is necessary during childhood and adolescence for growth. In addition, eggs contain an abundant supply of minerals, such as iron and phosphorus, which are vital for maintaining good physiological condition. On the other hand, eggs are low in calcium, which is contained in the shell, and have little or no vitamin C. Eggs are quite valuable in weight-reduction diets. Losing weight involves reducing calories while maintaining a well-balanced diet. An egg supplies good nutrition and has only ~80 calories. With food costs consistently on the rise, primarily for high-protein foods, shoppers are constantly searching for ways to decrease their expenses. One option is to include more eggs in the diet given that it is difficult to buy any other high-protein food—meat or fish—for such a low cost. In the food industry as well as in the kitchen, eggs are considered a multifunctional ingredient since they can be used for emulsification, foaming, gelling, and thickening, among other functions.

On the other hand, eggs are the most costly ingredients in some cakes. In yellow cakes, eggs are a major source of cholesterol. The use of vegetable proteins to substitute, either partially or completely, eggs in cake formulations is important, particularly for people with specific dietary needs or restrictions. The exceptional emulsifying, foaming, and heat coagulation properties of egg proteins are important for cake volume and texture. Therefore, it is particularly difficult to replace eggs effectively with a dissimilar source of proteins, even with the use of several types of additives, such as hydrocolloids, in cakes. A combination of lupine protein, emulsifiers, and xanthan gum (**Chapter 15**) was successful in improving the internal structure of crumbs, reducing shrinkage, and increasing cake height. In angel food cakes, interactions with other ingredients, in addition to intrinsic properties of individual proteins, are responsible for functional differences between egg white protein and whey protein isolate (**Supplemental G**).

Egg Yolk Emulsions

An emulsion is a mixture of two or more liquids which are normally immiscible. In an emulsion, one liquid (the dispersed phase) is dispersed in the other (the continuous phase). Examples of emulsions include vinaigrettes, milk, and cutting fluid for metal working. Yolk has emulsifying properties, that is, it has the ability to reduce interfacial tension between oil and water. In general, the more active the emulsifying agent, the more the interfacial tension is lowered. The plasma provides the yolk's emulsification ability. It has been shown that LDLs are better emulsifiers than bovine serum albumin and casein. Under certain circumstances, high-density lipoproteins (HDLs) have been suggested to be more efficient than LDLs for stabilization of oil-in-water emulsions; nevertheless the prevalent role of LDL in yolk emulsions has been confirmed. The emulsifying properties of LDL depend on the integrity of their structure. The interactions between apoproteins and lipids in the LDL particles are essential to transporting the surfactants in a soluble form to the interface and then releasing them at the interface. Another study dealt with the stability of oil-in-water emulsions prepared with egg white protein at pH 3.8 and containing 150 mM NaCl. Egg white emulsions exhibited higher stability against creaming than yolk emulsions, irrespective of the presence or absence of xanthan gum (**Chapter 15**). At relatively low xanthan gum contents, the emulsions exhibited higher stability against creaming compared to the respective control emulsions, probably due to the formation of a continuous droplet aggregate network structure.

Egg White Foams

Foam is defined as a substance formed by trapping many gaseous bubbles in a liquid or solid. It is a complex system consisting of polydispersed gas bubbles separated by draining films. The ability of a protein solution (e.g., egg white) to foam depends on the protein's structure and conformation. These factors depend on, among many things, pH and ionic strength. The mechanism underlying the formation of globular protein foams consists of three steps, namely: the diffusion of proteins toward the air–solution interface, confirmation of changes in the adsorbed proteins, and irreversible rearrangement of the protein film. The stability of the foam depends on protein associations at the air–solution interface in forming a continuous intermolecular network. As stated, the interfacial properties of egg white proteins are responsible for its successful foaming properties. Ovalbumin forms a single layer at the air–water interface; lysozyme creates films that are much thicker than the protein monolayer. Nevertheless, ovalbumin foaming properties are superior to those of lysozyme, due to the weak foaming capacity of the latter in its native state at pH 7.0. When egg white, which can be considered a solution of efficient surfactants, is compared with other proteins, it usually provides the best foaming properties. Since egg white is composed of many proteins that have different surface tensions and foamabilities, it is difficult to predict foaming properties of particular mixtures of egg white proteins. The foaming properties of isolated egg white proteins are lower than those of egg white as a whole, confirming the existence of interactions between proteins. Foam stability decreases when egg whites liquefy during storage. Moreover, egg white's foaming properties increase upon addition of sucrose and sodium chloride. A combination of physical measurements and artificial vision has enabled obtaining an apparent and applicable evaluation of the foaming properties of protein-based foams and illustrating their modifications due to changes in protein conformation. Controlled thermal treatment of solutions drastically increased the foaming quality of lysozyme and improved that of ovalbumin, both of which gave very stable foams after being heated to 90°C.

Gels

Proteins are responsible for the gelling properties of whole egg, as well as of egg white and yolk. Gelation occurs when the equilibrium between Van der Waals and electrostatic interactions is disrupted. Upon heating treatments, the protein structure is modified. Heat-induced gelation of globular proteins is a two-step process. The first stage includes unfolding of native proteins, disruption of their secondary and tertiary structures, and production of denatured proteins exposing their inner hydrophobic regions; this is followed by unfolding and interaction of the denaturated proteins to form high-molecular-weight aggregates that can further interact to form a gel. The gel properties depend on protein concentration, ionic strength, pH, and presence of sucrose. In heat-induced gels, hydrophobic and electrostatic interactions are involved. Nevertheless, highly energetic interactions are also observed.

Egg Yolk Gels

LDLs are responsible for yolk gelation. Yolk jellifies upon freeze-thawing or heat treatment. Freeze-thawing gelation occurs at temperatures below −6°C. Such gelation is undesirable since the yolk is difficult to deal with. Gelation rate is influenced by the rate and temperature of the freezing and thawing. Rapid freezing results in less gelation than a slow process. Freeze-thawing gelation is partially reversible by heating for 1 h at 45 to 55°C. LDLs at 4% (w/v) start to denature at 70°C and form gels at 75°C. If such a solution is heated at 80°C for 5 min, it forms more stable gels than ovalbumin or bovine serum albumin. LDLs form heat-induced gels at pH 4 to 9. Between pH 6 and 9, LDL solutions form opaque gels, whereas they form translucent gels at extreme pH (i.e., 4–6 and 8–9). The mechanism underlying yolk gelation involves unfolding of the proteins during heating, exposure of functional groups, and attraction among groups through hydrophobic bonds to form a gel. The critical concentration for heat gelation is 12–28 mg protein/mL for plasma and 26–120 mg protein/mL for granules. For a yolk solution, the critical concentration was found to be 16 to 39 mg protein/mL.

Egg White Gels

Egg white proteins coagulate upon heating, except for ovomucin and ovomucoid. The denaturation temperatures at pH 7.0 for ovalbumin, lysozyme, and ovotransferrin are 84.5, 74, and 65°C, respectively. Ovotransferrin is the most sensitive of these proteins and as such is regarded as a gelation initiator and limiting factor in the jellification. Thus ovotransferrin elimination has been suggested to improve egg white gelling properties. Ovotransferrin is more stable when metal ions are bound to it and therefore gelation temperature can be increased by adding Fe^{3+} or Al^{3+}. Heat denaturation induces an increase in the hydrophobicity of the protein surface. At high ionic strength, random aggregates of slightly denatured proteins appear, resulting in opaque gels having low rigidity, elasticity, and water-retention capacity. At low ionic strength, aggregation is delayed due to high electrostatic repulsions and denaturation is favored. Further aggregation involves hydrophobic regions and induces linear polymeric aggregates. Under high ionic strength, these aggregates can interact to form gels. In other words, a two-step heating process is recommended to produce translucent, firm, elastic egg white gels with high water-retention capacity. The gelling properties of egg white can be improved by heating egg white powder at 80°C for up to 10 days. This process expands the flexibility of the protein and the exposure of reactive groups that can further interact to strengthen the gel formed when the formerly dry heated egg white is solubilized and heated in solution.

Recipes with Eggs

Plain Omelette (Figure 7.2)

Ingredients

(Serves 5)
500 g (10) eggs
125 mL (½ cup) milk
50 g (a little over 4 Tbsp) butter or vegetable oil
Salt, pepper

Preparation

1. Beat eggs in a bowl, add milk, season with salt and pepper, and mix well with a whisk or chopsticks (**Hint 1**).
2. Heat butter in a frying pan (20 cm diameter), and pour in the beaten eggs (**1**) (**Hint 2**) (**Figure 7.2A**). When eggs begin to thicken, reduce heat and stir with a spatula or chopsticks a few times around the side of the pan.

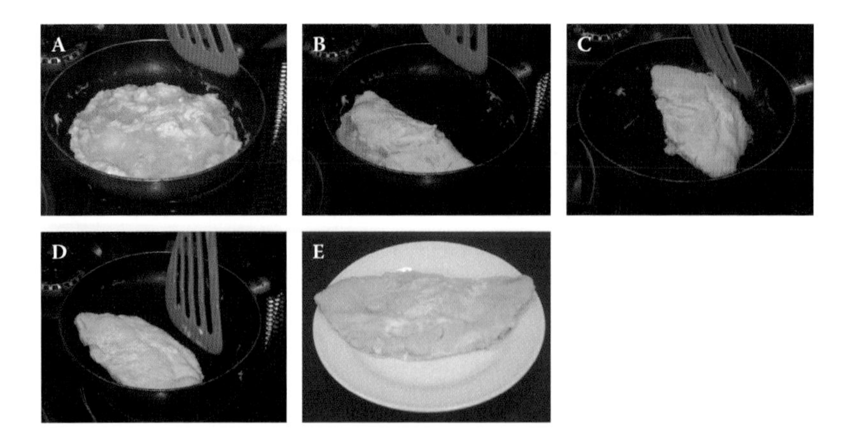

FIGURE 7.2 Plain omelette. (**A**) Pour the beaten eggs with other additives on a heated and buttered frying pan. (**B**) Fold omelette in half. (**C**) Turn it over. (**D**) Move it to far end of pan. (**E**) Turn omelette out onto a plate.

3. Fold omelette in half, away from you (**Figure 7.2B**), turn it over toward you (**Figure 7.2C**), move it to far end of pan (**Figure 7.2D**), and turn again onto a plate to serve (**Figure 7.2E**).

pH of egg dispersion = 7.41

pH of omelette = 7.68

Hints

1. Beating incorporates air, which produces a fluffy omelette. Adding milk or heavy cream leads to an omelette with a smooth surface, but less fluffiness. Do not add too much milk.
2. To cook the omelette, higher heat is preferable to lower heat. If it is difficult for you to make the omelette at high heat, medium heat is acceptable.

Omelettes can hold an unlimited variety of additions: tomato, cheese, many kinds of vegetables, and many types of sauces.

Japanese Rolled Omelette (*Sushi* Egg) (Figure 7.3)

Ingredients

(Serves 5)

250 g (5) eggs

A:
- 75 mL (⅓ cup) Japanese broth
- 7.5 mL (1½ tsp) light soy sauce
- 10 g (2 tsp) white sugar
- 1.3 g (1/5 tsp) salt

250 g Japanese white radish

5 buds of Japanese pepper†

Vegetable oil

†YOU CAN ALSO USE GREEN BEEFSTEAK PLANT OR *DAIKON* CRESS.

Preparation

1. Beat eggs, add **ingredients A**, and stir gently so that the egg yolks and egg whites are lightly mixed.
2. Heat vegetable oil in a square frying pan, coating all surfaces (**Hint 1**).
3. Pour one-third of the egg mixture to coat the pan (**Figure 7.3A**), and when the surface of the mixture is half cooked (**Hint 2**), fold the far end of the cooked egg with chopsticks, and roll toward you (**Figure 7.3B,C**).

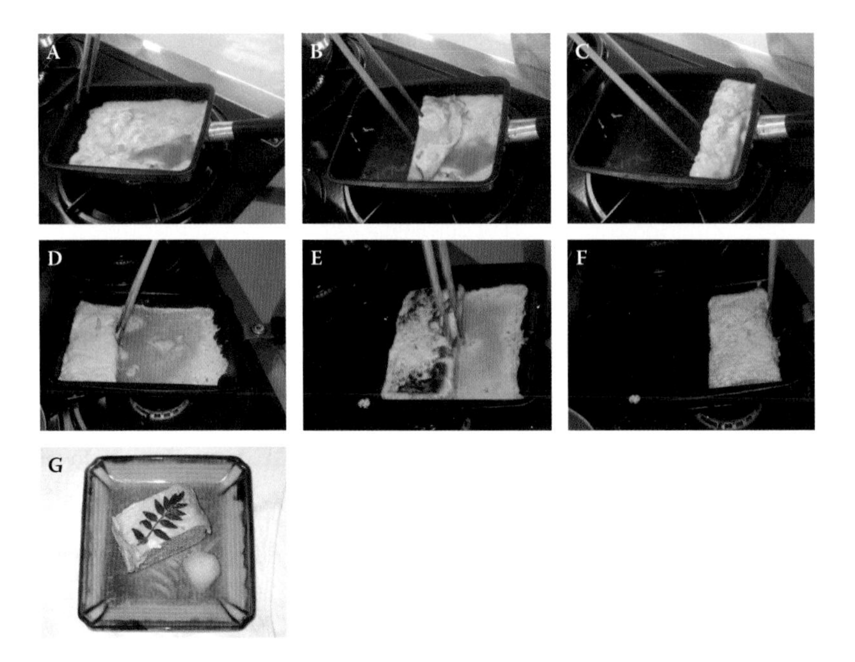

FIGURE 7.3 Japanese rolled omelette (*sushi* egg). (**A**) Pour one-third of egg mixture to coat heated oiled pan. (**B**) and (**C**) Fold far end of cooked egg and roll toward you. (**D**) Add vegetable oil to the pan and pour more of the egg mixture while lifting rolled omelette. (**E**) and (**F**) Upon setting, roll and repeat for third omelette. (**G**) Cut and garnish.

4. Add vegetable oil to the pan, and pour more of the egg mixture (**1**) while lifting the rolled omelette (**3**) (**Figure 7.3D**). When this mixture has set, roll toward you, and repeat again (**Figure 7.3E,F**).
5. Remove the rolled omelette from the pan, wrap in a bamboo mat or wax paper, and shape into a rectangular block.
6. Grate Japanese white radish.
7. Cut the rolled omelette into 2 cm thick pieces, and garnish with grated white radish and Japanese pepper buds (**Figure 7.3G**).

pH of egg dispersion = 7.31

pH of Japanese rolled omelet = 7.49

Hints

1. This is important to prevent the egg dispersion from sticking to the pan. If you don't have a square frying pan, you can use a round one.
2. The inclusion of Japanese broth in the egg mixture leads to a juicy rolled omelette. Small air bubbles in the mixture also contribute to its texture.

This recipe is very popular in Japan. It is a favorite for lunch boxes although it is popular as a *sushi* component in other countries.

Crabmeat and Egg Drop Soup

Ingredients

(Serves 5)
50 g canned crabmeat
40 g long onion
5 g dried Judas ear (**Figure 7.4A**)
100 g (2) eggs
750 mL (3 cups) Chinese soup stock
10 g (a little over 1 tsp) potato starch or cornstarch
Salt, soy sauce

Preparation (Figure 7.5)

1. Tear crabmeat into pieces with hands, and remove sinew (**Figure 7.5A**).
2. Soften dried Judas ear in water (**Figure 7.4B**), and slice thinly (**Figure 7.5A**).
3. Slice long onion thinly (**Figure 7.5A**).
4. Pour Chinese soup stock in a pan, heat to boiling, and season with salt and soy sauce. Then add torn crabmeat (**1**), sliced Judas ear (**2**), and sliced long onion (**3**) (**Figure 7.5B**).
5. Mix together potato starch and water, and add to the soup (**Hint 1**). Raise heat, pour in beaten egg (**Figure 7.5C**) (**Hint 2**), cover pan and turn off heat.
6. Serve soup (**Figure 7.5D**).

pH = 6.36

Hints

1. If the temperature of the soup is too low, the potato starch will settle at the bottom of the pan because the starch partly gelatinizes. Pour starch–water dispersion into the soup and mix quickly at high temperature.
2. Add the beaten eggs to the soup at higher temperature to make a clear soup.

This soup is made year round. Different ingredients can be used to adapt it to the season.

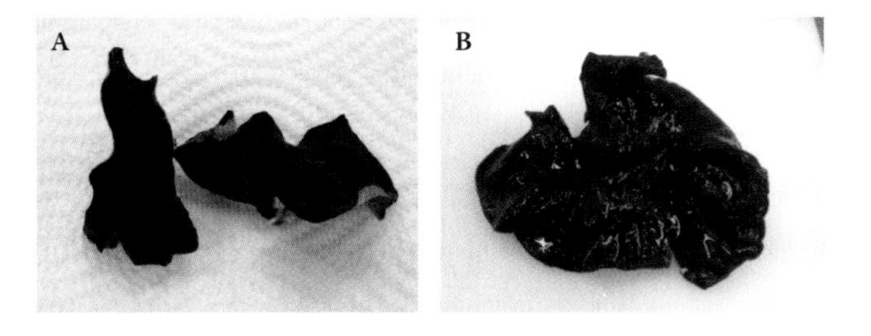

FIGURE 7.4 (**A**) Dried Judas ear. (**B**) Judas ear softened in water.

FIGURE 7.5 Crabmeat and egg drop soup. (**A**) Tear crabmeat, slice long onion and Judas ear thinly. (**B**) Add crabmeat, Judas ear, and sliced long onion to Chinese soup stock in a pan. (**C**) Pour beaten egg around the edges of the pan. (**D**) Pour the soup into a dish and serve.

Chawan-Mushi (Savory Egg Custard) (Figure 7.6)

Ingredients

(Serves 5)

150 g (3) eggs

450 mL (1⅘ cups) Japanese broth

A:
- Salt
- 11 mL (a little over 2 tsp) soy sauce
- 11 mL (a little over 2 tsp) *mirin*†

1 chicken breast fillet (75–100 g)

B:
- 6 mL (a little over 1 tsp) light soy sauce
- 5 mL (1 tsp) *sake*

5 shrimps

C:
- Salt
- 5 mL (1 tsp) *sake*

5 fresh *shiitake* mushrooms

5 slices of *kamaboko*‡

5 field peas

Yuzu†‡

†**JAPANESE SWEET *SAKE***

‡**FISH CAKE (SEE CHAPTER 6)**

†‡**JAPANESE CITRUS FRUIT**

FIGURE 7.6 *Chawan-mushi* (savory egg custard). (**A**) Beat eggs, add Japanese soup stock, stir, and strain. (**B**) Put cut chicken, shrimps, *shiitake* mushrooms, and *kamaboko* into serving cups. (**C**) Pour the egg mixture over ingredients in the serving cups. (**D**) Garnish with boiled field peas. (**E**) Remove cups from steamer, garnish, and serve.

Preparation

1. Remove the skin and sinew from the chicken breast, slice thinly, and sprinkle with light soy sauce and *sake* (**Ingredients B**).
2. Devein the shrimps, and add a pinch of salt and *sake* (**Ingredients C**).
3. De-string field peas, and boil them with salt.
4. Heat Japanese broth, and season with salt, soy sauce, and *mirin* (**Ingredients A**).
5. Beat eggs in a bowl with chopsticks or a fork, add Japanese soup stock (**4**), stir gently and then strain (**Figure 7.6A**).
6. Put sliced chicken (**1**), shrimps (**2**), *shiitake* mushrooms, and *kamaboko* into serving cups (**Figure 7.6B**), pour the egg mixture (**5**) over them (**Figure 7.6C**), and garnish with boiled field peas (**3**) (**Figure 7.6D**).
7. Place the cups with lids or covered with aluminum foil in a steamer, steam for 5 min at high heat (100°C), and continue to steam for 10–15 minutes at 80–90°C (**Hint 1**).
8. When the surface of the egg custard (**7**) has set, remove the cups from the steamer, and garnish with cut *yuzu* (**Figure 7.6E**).

pH = 7.23

Hint

1. When *chawan-mushi* is steamed at 100°C for a long time, it becomes spongy because the egg overcooks. Therefore, *chawan-mushi* should be heated with the steamer lid open.

This recipe is representative of typical Japanese cuisine; it is often eaten with *sushi*. Different kinds of ingredients can be added, depending on the season: white fish, sea eel, spinach, other mushrooms, carrots, bamboo shoots, lily roots, ginkgo nuts, trefoil, and so on.

Caramel Custard (Figure 7.7)

Ingredients

(Serves 5)
125 g (2–3) eggs
40 g (¼ cup) white sugar
300 mL (1⅕ cup) milk
1 vanilla pod or vanilla flavoring
Butter
A { 50 g (a little under ⅓ cup) white sugar
   25 mL (1⅔ Tbsp) water
25 mL (1⅔ Tbsp) hot water

Preparation

1. Spread butter in a large mold or 5 small molds.
2. Make caramel sauce using **Ingredients A**. Heat 50 g white sugar and water in a pan.
3. When the sugar solution becomes brown and thickens (**Figure 7.7A**), turn off the heat, add hot water, mix well (**Figure 7.7B**) (**Hint 1**), and pour caramel sauce into the greased mold.

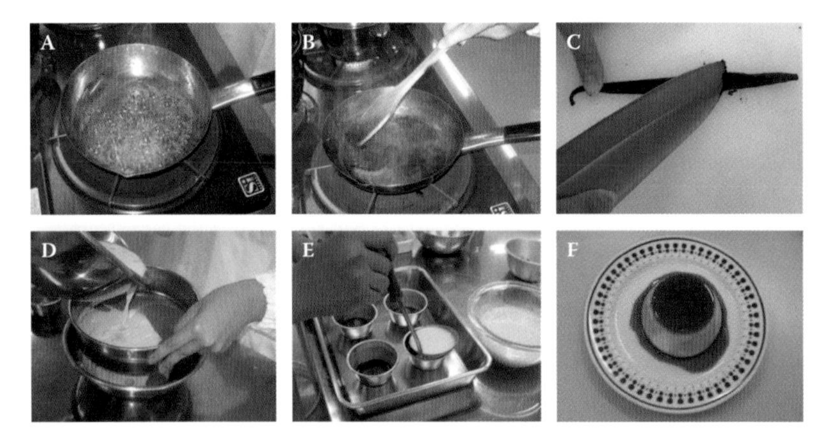

FIGURE 7.7 Caramel custard. (**A**) Thicken caramel sauce. (**B**) Add hot water and mix. (**C**) Split open a vanilla pod and scrape out beans. (**D**) Strain egg mixture. (**E**) Pour strained egg mixture into mold. (**F**) Remove pudding from mold and serve.

4. Preheat oven to 160°C (**Hint 2**).
5. Beat eggs in a bowl, add 40 g white sugar, and mix well.
6. Heat milk to 60°C, add to egg mixture (**5**); split open vanilla pod (**Figure 7.7C**) and add beans to mixture; mix well and strain (**Figure 7.7D**).
7. Pour the strained egg mixture into the mold (**3**) (**Figure 7.7E**).
8. Put water in a baking pan, put the mold in the baking pan, and heat for 20 minutes at 160°C until the egg mixture has set (**Hint 3**).
9. Remove the pudding from the mold (**Figure 7.7F**).

$$pH = 6.89$$

Hints

1. Add hot water carefully, without splattering, as the sauce is very hot and will stick to the skin.
2. You can make steamed caramel custard in a steamer using the same method as with *chawan-mushi.*
3. Insert a bamboo skewer or a toothpick to check that the custard cream has set. When the bamboo skewer comes out clean, the custard is ready.

Caramel custard is widely eaten all over the world, with many variations depending on the country. *Crème brûlée* is one of the variations of caramel custard (see below).

This recipe is popular with everyone, from little children to the elderly, because of its smooth texture and ease of swallowing.

Crème Brûlée (Figure 7.8)

Ingredients

(Serves 5–8)
4 egg yolks
35–70 g (⅕–⅖ cup) white sugar
400 g (1⅗ cups) heavy cream (40% or more milk fat)
100 mL (⅖ cup) milk
Vanilla beans, cassonade (brown sugar), rum

Preparation

1. Heat heavy cream and milk to boiling (**Hint 1**).
2. Strip vanilla beans from the pod (**Figure 7.7C**).
3. Whip egg yolks and white sugar well until it becomes whitish (**Hint 2**), and gradually add the boiled cream and milk (**1**) (**Figure 7.8A**); then add vanilla beans and rum.

FIGURE 7.8 *Crème brûlée.* (**A**) Gradually add boiled cream and milk to whipped egg yolk and white. (**B**) Pour cream dispersion into plates, and place in a hot water bath. (**C**) Cool the heated cream. (**D**) and (**E**) Sprinkle cassonade evenly over surface and torch.

4. Preheat oven to 150°C. Pour cream dispersion (**3**) into heat-resistant plates, place in a hot water bath (**Figure 7.8B**), and bake for 30 minutes.
5. Cool the heated cream down to 10°C in the refrigerator or an ice-water bath (**Figure 7.8C**), sprinkle cassonade evenly, and heat the cassonade with a gas torch until the color changes (**Figure 7.8D,E**).

$$\text{pH} = 6.28$$

Hints

1. Using more heavy cream than milk will make a richer *crème brûlée.*
2. If you do not whip egg yolk and white sugar well, you will get caramel custard.

This recipe is also known as burnt cream, crema catalana, or Trinity cream, and it is a traditional dessert in Europe.

Cream Puffs (Figure 7.9)

Ingredients

(Serves 4)

A {
- 40 g butter
- 100 mL (⅔ cup) water
- 20 g (a little over 2 Tbsp) weak flour
- 20 g (a little over 2 Tbsp) strong flour
- 2 eggs
- 2 egg yolks

FIGURE 7.9 Cream puffs. (**A**) Put butter and water in a pan and heat until boiling. (**B**) Heat until a thin skin form on the bottom of the pan. (**C**) Pipe the batter onto the baking paper. (**D**) Mix ingredients for cream custard. (**E**) Heat until thickened. (**F**) Make a cut in each baked choux pastry. (**G**) Pipe the cream custard into choux pastries. (**H**) Dust product with powdered sugar.

B {
240 mL (a little under 1 cup) milk
60 g (a little over 1/4 cup) granulated sugar
20 g (a little over 2 Tbsp) weak flour†
}

Vanilla, powdered sugar

†FOR A STIFFER CREAM, USE CORNSTARCH INSTEAD.

Preparation

1. Make choux pastry using **ingredients A**. Cut butter into small pieces, sieve together weak and strong flours, beat egg, and preheat oven to 200°C.
2. Heat cut butter (**1**) and water in pan (**Figure 7.9A**) to boiling; remove from heat and mix in the sieved flour (**1**) with a spatula. Heat until a thin skin forms on the bottom of the pan (80°C) (**Figure 7.9B**), and turn off the heat.
3. Cool the batter (**2**) in a bowl to 60°C; gradually add the beaten egg (**1**), stirring well with a spatula.
4. Spread baking paper on baking tray and place batter into a piping bag; pipe the batter onto the baking paper (**Figure 7.9C**), and spray with water.
5. Bake for 15 minutes at 200°C, and continue to bake for 8 minutes at 180°C (**Hint 1**).

6. Make cream custard using **ingredients B**. Sieve weak flour, and heat 225 g milk to 40–50°C.
7. Put egg yolk and 15 g (1 Tbsp) milk in a pan and mix, add granulated sugar to the mixture, and mix well.
8. Add the sieved weak flour and the heated milk (**6**) to the mixture (**7**) (**Figure 7.9D**), and heat until thickened (**Figure 7.9E**).
9. Remove from heat, mix in vanilla, and place the cream custard into a piping bag.
10. Make a cut in each baked choux pastry (**Figure 7.9F**) and pipe the cream custard into the pastry (**Figure 7.9G**).
11. Dust with powdered sugar (**Figure 7.9H**).

pH of choux pastry = 7.25

pH of cream custard = 6.34

Hint

1. You must not open the oven door while baking. If choux pastry does not dry enough while baking, it deflates.

This recipe is called "*choux à la crème*" in France. It has many variations, such as choux pastry with whipped cream, chocolate cream or ice cream, cream puff garnished with chocolate sauce, éclair or choux pastry glazed with caramel, and croquembouche.

Meringues (Figure 7.10)

Ingredients

(Serves 5)

2 egg whites

48 g (a little over ⅓ cup) powdered sugar

5 mL (1 tsp) vinegar

10 mL (2 tsp) hot water

Preparation

1. Mix powdered sugar and hot water in a bowl until sugar dissolves, then add vinegar (**Hint 1**), and mix (**Figure 7.10A**).
2. Add egg white to the mixture (**Figure 7.10B**), and whip with a hand mixer until peaks form (**Figure 7.10C**) (**Hint 2**).
3. Preheat oven to 100°C. Spread baking paper on a baking tray, and pipe (**Figure 7.10D**), or spoon the meringues onto it.
4. Bake meringues for 30 minutes at 120°C (**Figure 7.10E**), drying well.

pH = 5.95

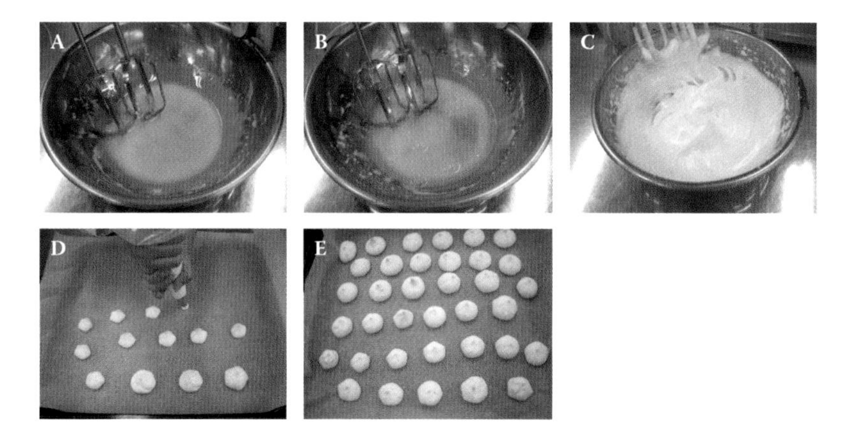

FIGURE 7.10 Meringues. (**A**) Mix powdered sugar and hot water in a bowl, then add vinegar. (**B**) Add egg white to the mixture. (**C**) Whip until peaks form. (**D**) Spread baking paper on a baking tray, and pipe mixture. (**E**) Bake meringues to dryness.

Hints

1. It is easier to whip egg whites at lower pH because the whites are more easily denatured; the isoelectric point of egg white is at around pH 5.0.
2. As it is harder to whip up egg whites in the presence of white sugar, use an electric blender.

This recipe is called *French meringue* and is used for baking sweets. *Italian meringue* is made of egg white and hot sugar syrup (see marshmallows in **Chapter 9**), and is used for fresh sweets.

Tips for the Amateur Cook and Professional Chef

- Refrigerated eggs are easier to separate than eggs at room temperature.
- Egg whites whip to their maximum volume at room temperature.
- Egg whites whip up better in a copper bowl.
- Even a speck of yolk makes it difficult to whip egg whites.
- Egg whites should be beaten slowly at the beginning of the whipping process.
- Sugar or cream of tartar (or other ingredients) should be added to whipped egg whites midway through the whipping process.

REFERENCES AND FURTHER READING

Anton, M. 2009. Egg proteins. In *Handbook of hydrocolloids,* ed. G. O. Phillips and P. A. Williams, 359–76. Boca Raton: CRC, and Oxford: Woodhead Publishing Limited.

Anton, M. and G. Gandemer. 1997. Composition, solubility and emulsifying properties of granules and plasma of hen egg yolk. *J. Food Sci.* 62:484–7.

Arozarena, I., Bertholo, H., Bunger, J. E. A., and I. de Sousa. 2001. Study of the total replacement of egg by white lupine protein, emulsifiers and xanthan gum in yellow cakes. *Eur. Food Res. Technol.* 213:312–6.

Berry, T. K., Yang, X., and E. A. Foegeding. 2009. Foams prepared from whey protein isolate and egg white protein. 2. Changes associated with angel food cake functionality. *J. Food Sci.* 74:E269–77.

Burley, R. W. and W. H. Cook. 1961. Isolation and composition of avian egg yolk granules and their constituents α- and β-lipovitellins. *Can. J. Biochem. Physiol.* 39:1295–307.

Chang, C. M., Powrie, W. D., and O. Fennema. 1977. Microstructure of egg yolk. *J. Food Sci.* 42:1193–200.

Clark, A. H., Kavanagh, G. M., and S. B. Ross-Murphy. 2001. Globular protein gelation—theory and experiment. *Food Hydrocolloids* 15:383–400.

Croguennec, T., Renault, A., Beaufils, S., Dubois, J.-J., and S. Fezennec. 2007. Interfacial properties of heat-treated ovalbumin. *J. Colloids Interfaces Sci.* 315:627–36.

Cunningham, F. E. and H. Lineweaver. 1965. Stabilization of egg white proteins to pasteurizing temperatures above 60°C. *Food Technol.* 19:136–41.

Davis, J. P. and E. A. Foegeding. 2007. Comparisons of the foaming and interfacial properties of whey protein isolate and egg white proteins. *Colloids Surf. B: Biointerfaces* 54:200–10.

Donovan, J. W., Mapes, C. J., Davis, J., and J. Garibaldi. 1975. A differential scanning calorimetric study of the stability of egg white to heat denaturation. *J. Sci. Food Chem.* 26:73–83.

Drakos, A. and V. Kiosseoglou. 2006. Stability of acidic egg white protein emulsions containing xanthan gum. *J. Agric. Food Chem.* 54:10164–9.

Dyer-Hurdon, J. N. and I. A. Nnanna. 1993. Cholesterol content and functionality of plasma and granules fractionated from egg yolk. *J. Food Sci.* 58:1277–81.

Foegeding, E. A., Luck, P. J., and J. P. Davis. 2006. Factors determining the physical properties of protein foams. *Food Hydrocolloids* 20:284–92.

Hagolle, N., Relkin, P., Popineau, Y., and D. Bertrand. 2000. Study of the stability of egg white protein-based foams: Effect of heating protein solution. *J. Sci. Food Agric.* 80:1245–52.

Kato, A., Ibrahim, H. R., Watanabe, H., Honma, K., and K. Kobayashi. 1989. New approach to improve the gelling and surface functional properties of dried egg white by heating in dry state. *J. Agric. Food Chem.* 37:433–7.

Katz, S. H. and W. W. Weaver. 2003. *Encyclopedia of food and culture*, vol. 1, 558. New York: Charles Scribner's Sons.

Kojima, E. and R. Nakamura. 1985. Heat gelling properties of hen's egg yolk low density lipoprotein (LDL) in the presence of other protein. *J. Food Sci.* 50:63–6.

Kudryashova, E. V., Meinders, M. B. J., Visser, A. J. W. G., Van Hoek, A., and H. H. J. De Jongh. 2003. Structure and dynamics of egg white ovalbumin adsorbed at the air-water interface. *Eur. Biophys. J.* 32:553–62.

Kurakawa, H., Mikami, B., and M. Hirose. 1995. Crystal structure of diferric hen ovotransferrin at 2.4 Å resolution. *J. Mol. Biol.* 254:196–207.

Li-Chan, E. C. Y., Powrie, W. D., and S. Nakai. 1995. The chemistry of eggs and egg products. In *Egg science and technology,* 4th ed., ed W. J. Stadelman and O. J. Cotterill, 105–76. New York: Food Product Press.

Lin, L. N., Mason, A. B., Woodworth, R. C., and J. F. Brandts. 1994. Calorimetric studies of serum transferring and ovotransferrin. Estimates of domain interactions and study of the kinetic complexities of ferric ion binding. *Biochemistry* 33:1881–8.

Mann, K. 2007. The chicken egg white proteome. *Proteomics* 7:3558–68.

Martinet, V., Saulnier, P., Beaumal, V., Couthaudon, J. L., and M. Anton. 2003. Surface properties of hen egg yolk low-density lipoproteins spread at the air-water interface. *Colloids Surf. B: Biointerfaces* 31:185–94.

McBee L. E. and O. J. Cotterill. 1979. Ion-exchange chromatography and electrophoresis of egg yolk proteins. *J. Food Sci.* 44:657–66.

McRitchie, F. 1991. Air/water interface studies of protein. *Anal. Chim. Acta* 249:241–5.

Mine, Y. 1996. Effect of pH during the dry heating on the gelling properties of egg white proteins. *Food Res. Int.* 29:155–61.

Mine, Y. 1997. Effect of dry heat and mild alkaline treatment on functional properties of egg white proteins. *J. Agric. Food Chem.* 45:2924–8.

Mizutani, R. and R. Nakamura. 1984. Emulsifying properties of egg yolk low density lipoprotein (LDL): Comparison with bovine serum albumin and egg lecithin. *Lebensmittel Technol.* 17:213–6.

Nakamura, R. 1963. Studies on the foaming property of chicken egg white. Spread monolayer of the protein fraction of the chicken egg white. *Agric. Biol. Chem.* 27:427.

Raikos, V., Campbell, L., and S. R. Euston. 2007. Effects of sucrose and sodium chloride on foaming properties of egg white proteins. *Food Res. Int.* 40:347–55.

Raikos, V., Campbell, L., and S. R. Euston. 2007. Rheology and texture of hen's egg protein heat-set gels as affected by pH and the addition of sugar and/or salt. *Food Hydrocolloids* 21:237–44.

Saari, A., Powrie, W. D., and O. Fennema. 1964. Isolation and characterization of low-density lipoproteins in native egg yolk plasma. *J. Food Sci.* 29:307–15.

Smith, P. and C. Daniel. 1975. *The chicken book*, 11–12. Athens: University of Georgia Press.

Totosaus, A., Montejano, J. G., Salazar, J. A., and I. Guerrero. 2002. A review of physical and chemical protein-gel induction. *Int. J. Food Sci. Technol.* 37:589–601.

Townsend, A. and S. Nakai. 1983. Relationship between hydrophobicity and foaming characteristics of food proteins. *J. Food Sci.* 48:588–94.

Yamasaki, M., Takhashi, N., and M. Hirose. 2003. Crystal structure of S-ovalbumin as a non-loop-inserted thermostabilized serpin form. *J. Biol. Chem.* 278:35524–30.

8

Galactomannans

Introduction

Different galactomannans, consisting of the hydrocolloids fenugreek, guar gum, tara gum, and locust bean gum (LBG), are generally used for thickening purposes.

Locust Bean Gum: Sources, Manufacturing, and Legislation

The tree *Ceratonia siliqua*, commonly known as Carob tree or St. John's bread, is found in Mediterranean regions. It is ~10–15 m in height, with roots penetrating the soil to depths of 18–27 m. It has brownish, 10- to 20-cm long pods containing ~10-mm long seeds (each fruit contains 10–15 seeds) that weigh ~0.2 g each. The tree flourishes on rocky, semiarid soil. It yields fruit for 5 years after budding and reaches maturity at the age of 50. The carob fruit is eaten raw. In addition, its sugar contents can be concentrated into a viscous fluid or solid. The fruit can also serve as animal feed. In northern Africa, carob fruit supplements the poor natives' diet, whereas in Europe it is sometimes roasted as a substitute for coffee. In the 1920s, wheat and carob flour were mixed to prepare unique breads. Each pod contains 10 to 15 seeds that include ~38% galactomannan. Seeds are composed of 30–33% husk, 23–25% germ, and the remainder endosperm. The seeds have a dark coat, which is removed from the endosperm prior to pulverizing the gum.

Commercial production of LBG started in the 1920s, and today it is marketed globally under multiple trade names. Carob pods are gathered by hand. The kernels are removed from the pod, the husks are eliminated, the seeds are split lengthwise and the endosperm is separated from the germ. Following germ separation, the endosperm is passed through a sieve, graded, packaged, and marketed as LBG or carob gum. The gum is sold in a variety of particle sizes, and can include delicately pulverized portions of the dark-brown testa (seed coat), which emerge as small dark pieces in the powder; the higher the inclusion of the specks, the lower the value of the product. As a result of the presence of undissolved fat and proteins, the gum solution appears opaque, although its appearance can be improved with an alcohol wash. If a clear product is desired (e.g., for fruit jellies), a more costly, alcohol-precipitated polysaccharide is employed. To achieve good solubility in water, the dispersion must be heated, or hot solutions of the gum should be dried after mixing with sugars to prevent recrystallization of the polysaccharide. LBG is classified as a generally regarded as safe (GRAS) food ingredient by the Food and Drug Administration (FDA). The highest allowable concentration in the United States is 0.80%, in cheeses. The gum's inclusion in an item for consumption must be specified in that item's list of ingredients.

Guar Gum: Sources, Processing, and Regulatory Status

Guar gum is extracted from two leguminous (guar) plants: *Cyamopsis tetragonolobus* and *Cyamopsis psoraloides*, both distributed in northwest India and Pakistan. The guar pods are utilized there for animal feed and human consumption. The guar plant is planted following the monsoons in June/July and cropped in December at ~1 m in height. The plants are drought-resistant and can develop in semiarid areas. The pods are green, and each encloses six to nine pea-shaped seeds which are ~2–4 mm in diameter and weigh about 35 mg apiece, of which ~36% is galactomannan. The guar plant was introduced to the United States at the start of the 20th century, when it was predestined for planting in the semiarid regions of the Southwest. In fact, nearly all of the guar plants in the United States are grown in the dry areas of Texas. From 1945, when there was a scarcity of LBG as a result of the Second World War, its replacement by guar gum was explored. The first of such business enterprises were run by General Mills Inc. and Stein, Hall & Co. Inc.

The pods are gathered from the plant by hand, except in Texas where automatic harvesters are used. Manufacturing involves disjointing the endosperm from the germ and testa, and pulverizing to produce the gum. The gum is obtainable in a large variety of sizes, which fluctuate in their solubilization rates. Thermally degraded gums with decreased viscosities are commercially available. Steamed powders show an augmented dissolution rate and decrease in their characteristic tastes. The two main commercial grades of guar gum are food and industrial. For industrial purposes, pulverized endosperm is employed, with minute concentrations of hull and germ due to inadequate purification. The industrial grades are produced with chemical additives, for example, carboxymethyl, hydroxyalkyl, and quaternary amine derivatives, to control useful properties such as viscosity, solubility, and swelling. Guar gum is classified as GRAS by the FDA and by additional worldwide regulatory agencies. The maximal permissible concentration of guar gum in foods is 2% by weight as found in vegetable products and in fats and oils. The gum's presence in foods has to be listed on the label.

Tara Gum

Tara gum is derived from the tara bush, *Caesalpinia spinosa*, which is indigenous to Ecuador and Peru and is grown in Kenya. The reddish tara pods contain seeds which are about 10 mm long and weigh ~0.25 g each, of which ~18% is galactomannan. The ratio of mannose to galactose in tara gum is 3:1 (in LBG this ratio is 4:1 and in guar gum 2:1). Tara gum (E417) is manufactured as a white to white-yellow odorless powder. It is soluble in water and insoluble in ethanol. To form a gel, addition of small amounts of sodium borate to an aqueous solution of the sample is required. As for the purity of the gum, its loss on drying should not exceed 15%, and it should not have more than 1.5% ash, 2% acid insoluble matter, 3.5% protein, 3 mg/kg arsenic, and 5 mg/kg lead; the level of starch should be undetectable.

Fenugreek Gum

The name *fenugreek* is from the Latin for "Greek hay." India is the world's main producer of fenugreek gum. Fenugreek seeds can be employed in the preparation of pickles, vegetable dishes, daals, and spice mixes. The seed is used as a natural herbal medicine in the treatment of diabetes. Yemenite Jews use fenugreek gum to produce *Hilba*, a foamy sauce reminiscent of curry. In India, where a great deal of fenugreek gum is consumed, there is a low incidence rate of arthritis. Fenugreek seeds are a rich source of galactomannans and saponins. Other bioactive constituents of fenugreek gum are volatile oils and alkaloids. Aside from in meals or food preparations, fenugreek gum can be consumed in encapsulated form and it is sometimes prescribed in alternative medicine as a dietary supplement for the control of hypercholesterolemia and diabetes. It is also commonly believed that, due to its high level of dietary fiber, a few fenugreek seeds taken with warm water before bedtime can help prevent constipation. The level of substitution has a strong impact on the property of the gum, as polymannan can associate with itself and cross-link. Consequently, ivory nut mannan is fully insoluble whereas fenugreek gum shows the properties of a simple random coil. In fenugreek gum, the galactomannan substitution level is 1 galactose/1 mannose.

Galactomannan Structure

Galactomannans are linear polysaccharides based on a backbone of β(1,4)-linked D-mannose residues. Single α-D-galactose residues are connected to the chain by C-1 via a glycosidic bond to C-6 of mannose. Gum from a single source encloses molecules with dissimilar degrees of polymerization. The degree of galactose substitution changes from one botanical source to the next, as well as between molecular species of a particular type of gum. These dissimilarities confer the molecules with different properties and this can be used to fractionate the gum. LBG has a "block" structure, that is, the branching units come together mostly in blocks of ~25 residues ("hairy" region), followed by even longer blocks of unsubstituted β(1,6)-D-mannopyranosyl units ("smooth" region) that are vital in the arrangement of interchain connections. Given that separation of the testa and endosperm is limited, impurities can be located in the commercially sold gum. The most important amino acids present in the proteinaceous constituents of LBG in diminishing order of quantity are glutamic and aspartic acids, glycine, arginine, alanine, and serine. In contrast to LBG, guar gum has a regular, alternating structure. Schematically, it can be characterized by its regular, twofold conformation. The unsubstituted D-mannopyranosyl units make up the so-called "smooth" side, while the substituted D-galactopyranosyl units comprise the "hairy" side. This conformation explains guar gum's useful properties.

Commercial applications can be improved and expanded using custom-made guar gum in which modifications are initiated in its chain structure. Properties of the carboxylated polyelectrolyte attained from guar gum were explored. The charged macromolecules formed from native guar gum demonstrated all the distinctiveness of a polyelectrolyte. Viscometry results showed that carboxylated guar gum has markedly elevated viscosity in a medium with low salt content relative to the native polymer, improving its viscosity-forming properties. The average molecular weights of LBG and guar gum are comparable, 1.94 × 106 and 1.9 × 106, respectively, while the number of average molecular weights are ~80,000 for LBG and ~250,000 for guar gum. The substitution ratio is not adequate to completely illustrate the molecules, but provides a suggestion of potential interactions with additional polysaccharides. To provide an unambiguous look at the gum, additional information on the substitution pattern is required. Analogous to LBG, an endogenous constituent of the gum is its proteinaceous substance, an important fraction of the seed endosperm, which stays joined following pulverization and purification.

Gum Solution Properties

Marketable gums displaying high substitution ratios (e.g., guar gum) are more highly hydrated in cold water than gums with limited substitutions (such as LBG), because the presence of side chains interferes with the configuration of stable crystalline regions and supports water penetration, thereby improving solubility. This is reflected in the viscosity of hot- and cold-produced solutions. A number of LBG derivatives have been produced to achieve enhanced rates of swelling and hydration via etherification, and these are used for nonfood purposes. To achieve full viscosity in the shortest amount of time, a suitable mixer, fine powder, and higher temperatures are adopted. Safety measures should be taken to keep the temperature below 80°C, thereby circumventing thermal degradation. The high viscosities attained with seed gums serve in their everyday exploitation. Water is employed exclusively as a solvent in food applications, but LBG also dissolves in low-molecular-weight alcohols. Actually, the viscosity of LBG or carob gum remains unmodified at pH 3.5–11.0.

Following dissolution in water, galactomannans acquire a random coil conformation. Subsequent to separation by the solvent, a linear increase in viscosity with concentration is detected. While the concentration increases, mutual entanglements as a consequence of random contact flanked by polysaccharide chains take place. This causes an exponential amplification in solution viscosity with concentration, that is, the longer the molecule and the more extended the conformation in solution, the higher the possibility of former, stronger entanglements. LBG concentrations under 0.5% do not display any substantial augmentation in viscosity. Nevertheless, over this level, viscosity increases exponentially. Characteristically, a 1% solution of LBG at ambient temperature has a viscosity of 2.4–3.2 × 103 mPa·s. In solution, galactomannans exist as disordered mobile coils. Galactomannans have been shown to depart from characteristic random-coil features and display intermolecular associations between unsubstituted regions of glycan chains. Hyperentanglement was studied by evaluating solution characteristics in strong alkali and at neutral pH. Minute decreases in viscosity values for guar gum were consistent with a lower content of unsubstituted sequences able of forming intermolecular associations. To put it simply, topological entanglement in solutions of galactomannans is enhanced by alkali-labile noncovalent associations. These associations give rise to departures from the general form of concentration, a dependence that is reflected by disordered polysaccharides.

To utilize galactomannans efficiently as a viscosity former, a number of fundamental concepts need to be taken into account. Its concentration should be within the entanglement domain. For galactomannans, this happens when the space occupancy or overlap factor is somewhat higher than unity. Theory predicts that at higher concentrations, the zero-shear viscosity will increase with the third power of concentration, such that doubling the concentration will result in an eightfold increase in viscosity. For guar gum, the dependence on concentration was found to be proportional to the fifth power, perhaps owing to the association with mannan chains in solution. Galactomannans (which form extended random coils in solution) exhibit pseudoplastic flow behavior, in which viscosity declines with shear. Viscous solutions appear to thin upon mastication, offering an enjoyable, light mouth feel. Shear thinning, a consequence of orientation of the extended molecules under the shear gradient, indicates that the molecules are aligning themselves parallel to the direction of flow. At higher gum concentrations, the overlap factor is greater than unity, and shear causes disentanglement. While the speed of the shear increases with respect to the speed of re-entanglement, the viscosity drops, but as the shear stops, it is followed by re-entanglement and revitalization of the unique low-shear viscosity. The decrease in viscosity that is amplified throughout the heating of galactomannan solutions of processed foods, which enclose these thickening agents is a reversible procedure.

Gelation and Interactions of Galactomannans

LBG solutions do not gel. Nevertheless, weak cohesive gels can be structured, even at concentrations of ~0.5%, upon freezing and thawing. The gels dissociate at temperatures of 50–55°C. At LBG concentrations of 0.75% or 1.0%, freeze-thaw gel collapse occurs at 60–65°C, or 64–67°C, respectively. As an explanation, it was proposed that LBG's ability to gel in its native form is a result of its "block" conformation, which allows accumulation of consecutive "smooth" regions to form junction zones. The "hairy" regions are accountable for the network's dispersibility by means of hydrogen bonding with water molecules. This same model clarifies the inability of native guar gum to gel. The lower the concentration of β(1,6)-D-galactopyranosyl, the higher the probability of generating a firm rubbery gel, which can be melted by retortion. If more than one cycle of freeze thawing is applied to these gels, syneresis is more rapid, causing them to lose up to 50% of their water content. Weak-cohesive carob gels with a very low gum concentration (~0.2%) can be shaped into 50% aqueous ethylene glycol.

Due to its alternating chemical structure, which sterically impedes the formation of interchain junction zones, guar gum does not create gels under characteristic food-system conditions. When cations, for instance, borates, Ca^{2+}, Al^{3+}, are added to the gum under alkaline conditions (pH 7.5–10.5), a variety of irreversible gels of changeable consistencies are manufactured. Such gel creation is the consequence of complex formation flanked by the cross-linking agent and the cis-hydroxyl groups. Aside from the addition of cross-linking agents, there are three other methods of inducing guar gum gelation, specifically: reducing the water activity of the system, freeze-thaw treatments, and a reduction in galactose content. Decreased water activity can be accomplished by adding sucrose, or additional hydrophilic molecules to the system, consequently creating competition for the available water between the sucrose and the gum; this supports interchain binding. Freezing also supports interchain binding of the guar-gum molecules, as a consequence of increasing the useful hydrocolloid concentration in the residual unfrozen solution. Upon thawing, the junction zones dissociate and produce a dispersed hydrated gum. Interactions of LBG and xanthan gum (**Chapter 15**) synergistic gelation of LBG or konjac mannan (**Chapter 12**) with xanthan gum, and strong elastic gels that are achieved via blends of LBG and carrageenan (**Chapter 4**) are detailed in **Chapter 16.**

Stability

While galactomannans are employed as viscosity formers in processed foods, heat processing at 120°C for 10 minutes under neutral conditions results in an ~10% reduction in viscosity. This situation can be improved by adding trace elements, for example, sodium sulfite and propyl gallate, suggesting that the principal mechanism is reductive-oxidative depolymerization. However, this mechanism only applies at pH > 4.5. At low acidic pH, hydrolysis of the glycoside bond results in a ~90% drop in viscosity, while the solution is heated to 120°C for 10 minutes. Consequently, the buffering of processed systems, in addition to a reduction in processing temperature, can decrease degradation, and hydration can be delayed by means of crudely ground galactomannan powder. Galactomannans are usually stable to shear forces except under the extreme conditions employed throughout processing, for instance, those in high-pressure homogenization.

Food Applications

Galactomannans are employed for their intrinsic properties, the most extensively utilized being viscosity formation in aqueous solutions. Supplementary sought-after properties involve their synergistic interaction with additional gums and their potential to decrease syneresis. Intended for viscosity-forming functions, low concentrations of 0.5–1.0% are utilized. The galactomannans generate a light consistency, which improves following shearing; a few of them are cold soluble, and viscosity is dependent only on temperature. Enzymatic attack is uncommon given that the applicable enzymes are rarely found in foods and in subsequent eating they are not degraded, so minute or no energy is obtainable from their intake.

Galactomannans are frequently used to stabilize ice cream, compared to the limited use of seed gums in other milk-based products. In an ice-cream mix containing either milk fat or fat from a different source, nonfat milk solids, and added sugars, 0.5% mono/diglyceride is added to destabilize the protein layer around the emulsified fat globules, with 0.3% high-viscosity LBG or a blend of guar gum–LBG. Then, throughout cooling and whipping, incomplete churning of the fat occurs and the free liquid fat with fat crystal structures forms a stabilizing layer around air pockets inside the mass. The function of the galactomannans in ice cream is to slow growth of the bigger ice crystals, by uniting the fluid water to the gum and consequently preventing its mass transfer, and to increase viscosity, yielding a product with a creamier mouthfeel. Ice-crystal size in soft, flavored ices is decreased by adjoining

~0.1% galactomannan. Lactose or ice crystals greater than 20 μm in diameter can be distinguished sensorially and are perceived as sandiness. This sensorial inspection points to large temperature fluctuations throughout storage, and poor product formulation. As formerly mentioned, guar gum mixed with LBG or carrageenan (**Chapter 4**) delays the creation of large lactose and ice crystals. LBG and guar gum assist in maintaining smooth textures in ice cream by slowing down ice-crystal growth throughout constant and fluctuating temperatures. LBG and guar gum were dissolved in sucrose solutions with or without milk solids without fat, fat, and/or emulsifier. Solutions were temperature-cycled at subzero temperatures. LBG solutions developed weak gel structures with temperature cycling, particularly in the presence of milk solids without fat, but guar gum solutions did not. Fat droplets interfered with the formation of weak LBG gel networks while emulsifiers did not change the rheological properties of the emulsions. The ability of a polysaccharide to cryo-gel with both temperature cycling and protein/stabilizer incompatibility leading to phase separation helped create elastic structures.

Galactomannans were used in the development of edible films/coatings for food applications. LBG or guar gum were introduced into traditional wax formulations of two easy-peeler citrus fruit cultivars. Both galactomannans reduced weight loss of the fruit during respiration in a manner similar to a wax-based coating without gum. The incorporation of guar gum or LBG into the wax coatings decreased their permeability to oxygen and carbon dioxide. The LBG–wax coating produced the juice with the best taste.

Nearly all crude galactomannans include proteinaceous substances. Crude LBG and guar gum decreased the surface tension of water, and adsorbed to oil–water interfaces, decreasing their interfacial tensions. Coalescence and flocculation were reduced by finding the best ratios of gum to oil for complete droplet coverage. In fermented milk items for consumption, the inclusion of a gum mixture consisted of CMC (**Chapter 5**), galactomannan, and gelatin (**Chapter 9**), employed to overcome structural loss when the fresh cheese is cut by standard processing. In milkshakes, a lesser amount of galactomannan (0.1%) was employed to increase viscosity and convey a creamier mouthfeel to the manufactured product. Galactomannans are also used in milk-fortified fruit-juice drinks where the proportion of milk is low, to hold up sedimentation of the casein (**Supplemental G**) micelles and guarantee approval of the product.

In addition to their utilization in gelled desserts based on carrageenans (**Chapter 4**), galactomannans are beneficial in modifying gel consistency and avoiding syneresis. A dessert with an indistinguishable chiffon-type gel can be created by incorporating gelatin, LBG, soy protein (**Supplemental K**), sugar, and organic acids. In mayonnaises whose oil content is below ~60%, rheological properties can be altered by substituting the oil with chemically modified starch (**Chapter 14**) and including a stabilizer, for instance, guar gum, with a small addition of LBG and xanthan gum (**Chapter 15**) to the formulation. Guar gum is utilized in meat sauces and barbecue, and in a variety of salad dressings to prevent phase separation and convey a pleasing mouthfeel. If pieces of fish, meat, or other constituents are added to the mayonnaise, then it should be constructed such that it takes up the liquids and leaves the pieces undamaged. This can be accomplished by increasing the inherent galactomannan content. If fresh fruits and vegetables are included in the mayonnaise, amylases may be liberated, causing starch hydrolysis. In this case, the starch should be replaced with a blend of high-viscosity guar gum and xanthan gum, both at concentrations of not more than 0.7%.

Guar gum, LBG, and fenugreek gum could be incorporated at up to 15% in a pea–rice blend to develop nutritious, organoleptically acceptable, extruded snack products with low glycemic index. Galactomannans are also used in ketchups and dressings to increase viscosity and reduce syneresis. In sterilized soups and sauces, galactomannans are utilized as viscosity formers in blends with xanthan gum at concentrations of ~0.2–0.5%. The effect of various levels of replacement of a fenugreek polysaccharide product (0.1–0.9%) on 5% wt/wt corn starch (**Chapter 14**) cream soups was investigated. The apparent measured viscosity of the cooked soups increased gradually at all fenugreek levels and revealed shear-thinning behavior. In terms of creaminess, soups containing fenugreek gum scored higher than those without it. In terms of taste, soups with no fenugreek gum or with a high level of it (0.9% fenugreek gum) scored lowest. Soup with 0.5% fenugreek gum substitution was found to have the best taste and overall acceptability.

In deep-frozen foods, the addition of cold-soluble galactomannans improve freeze-thaw resistance, presumably by sterically hindering the formation of aggregates in the interstitial fluid phase. Other applications are to prevent weeping in sausages, as a suspension agent in pumped meat blends, as a binder and phosphate replacer in fish fillets, as a gluten replacer in low-calorie jams and spreads, and in baked goods. In sausages, salami, and bologna, LBG is included to advance comminution of constituents, to improve yield as a result of the binding of free water, to extrude mixtures easily, and to prevent phase separation during cooking, smoking and storage. The addition of LBG or carob gum to pet foods influences the texture of the finished product. For instance, a viscous gravy sauce with gloss can be produced. A similar result can be obtained with canned meat, where a viscous stew is desired. Guar gum can be added at a concentration of ~0.5% of the total batch weight to canned meat products, resulting in cleaner mixing throughout meat cooking, easier pumping of the cooked product, cleaner technological processing, and improved control of the practice. In stuffed meat products, guar gum offers quick binding of free water for the duration of comminution, improved stuffing into casings, elimination of fat and free-water separation, and migration during cooking, smoking and storage, and improved firmness upon cooling.

In addition, LBG can assist in producing the desired baked textures. The addition of LBG to wheat flours results in a softer, tastier item for consumption with a longer shelf life. Staling is slowed, crumbliness is diminished, and the number of eggs used for the manufacture of biscuits, rolls, and cakes is reduced. The strength and flexibility of baked tortillas can be improved by inserting a combination of LBG or carob gum and guar gums. In producing dry mixtures for baking, the addition of LBG or carob gum and guar gum at a maximum concentration of 0.15% improves mixing and the resultant mix's characteristics. Guar gum added to bakery good mixes imparts shorter batter mixing times, less crumbling of the finished product, improved ingredient mixing, reduced moisture loss during storage, and the ability to freeze the finished product. Reconstituted flours, consisting of rice, potato or tapioca starch (**Chapter 14**), vital wheat gluten, lecithin, and ethoxylated mono/diglycerides, plus xanthan gum, guar gum or cellulose gum (**Chapter 5**), were made into batters which were assessed for specific gravity and viscosity. On the whole, reconstituted rice flours with guar or xanthan gums produced cakes that were most similar to control cakes.

The addition of 1% guar gum decreases the quantity of oil and fat absorbed during frying in deep fat. This is due to the construction of a shielding film, resulting in better-moistened products. Guar gum has been integrated into breads, oatcakes, and biscuits to yield baked items with a high soluble fiber content, which can be exploited to decrease the levels of serum cholesterol and low-density lipoproteins (LDLs). Another report dealt with biscuits that were prepared from blends containing different proportions of raw, soaked, and germinated fenugreek seed flour. Fenugreek flour could be incorporated up to 10% into the formulation of biscuits without affecting their overall quality. Physical, sensory, and nutritional characteristics revealed that among all composite fenugreek flour biscuits, those containing 10% germinated fenugreek flour were the best. Use of fenugreek gum, a food with demonstrated efficacy in lowering blood sugar, is limited by its bitter taste and strong flavor, and therefore a special fenugreek-gum bread formula was developed and tested for its taste acceptability and its effect on carbohydrate metabolism. There was no statistically significant difference in proximate composition, color, firmness, texture, or flavor intensity between the fenugreek gum and wheat breads. The bread retained fenugreek gum's functional property of reducing insulin resistance. Thus acceptable baked products could be prepared with added fenugreek gum, which would reduce insulin resistance and treat type 2 diabetes. Over the years, guar gum has been employed in a small number of manufactured goods, as a foam stabilizer in foam-dried coffee concentrate, to manufacture a simulated chocolate–chiffon dessert, and to improve mechanical properties, water absorption and puffing characteristics of heat-extruded corn grits.

Recipes with Galactomannans

Brown Sugar Sherbet (Figure 8.1)

Ingredients

(Serves 5)

250 mL (1 cup) water

250 g (2¼ cups) brown sugar or cane sugar

50 g (2½ Tbsp) glucose

3 g (2½ tsp) locust bean gum (LBG) powder

250 mL (1 cup) lemon juice

Preparation

1. Heat water, brown sugar, glucose, and lemon juice in a pan to boiling while stirring with a whisk (**Figure 8.1A**).
2. Whisk in LBG (**Figure 8.1B**).
3. Strain the solution (**Figure 8.1C**) (**Hint 1**), pour into a shallow container, and cool to around 10°C in a refrigerator or ice-water bath (**Figure 8.1D**) (**Hint 2**).

FIGURE 8.1 Brown sugar sherbet. (**A**) Heat, boil, and stir water, brown sugar, glucose, and lemon juice. (**B**) Stir in LBG. (**C**) Strain solution. (**D**) Cool. (**E**) Place frozen solution in a blender. (**F**) Introduce air into the frozen mass. (**G**) Serve.

4. Freeze the solution.
5. Place the frozen solution in a blender (**Figure 8.1E**), and mix until it contains air (**Figure 8.1F**).
6. Freeze the mixed sherbet, and serve (**Figure 8.1G**).

$$pH = 2.85$$

Hints

1. Straining eliminates lumps.
2. The solution is first cooled so that foods next to it in the freezer will not thaw.

This recipe is very sour, and good for hot summer days. If you do not like sour sherbet, the amount of lemon juice and brown sugar can be adjusted accordingly.

Raspberry Sauce (Figure 8.2)

Ingredients

(Makes 300 g)
0.4 g (⅖ tsp) locust bean gum (LBG) powder
2 g (⅚ tsp) agar–agar powder

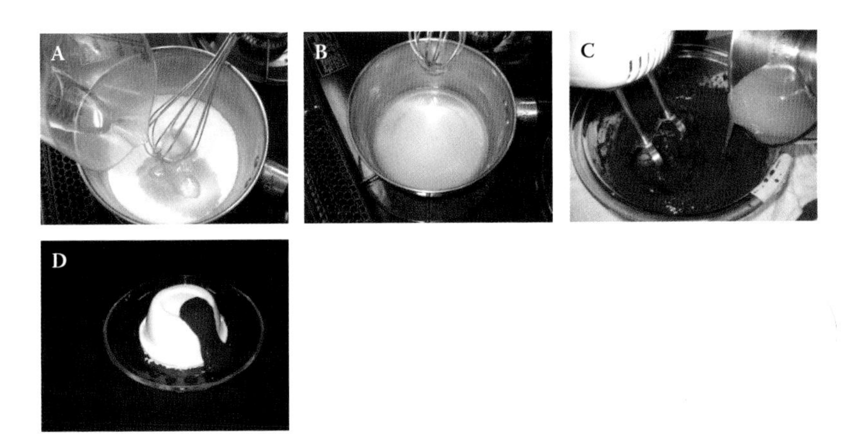

FIGURE 8.2 Raspberry sauce. (**A**) Mix water with LBG, agar–agar, and granulated sugar. (**B**) Heat the solution until ingredients dissolve. (**C**) Add the heated solution to raspberry purée while mixing. (**D**) Serving suggestion: use as garnish.

50 g (a little over ¼ cup) granulated sugar
25 mL (2⅔ Tbsp) water
250 g (1⅕ cups) raspberry purée†

†YOU CAN ALSO USE CRANBERRY PURÉE.

Preparation

1. Mix LBG, agar–agar, and granulated sugar in a pan, and whisk in water (**Figure 8.2A**).
2. Heat solution (to 90°C) until all components are dissolved (**Figure 8.2B**).
3. Whisk raspberry purée with a hand mixer, and add the heated solution to it with stirring with a hand mixer (**Figure 8.2C**).
4. Cool the jam in the refrigerator or in an ice-water bath, and use to garnish desserts (**Figure 8.2D**) (**Hint 1**).

$$pH = 2.95$$

Hint

1. This is a good accompaniment for ice cream, cheese cake, milk jelly, panna cotta, yoghurt, and so on.

The texture of this recipe is between a jelly and a jam. It has a sour taste, and the sauce is therefore well suited to sweets.

Banana and Nuts Ice Cream (Figure 8.3)

Ingredients

(Serves 5)
60 g roasted pecans or walnuts
500 mL (2 cups) cold water (5°C)
1 ripened banana (150–200 g) (**Hint 1**)
1.2 g (⅓ tsp) guar gum powder
85 g (4 Tbsp) honey
1 g (½ tsp) cinnamon

Preparation

1. Finely grind pecans or walnuts in a blender (**Figure 8.3A**) (**Hint 2**).
2. Add 350 g cold water to the ground nuts and blend until smooth (**Figure 8.3B**).
3. Add banana, guar gum, honey, and cinnamon to the nut mixture (**Figure 8.3C**), and blend until smooth.
4. Pour the mixture into an ice-cream maker (**Figure 8.3D**), add 150 g cold water, and follow the manufacturer's instructions. Serve garnished with nuts (**Figure 8.3E**).

pH = 5.37

Hints

1. Thicker ice cream can be made using more ripened bananas.
2. For a crunchier texture, chop nuts roughly with a knife.

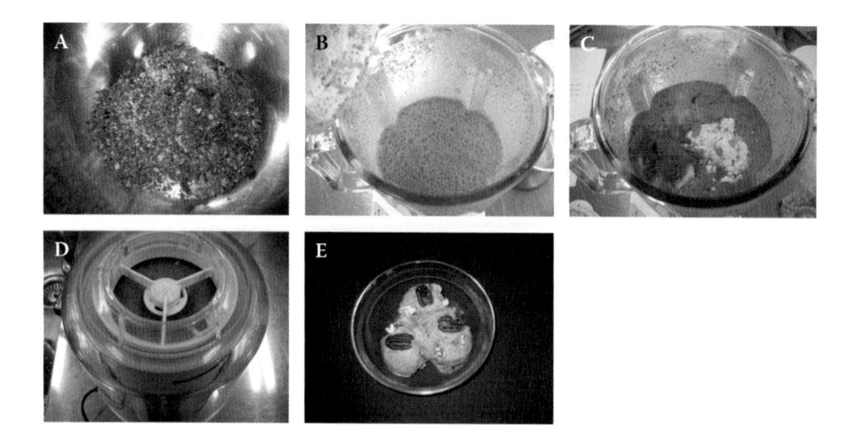

FIGURE 8.3 Banana and nuts ice cream. (**A**) Grind pecans or walnuts. (**B**) Add cold water to the ground nuts, continue to blend the mass. (**C**) Add banana, guar gum, honey, and cinnamon to the nut mixture and blend again. (**D**) Pour the mixture into an ice-cream maker. (**E**) Garnish with nuts.

Frozen Fruit Drink (Figure 8.4)

Ingredients

(Serves 5)
500 g (2 cups) low fat milk
200 g frozen fruit purée†
120 g (½ cup) fruit juice
1.5 g (½ tsp) guar gum powder
50 g (2⅓ Tbsp) corn syrup
200 g ice cubes (**Hint 1**)
Vanilla flavoring

†THE AMOUNT OF FRUIT REQUIRED TO MAKE 200 G OF PURÉE DEPENDS ON THE FRUIT, AND IS ABOUT 1 CUP.

Preparation

1. Place all ingredients in a blender (**Figure 8.4A**), and blend for 2 minutes (**Figure 8.4B**).
2. Pour into glasses to serve.

pH of raspberry drink = 4.03

Hint

1. Use bigger ice cubes to enjoy the texture of crushed ice.

This recipe is very easy to make. Drink it cold.

FIGURE 8.4 Frozen fruit drink. (**A**) Place all ingredients in a blender. (**B**) Blend for 2 minutes.

Empanada Dough (Figure 8.5)

Ingredients

(Serves 4)
6 g (2 tsp) guar gum powder
110 g (1 cup) sorghum blend powder
65 g (¾ cup) tapioca starch
30 g (¼ cup) sweet rice flour
9 g (1 Tbsp) white sugar
3 g (½ tsp) salt
2 g (½ tsp) baking soda
120 g (¾ cup) shortening
70 mL (4⅔ Tbsp) water
Rice flour

Preparation

1. Put guar gum, sorghum blend, tapioca starch, sweet rice flour, white sugar, salt, baking soda, and shortening in a food processor (**Figure 8.5A**), and mix at low speed until smooth (**Figure 8.5B**) (**Hint 1**).

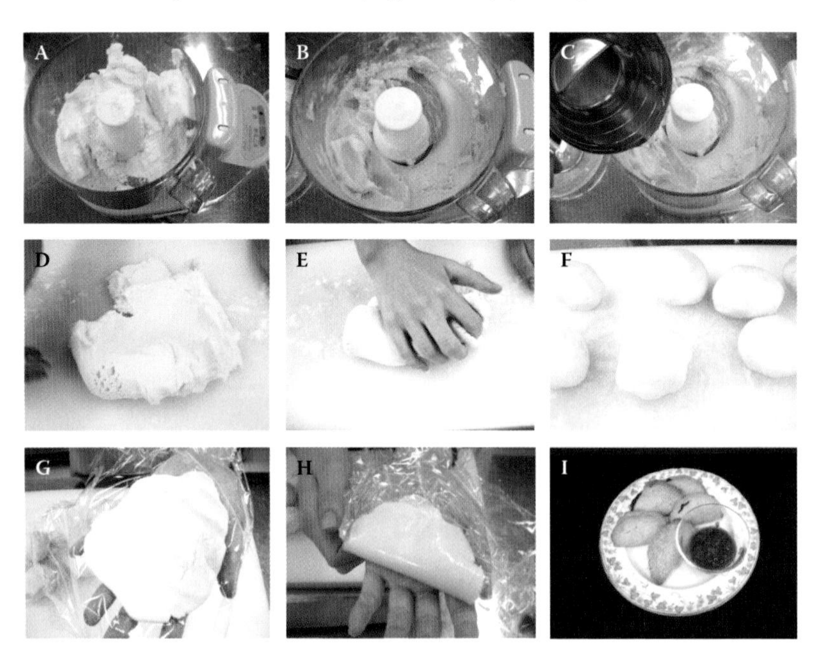

FIGURE 8.5 Empanada dough. (**A**) Put guar gum, sorghum blend, tapioca starch, sweet rice flour, white sugar, salt, baking soda, and shortening in a food processer. (**B**) Mix until it becomes smooth. (**C**) Add water and continue to mix. (**D**) Remove dough. (**E**) Knead. (**F**) Divide the dough. (**G**) Roll the dough pieces. (**H**) Add your favorite filling and fold. (**I**) Bake and serve.

2. Add water to the mixture (**Figure 8.5C**), and mix at low speed for a few min. (Figure 8.5D). Sprinkle rice flour on a cutting board, and knead the mixture by hand (**Figure 8.5E**) until it becomes flexible.
3. Divide dough into 10 pieces (**Figure 8.5F**), and roll each out into a 15 cm diameter circle in the palm of your hand covered with plastic wrap (**Figure 8.5G**) (**Hint 2**); add your favorite filling, such as meat, fish, shrimp, cheese, tomato sauce, apple filling, and so on, and fold the dough in two over the filling (**Figure 8.5H**).
4. Preheat the oven to 190°C, bake empanadas for 20 minutes, and continue to bake at 220°C for 5 minutes until they brown.
5. Serve hot empanada with salsa (optional) (**Figure 8.5I**).

pH of dough = 5.62

Hints

1. Using a food processor produces a smooth dough.
2. The dough is sticky but brittle, and using plastic wrap helps form the dough.

Empanada is eaten in Latin America, Western Europe, and parts of Southeast Asia. It generally consists of stuffed bread or pastry made of wheat flour. However, this recipe is made of sorghum, tapioca, and rice flour, and it is gluten-free.

Loukoums (Figure 8.6)

Ingredients

(Serves 5)
1.4 g (a little over ½ tsp) agar–agar powder
0.6 g (⅕ tsp) tara gum powder
150 g (⅗ cup) fruit juice
100 g (¾ cup) white sugar
70 g (3⅓ cups) corn syrup
15 mL (1 Tbsp) lemon juice
8 g (2 tsp) gum arabic powder
50 mL (⅕ cup) water
20 g (1 Tbsp) glucose

Preparation

1. Put 100 g fruit juice in a pan, and whisk in agar–agar and tara gum (**Figure 8.6A**).
2. Heat the gum mixture until it dissolves (**Figure 8.6B**) and remove from heat.
3. Mix white sugar, corn syrup, and 50 g fruit juice in another pan, heat it with whisking until it boils, and remove from heat.

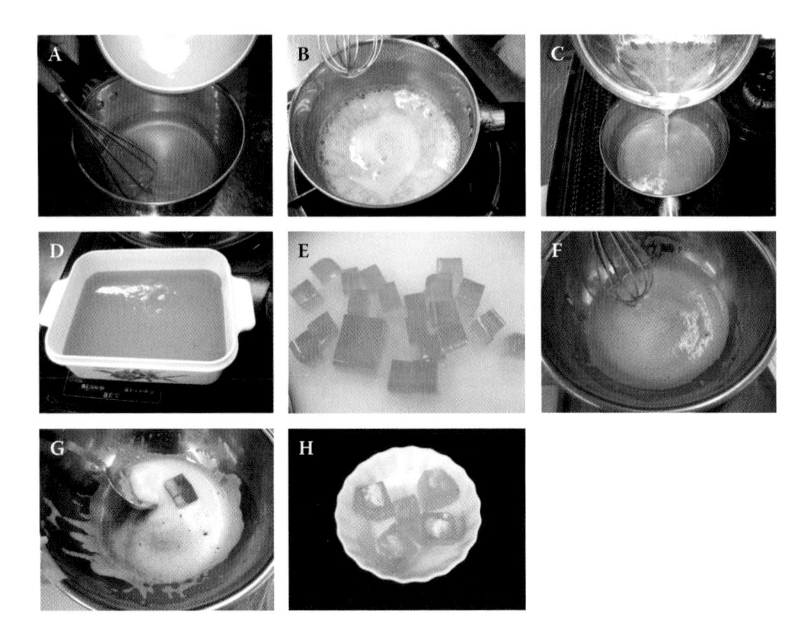

FIGURE 8.6 *Loukoums*. (**A**) Stir together fruit juice, agar–agar, and tara gum. (**B**) Heat until dissolution. (**C**) Cool gum and sugar solution and add lemon juice. (**D**) Pour solution into a greased mold. (**E**) Cut jelly into cubes. (**F**) Mix gum arabic into water. (**G**) Dip the cut jellies into the gum arabic solution. Sprinkle with glucose powder and dry.

4. Mix the gum solution (**2**) and sugar solution (**3**), cool to 40–45°C (**Hint 1**), add lemon juice and mix well (**Figure 8.6C**).
5. Spread vegetable oil on a rectangular mold (**Hint 2**), and pour the solution (**4**) into it (**Figure 8.6D**).
6. Leave the solution to dry for at least 4 hours (preferably overnight), then cut it into cubes (**Figure 8.6E**).
7. Put water in a bowl, and whisk in gum arabic (**Figure 8.6F**).
8. Dip the cubed jellies (**6**) into the gum arabic solution (**7**) (**Figure 8.6G**) with using a spoon, and sprinkle with glucose (**Figure 8.6H**).
9. Dry the *loukoums* for a few days.

$$\text{pH} = 3.05$$

$$\text{pH of gum arabic solution} = 4.23$$

Hint

1. If the temperature of the dispersion drops below 40°C, gelation will occur, and lemon juice will not be mixed in.
2. Vegetable oil prevents *loukoums* from sticking to the mold.

Loukoums are traditional Turkish sweets. They are widely eaten in the Balkans, North Africa, the Middle East, and so forth. They can be stored for long periods.

Yoghurt Dressing (Figure 8.7)

Ingredients

(Makes ~1/2 cup)
0.2 g tara gum powder
20 mL (1⅓ Tbsp) lukewarm water (40°C)
15 g (1 Tbsp) white wine vinegar
2 g (⅓ tsp) salt
0.5 g (¼ tsp) black pepper
50 g (⅕ cup) yoghurt
12 g (1 Tbsp) olive oil

Preparation

1. Whisk tara gum into lukewarm water (**Hints 1** and **2**).
2. Mix in white wine vinegar, salt, black pepper, and yoghurt (**Figure 8.7A**).
3. Gradually add olive oil to the tara gum mixture, stirring with a whisk (**Figure 8.7B**) and mix well (**Figure 8.7C**).

pH = 3.62

Hints

1. A better solution is produced at 40°C.
2. A blender or food processor can be used instead.

This recipe produces a fresh and creamy type of dressing. It is good on green vegetables, tomatoes, smoked salmon, and so forth.

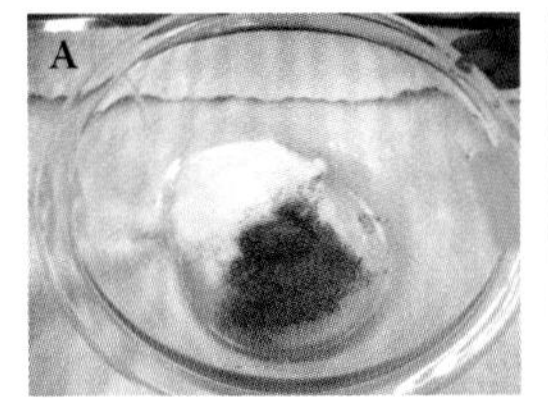

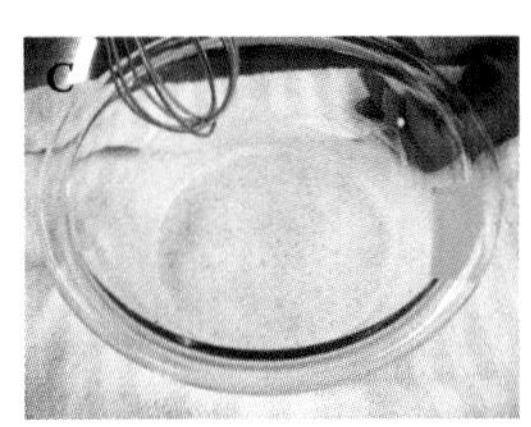

FIGURE 8.7 Yoghurt dressing. (**A**) Mix tara gum solution, white wine vinegar, salt, black pepper, and yoghurt in a bowl. (**B**) Gradually add olive oil while stirring. (**C**) Mix well.

Sharbat Hilba (Fenugreek Soup) (Figure 8.8)

Ingredients

(Serves 5)
900 mL (3⅗ cups) water
1 kg meat†
20 g (4 tsp) fenugreek seed
200 g (1–2) onion
4 g (½ Tbsp) bzaar‡
3 g (½ Tbsp) chili powder
20 g (1⅓ Tbsp) tomato purée
30 g (2½ Tbsp) vegetable oil
Salt, dry mint

†LARGE PIECES OF LAMB WITH BONES ARE RECOMMENDED

‡NORTH AFRICAN SPICE

Preparation

1. Crush fenugreek seeds lightly with a knife (**Figure 8.8A**), and slice onion thinly.
2. Pour water into a large pot, and add vegetable oil, meat, and sliced onion (**Figure 8.8B**), stirring with a spatula.
3. Heat and add tomato purée, bring to boil and then simmer for 90 minutes or until the meat is tender.

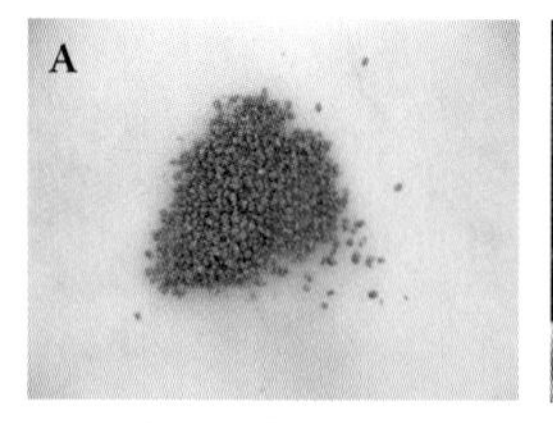

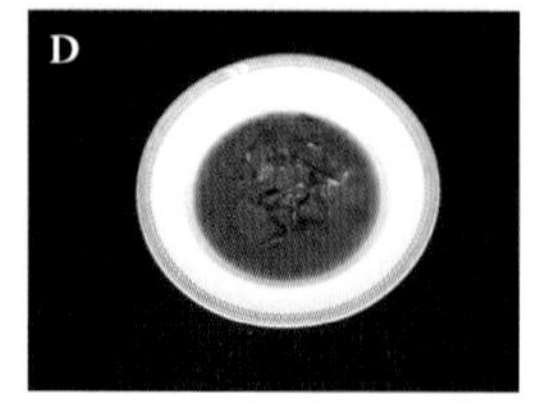

FIGURE 8.8 *Sharbat hilba* (fenugreek soup). (**A**) Crush fenugreek seeds. (**B**) Pour water into pot, add vegetable oil, meat, and sliced onion. (**C**) Add the crushed fenugreek seeds and additional spices into the soup. (**D**) Serve.

4. Add the crushed fenugreek seeds, bazzar, chili powder, and salt to the soup (**Figure 8.8C**) and simmer for 1 hour or until the soup has thickened slightly.
5. If you used large pieces of meat, remove the simmered meat from the pan, chop it, and return to the soup.
6. Serve in soup bowls (**Figure 8.8D**), and garnish with dry mint.

pH = 6.18

This recipe is for a traditional Libyan soup. Fenugreek seeds are widely used in Middle Eastern, African, and Indian cuisine.

Pear and Fenugreek Jam (Figure 8.9)

Ingredients

(Makes 300 g)
300 g (2) pears†
2 g (½ tsp) fenugreek powder
90–120 g (30–40%) maple syrup
15 mL (1 Tbsp) lemon juice

†YOU CAN USE PERSIMMON INSTEAD.

Preparation

1. Peel and core pears, and chop roughly.
2. Grind chopped pears in a food processor (**Figure 8.9A**) (**Hint 1**).
3. Place the ground pear, fenugreek powder, and maple syrup in a pan (**Figure 8.9B**), and cook on low heat with occasional stirring.
4. When the jam has thickened (**Figure 8.9C**), remove from heat and add lemon juice.
5. Pour jam into sterilized glass bottles (see **Chapter 13**).

pH = 3.53

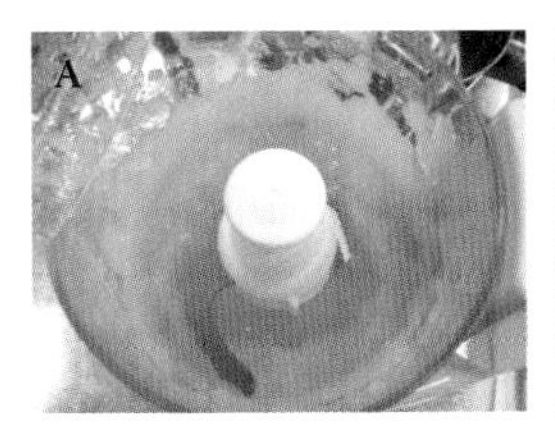

FIGURE 8.9 Pear and fenugreek jam. (**A**) Grind chopped pears. (**B**) Mix ground pears, fenugreek, gum powder, and maple syrup in a pan. (**C**) Upon thickening, remove from heat.

Hint

1. Leave pears chunky for a different texture.

This recipe differs from other jams. Enjoy the maple syrup and caramel flavor.

Tips for the Amateur Cook and Professional Chef

- LBG is used at ~0.6% in cheese spreads to obtain a highly homogeneous, fine product with a spreadable texture.
- To retain the cloudiness of citrus juice products, include up to 30 ppm LBG with up to 400 ppm sodium hexametaphosphate.
- Addition of 2.5–3.0 g of guar gum per 100 L milk increases the yield of curd solids in soft cheeses.
- Guar gum at a concentration of ~0.3% helps effectively stabilize ice cream.
- In cake mixes, guar gum at a concentration of ~0.10–0.15% of the total weight of the dry ingredients is used for greater moisture retention and prolonged shelf-life.
- The preferred level of guar gum in an icing mixture consists of 1 part guar gum to 250 parts sugar and 30 parts water.
- As a meat binder, 0.01 lb (4.536 g) guar gum to 10 lb (4.536 kg) free water is required.
- In canned meat and pet foods, addition of ~0.5% guar gum is used to prevent emulsion separation and reduce syneresis.
- Levels of 0.27 to 0.75% guar gum are used to stabilize fruit nectars, increasing their viscosity and smoothness.

REFERENCES AND FURTHER READING

Anderson, D. M. W. 1986. The amino acid components of some commercial gums. In *Gums and stabilisers for the food industry*, vol. 3, ed. G. O. Phillips, D. J. Wedlock and P. A. Williams, 79–86. London: Elsevier Applied Science Publishers.

Arbuckle, W. S. 1986. Effects of emulsifiers on protein-fat interactions in ice-cream mix during ageing. I. Quantitative analyses, In *Ice cream*, 84–94. Westport: Avi Publishing Co.

Block, H. W. 1961. Chocolate chiffon. U.S. Patent No. 2,983,617.

Cawley, R. W. 1964. The role of wheat flour pentosans in baking. II. Effect of added flour pentosans and other gums on gluten-starch loaves. *J. Sci. Food Agric.* 15:834–9.

Cerqueira, M. A., Bourbon, A. I., Pinheiro, A. C., Martins, J. T., Souza, B. W. S., Teixeira, J. A., and A. A. Vicente. 2011. Galactomannans use in the development of edible films/coatings for food applications. *Trends Food Sci. Technol.* 22:662–71.

Chen, S. and A. Nussinovitch. 2000. Galactomannans in disturbances of structured wax–hydrocolloid-based coatings of citrus fruit (easy-peelers). *Food Hydrocolloids* 14:561–8.

Chen, S. and A. Nussinovitch. 2001. Permeability and roughness determinations of wax-hydrocolloid coatings, and their limitations in determining citrus fruit overall quality. *Food Hydrocolloids* 15:127–37.

Dea, I. C. M. 1979. Interactions of ordered polysaccharide structures synergism, and freeze-thaw phenomena. In *Polysaccharides in food,* ed. J. M. V. Blanshard and J. R. Mitchell, 229–40. London: Butterworths.

Dea, I. C. M. 1990. Structure/function relationships of galactomannans and food grade cellulosics. In *Gums and stabilisers for the food industry*, vol. 5, ed. G. O. Phillips, D.J. Wedlock, and P. A. Williams, 373–82. Oxford: IRL Press.

Dea, I. C. M., Morris, E. R., Rees, D. A., Welsh, E. J., Barnes, H. A., and J. Price. 1977. Associations of like and unlike polysaccharides: Mechanism and specificity in galactomannans, interacting bacterial polysaccharides, and related systems. *Carbohydrate Res*. 57:249–72.

Doublier, J. L. and B. Launey. 1981. Rheology of galactomannan solutions. *J. Text. Studies*, 12:151–72.

Einstein, A. 1906. Eine neue Bestimmung der Molekuldimension. *Annale der Physik* 19:289–306.

Federal Register. 1987. *Guar gum*, Code of Federal Regulations, Title 21, 184.1339, 432–3. Washington DC: Office of the Federal Register.

Federal Register. 1987. *Locust (carob) bean gum*, Code of Federal Regulations, Title 21, 184.1343, 433. Washington DC: Office of the Federal Register.

Federal Register. 1987. *Food; designation of ingredients*, Code of Federal Regulations, Title 21, 101.4a, 14. Washington DC: Office of the Federal Register.

Fox, J. E. 1992. Seed gums. In *Thickening and gelling agents for food*, ed. A. Imeson, 153–70. Glasgow: Blackie Academic & Professional, an imprint of Chapman & Hall.

French, S. J. and M. A. Hill. 1985. High fibre foods: A comparison of some baked products containing guar and pectin. *J. Plant Foods* 6:101–09.

Frollini, E., Reed, W. F., Milas, M., and M. Rinaudo. 1995. Polyelectrolytes from polysaccharides: Selective oxidation of guar gum—A revisited reaction. *Carbohydr. Polym.* 27:129–35.

Garti, N. and D. Reichman. 1994. Surface properties and emulsification activity of galactomannans. *Food Hydrocolloids* 8:155–73.

Glicksman, M. 1969. *Gum technology in the food industry*, 130. New York: Academic Press.

Goldstein, A. M., Alter, E. N., and J. K. Seaman. 1973. Guar gum, In *Industrial gums*, eds. R. L. Whistler and J. N. BeMiller, 303–21. New York: Academic Press.

Gorton, L. 1984. Tortilla improvements—Less fragile, more firm. *Baker's Digest* 58:26.

Goycoolea, F. M., Richardson, R. K., Morris, E. R., and M. J. Gidley. 1995. Stoichiometry and conformation of xanthan in synergistic gelation with locust bean gum or konjac glucomannan. *Macromolecules* 28:8308–20.

Goycoolea, F. M., Richardson, R. K., Morris, E. R., and M. J. Gidley. 1995. Effect of locust bean gum and konjac glucomannan on the conformation and rheology of agarose and κ-carrageenan. *Biopolymers* 36:643–58.

Herald, C. T. 1986. Locust/carob bean gum. In *Food hydrocolloids*, vol. 3, ed. M. Glicksman, 161–70. Boca Raton: CRC Press.

Herald, C. T. 1986. Guar gum. In *Food hydrocolloids*, vol. 3, ed. M. Glicksman, 171–84. Boca Raton: CRC Press.

Hooda, S. and S. Jood. 2005. Organoleptic and nutritional evaluation of wheat biscuits supplemented with untreated and treated fenugreek flour. *Food Chem.* 90:427–35.

Jiang, J. X., Jian, H. L., Cristhian, C., Zhang, W. M., and R. C. Sun. 2011. Structural and thermal characterization of galactomannans from genus *Gleditsia* seeds as potential food gum substitutes. *J. Sci. Food Agric.* 91:732–7.

Lopeda Silva, J. A., and M. P. Goncalves. 1990. Studies on a purification method for locust bean gum by precipitation with isopropanol. *Food Hydrocolloids* 4:277–87.

Losso, J. N., Holliday, D. L., Finley, J. W., Martin, R. J., Rood, J. C., Yu, Y., and F. L. Greenway. 2009. Fenugreek bread: A treatment for diabetes mellitus. *J. Medicinal Food* 12:1046–9.

Mancuso, J. J. and L. Common, 1960. Chiffon. U.S. Patent No. 2,965,493.

Mitchell, J. R., Hill, S. E., Jumel, K., and S. E. Harding. 1992. The use of anti-oxidants to control viscosity and gel strength loss on heating of galactomannan systems. In *Gums and stabilisers for the food industry*, vol. 6, ed. G. O. Phillips, D. J. Wedlock and P. A. Williams, 303–10. Oxford: Oxford University Press.

Morris, E. R. and S. B. Ross-Murphy. 1981. Chain flexibility of polysaccharides and glycoproteins from viscosity measurements. *Tech. Carbohydr. Metab.* B310:1–46.

Nussinovitch, A. 1997. *Hydrocolloid applications: Gum technology in the food and other industries*. London: Blackie Academic & Professional.

Nussinovitch, A. 2003. *Water soluble polymer applications in foods*. Oxford, UK: Blackwell Publishing.

Nussinovitch, A. 2010. *Plant gum exudates of the world sources, distribution, properties, and applications*. Boca Raton: CRC Press.

Patmore, J. V., Goff, H. D., and S. Fernandes. 2003. Cryo-gelation of galactomannans in ice cream model systems. *Food Hydrocolloids* 17:161–9.

Ravindran, G., Carr, A., and A. Hardacre. 2011. A comparative study of the effects of three galactomannans on the functionality of extruded pea–rice blend. *Food Chem.* 124:1620–6.

Ravindran, G. and L. Matia-Merino. 2009. Starch–fenugreek (*Trigonella foenum-graecum* L.) polysaccharide interactions in pure and soup systems. *Food Hydrocolloids* 23:1047–53.

Rice, B. O. and M. K. Ndife. 1995. Effects of the addition of hydrocolloids on the baking performance of high ratio layer microwave cakes made from reconstituted flours. *IFT Annual Meeting, 1995, Conference Book*, 288.

Rol, F. 1973. Locust bean gum. In *Industrial gums: polysaccharides and their derivatives*, ed. R. L. Whistler and J. N. BeMiller, 323–37. New York: Academic Press.

Seaman, J. K. 1980. Locust bean gum. In *Handbook of water-soluble gums and resins*, ed. R. L. Davidson, 14.1–16. New York: McGraw-Hill Book, Co.

Tolstoguzov, V. B. 1991. Functional properties of food proteins and role of protein-polysaccharide interaction. *Food Hydrocolloids* 4:429–68.

9

Gelatin

Historical Background

The word *gelatin* is derived from the Latin verb *gelare* meaning "to congeal." The circumstances surrounding gelatin's discovery remain ambiguous. They are most likely related to the finding that boiling pieces of animal skin and bone produced a material that was fluid when hot, and solidified upon cooling. Extraction of this glue by cooking hides dates back to at least the time of the ancient Pharaohs of Egypt. A 3000-year-old stone carving from the ancient city of Thebes described the gluing of a thin covering of rare redwood to a flat yellow timber of sycamore. In the Roman era, evidence from Pliny described the cooking of glue from hides of bulls. Much later, in ~1600 A.D. (the Elizabethan period), both Bacon and Shakespeare referred to glue in their writings. A glue industry started in England in ~1700 and parallel manufacturing developed in the United States (US) in the following century. The first profitable fabrication of gelatin in the US was in Massachusetts in 1808, and the first production of edible gelatin is credited to Arney, who was granted a patent in 1846 for the preparation of powdered gelatin. Today, member companies of the Gelatin Manufacturers Institute of America produce an annual combined total of over 110 million pounds of gelatin. Gelatin is an important material, finding application in the food, pharmaceutical, and photographic industries as well as for diverse technical uses.

Definitions

Gelatin is classified in the US Pharmacopeia as "a product obtained by the partial hydrolysis of collagen derived from the skin, white connective tissue and bones of animals." Collagen, meaning a "glue-producing material," is a protein constituent of the connective tissues which serve as the main stress-bearing elements for all mammals and fishes. Collagen is one of a kind among proteins due to its notable amino acid composition. Collagen's distinctive structure is due to its high content of cyclic amino acids, that is, proline and hydroxyproline, in addition to high proportions of more common nonpolar amino acids such as glycine and alanine.

Manufacture and Sources

Manufactured gelatin stems from three basic sources and consists of two types of finished gelatin products. The sources include pig skins, cattle hides, and ossein (bone). In general, all gelatins are manufactured by one of the two following processes or their modifications.

Acid Processing

Depending on the pretreatment procedure, two commercial types of gelatin are available, type A and type B, obtained under acid and alkaline pretreatment conditions, respectively. Acid processing (type A) depends on the use of ossein and pig skins. In the US, vast use for this process includes the preparation of edible gelatin from frozen pig skins. Processing includes thawing, washing in cold water, and soaking in ~5% solutions of inorganic acids. This treatment ends with swelling and hydration of the skin devoid of substantial solubilization. For processing, different acids, such as hydrochloric, sulfuric, phosphoric and sulfurous are the most commonly employed. After soaking in acid for 10 to 30 hours, the supernatant acid is discarded and cold water is used to wash the excess acid and elevate the pH to ~4. Most of the noncollagenous proteins have isoelectric points in the vicinity of 4 to 5 pH units and consequently, they are readily coagulated and removed. The acid-conditioned skins are treated by a series of hot extractions with monitoring of temperatures and times. As a general rule, the gelatin prepared through the first extract exceeds the following extracts in gel strength, and the products are kept disjointedly and blended in line with requests. Vacuum evaporation is used to concentrate the dilute gelatin extract after it is pressure-filtered. The warm, concentrated solution is chilled to approximately its gelling point, and then refrigerated to pass through gelation upon further rapid chilling. The product, that is, the continuous gel sheet, is cut into pieces and dried at a controlled temperature and humidity to avoid melting and surface dehydration. At ~10% moisture, the product is returned to ambient temperatures and then passed through grinding or pulverizing for a final product. To obtain a product with the desired qualification and grade (related to gel strength and viscosity), different gelatin grades are blended.

Alkaline Processing

Via alkali processing (type-B gelatin), bones are first demineralized with dilute acid to eliminate the calcium salts, predominantly phosphate, and then both the ossein and cowhide are treated with an analogous method. The raw collagen stock is washed and systematically hydrated in cold water in large tanks. After drainage of the excess water, sufficient lime is added, followed by the addition of fresh water to get a saturated solution of calcium hydroxide. The hides are immersed in this solution for 3 to 12 weeks or even longer, to remove globulins, mucopolysaccharides, albumins, carotenes and other pigments. When this stage is completed, the lime is washed with water for a day or so. Residual base is neutralized with dilute acid until the collagen becomes limp and flaccid. The pH at this stage is 5 to 8, and the extraction of gelatin can begin, from preparations of successively higher temperatures. The best quality gelatin is obtained from the first extractions. The process includes filtration, concentration, and drying to get type-B gelatin powders. To control the marketable gelatin's quality, that is, to obtain gel strength and viscosity in the requested amounts, gelatins from different extractions can be blended.

Final Products

The differences between type A and B gelatin are not only in their source and extraction. Type-A gelatin has a moisture content of 8 to 12%, a pH of 3.8 to 5.5, an isoelectric point of 7.0 to 9.0, a gel strength (by Bloom test) equal to 50 to 300 g, a viscosity of 20 to 70 mPa·s and an ash content of 0.3% (mainly Na^+, Cl^-, SO_4^{2-}). For type-B gelatin, moisture content is 8–12%, pH 5.0–7.5, isoelectric point 4.7–5.1, gel strength 50–275 g, viscosity 2–75 mPa·s, ash content 0.5–2.0% (mainly Ca^{2+}, PO_4^{2-}, Cl^-).

Physical Properties

To solubilize gelatin, it should be hydrated in water then heated at ~71°C. Thus, a macromolecular colloidal dispersion is formed. For gelation, temperatures must be decreased to below 49°C. Gelatin is soluble in polyhydric alcohols but water is usually present as an accessory solvent. Gelatin is insoluble in alcohol, acetone, and nonpolar solvents.

Technical Data

Historically, the initial focus was on the strength of gelatin gels formed under standard conditions. This was followed by a determination of additional physical, chemical, and microbiological properties. Nevertheless, the commercial value of gelatins relies mainly on gel strength as reflected by Bloom test.

Bloom Strength

The rigidity of a gelatin gel formed under standard conditions is termed Bloom strength. The Bloom gelometer was developed and patented in 1925 by O. T. Bloom. The apparatus determines gel rigidity by reading the force in grams required to depress the surface of a 4-mm gelatin gel using a 12.7 mm diameter flat-bottomed cylindrical plunger. The gel is composed of 6.67% air-dried protein. It is prepared by introducing 7.5 g of gelatin into a Bloom jar and incorporating 105 g of distilled water. The mixture is allowed to stand until the gelatin swells, and is then heated in a water bath at 60°C with gentle swirling to obtain a homogeneous solution. Samples are then left for 15 min. at room temperature, followed by placing in a 10°C bath for 16 to 18 hr. before testing. Today, the Stevens-LFRA texture analyzer is replacing the Bloom gelometer. The relationship between gelatin concentration and its gel strength depends on the type and origin of the gelatin itself. In addition, the moisture content of gelatin samples can influence gel strength and therefore should be determined at or before the gel strength determination.

Chemical and Microbiological Properties

Powders, granules, or gelatin sheets should have a light amber to faintly yellow translucent hue. The supplier must inform the customer of the gelatin's Bloom strength.

Gelatins should swell upon immersion in water and jellify upon appropriate heating and cooling. Samples are subjected to several identification tests, as well as clarity and color intensity tests. The gelatin should contain less than 1 ppm arsenic and less than 50 ppm heavy metals. Total ash should be less than 2% and moisture content less than 15%. One gram of gelatin must be free of *Escherichia coli* and 10 g must be free of salmonella. Moreover, in most parts of the world, gelatin is produced with a total count of less than 10^3 CFU/g.

Food Uses and Applications

Gelatin has unique hydrocolloidal properties that enable its use in numerous applications. It has been estimated that about two-thirds of the gelatin produced in the US is used in food products such as desserts, marshmallows, confections, jellied meats, baked goods, and dairy products. The remaining third is used in pharmaceutical and specialized industrial applications.

Gelatin Desserts

A major use of gelatin is in the preparation of desserts. This is because of gelatin's unique ability to yield a product with a desirable texture. This product is a reversible gel that melts at mouth temperature. In 1845, Peter Cooper received a patent for "portable gelatin" which included all of the components for the manufacture of a clear concentrated or solidified jelly, except for the addition of hot water, prior to cooling the formulation for solidification. Of course, further advances in the technology of gelatin production have resulted in the development of many high-quality gelatin dessert mixes. Such mixes include gelatin, sugar, acid, buffer, salts, and color. The customary gelatin gel strengths for dessert products are from 175 to 300 g by Bloom test. Setting time and the rigidity of the prepared dessert depend on gel strength, concentration, viscosity, pH, and type of gelatin used. Gelatins with 220 to 250 g (Bloom test) at concentrations of 12 to 15% are used to decrease the setting or gelling times.

Food acids such as citric and tartaric acid are employed to convey sourness to the gels. In recent years, other acids, such as fumaric, adipic, and malic acids, have also been used to improve stability and performance. Buffer salts such as sodium bitartrate, tartrate, lactate, acetate, or citrate are also incorporated into the mix to maintain the desired pH of 3.0 to 4.7, which is required for the production of faster-setting, smoother gels. Gelatin desserts also include encapsulated powdered flavors, sugar (i.e., sucrose, dextrose, corn syrup solids, or mixtures), and for dietetic products, artificial sweeteners.

Confections

Foamed products make use of proteinaceous whipping agents such albumin and gelatin, and processed proteins such as casein (**Supplemental G**), soy albumin, and fish albumin. The good foaming ability of albumin and casein base materials contributes to the texture of the products. Gelatin is used in toppings, bakery fillings, confections, and related food products. Its use is based on its capacity to stabilize, for example, gas–solid and gas–liquid emulsions. Both the surface tension of the gelatin solution, which prevents coalescence of air bubbles, and its hydrophilic nature, which allows it to bind water in a semirigid structure, stabilize these foamed products. Gelatin is the single main ingredient in the manufacture of commercial marshmallows. Its success in marshmallow production is due to a reduction in the surface tension of the sugar syrup, an increase in the viscosity of the foam and finally, setting of the foam by gelation. The setting results from cooling, drying, and formation of a firm crust, due to the collapse of cells on the outside surfaces of individual pieces. Marshmallows contain ~1.5 to 2.5% gelatin and two sorts are generally manufactured: a soft type, and a harder-grained type. Gelatin can be hydrated in warm water for the preparation of aerated, gelatin-containing confections such as marshmallow. This process can be replaced by hydrating a dry blend of sucrose and gelatin in cold water prior to heating. This latter process is more efficient and is unbeatable for the manufacture of marshmallow pieces, marshmallow fillers, for example, for cookies and candies, dehydrated marshmallow bits, and spoonable marshmallow toppings. A mixture of gelatin and albumin is employed for the production of marshmallow fluff which is a marshmallow-type topping. The effects of gelatin concentration, Bloom strength, and origin on the quality and shelf life stability of marshmallows were studied. Gelatin A (150 bloom) with the highest viscosity, lowest density, and greatest amount of moisture loss produced the hardest marshmallows. Moisture loss was the main mechanism governing hardening. In addition, sugar crystallization had an influence on hardness. Gel network formation may contribute to hardness in gelatin B as there was an increase in hardness but no perceived change in water activity. Gelatin has been used in a lot of other confectionary products. For instance, in the production of caramels, gelatin was included to maintain moisture and chew aimed at a better textured product.

Ice-Cream Stabilizers, Dairy, Fish and Meat Products

Prior to World War II, gelatin was the oldest and most extensively used stabilizer for ice cream. Even though other hydrocolloids then took over the market from gelatin, at present there are many stabilizers that are composed of gelatin and other hydrocolloids. Gelatin binds a portion of the free water in ice cream, in addition to forming a 3D fibril network that slows down crystal formation, protects the ice cream from heat shock, and creates good texture. Gelatins of almost any Bloom strength will successfully stabilize ice creams but it is more economical to use higher Bloom strength materials. Ice crystal growth can be inhibited in an ice cream mix by gelatin hydrolysate produced by papain. The hydrolysate fraction containing peptides in the molecular weight range of about 2,000–5,000 displayed the highest inhibitory activity on ice crystal growth in an ice cream mix, while fractions enclosing peptides greater than 7,000 Da did not inhibit ice crystal growth.

Milk and gelatin are mixed and dried to create dried powdered dairy mixes. A few examples of such products are a frozen ready-to-use product based on butter fat cream, sugar, gelatin, and vanilla. Another example is a nutritious milk beverage combined with fruit juices in which inclusion of gelatin enables the successful combination of milk and acidic juices without curdling. Substantial microstructural differences between yoghurt made with and without gelatin at two levels of total milk solids were detected. Differences as a result of the presence of gelatin were also seen in acid heat-induced milk gels. The differences in firmness and water-retention properties could be attributed to the nature of the network of the different samples. The results revealed gelatin's suitability for improving the quality of milk products. Addition of commercial gelatin at 0.4% to milk yoghurt provided little syneresis of the product and yielded yoghurts with good eating quality.

Another interesting product combined butter with gelatin to achieve modified butter that does not require refrigeration. In the meat industry, a few examples of the use of gelatin in products are deboned hams, meat loaves, sausages, canned hams, and jellied tongues. Inside such products, gelatin both absorbs and holds the meat juices, which might separate during the cooking process and gives shape and firmness to the product. In processed hams, 1 part dry gelatin is added to 50 to 75 parts deboned ham. The hams are then cooked, cooled, and stored. Previous to putting into casings, hams are dipped in gelatin solutions to produce a pleasing form, to enclose the juices, and to preserve the ham's form for slicing. Sausages are usually composed of ~2 to 3% gelatin since the different types of cooked meats that served for their production are suspended in a gelatin solution. High-viscosity gelatins are used for glazes to permit their setting before the glaze runs off the product. Sometimes, combinations of gelatin, starch, sodium citrate, and flavoring are used. Modified collagens can also be used for the preparation of edible sausage casings.

Fish gelatin is a food additive obtained after hydrolysis of collagen from fish skin. The importance of fish gelatin as a food additive is on the rise due to its increased commercial availability. *Surimi* (**Chapter 6**) is washed minced fish meat used as the raw material for seafood analogs such as crabmeat substitutes. The most important attributes of *surimi* are gelling and whiteness. The effect of using fish gelatin as an additive in *surimi* to improve the mechanical and functional properties of gels was studied. Grade Alaska pollock *surimi* gels containing 7.5–15 g/kg of fish gelatin showed improved expressible moisture. However, gelatin added at 15 g/kg showed a disruptive effect on the mechanical properties. The whiteness attribute, as affected by increasing the fish gelatin, was detected by instruments but not observed by sensory panelists.

Baked Goods, Carriers, and Miscellaneous

Gelatin is used as a setting and stabilizing agent in baked goods. It can also be used as a foam-producing material. In chiffon-type pies, gelatins of high Bloom value are employed to reach a superior degree of consistency. In chiffon cheesecakes, gelatin combined with albumin is responsible for the setting property of the foam. In various types of icing, gelatin serves as a stabilizing agent. High Bloom gelatins at levels of 0.5 to 1.0% improve overrun of cream-type icings. Agar (**Chapter 2**) and alginate (**Chapter 3**) might also be used for an equal stabilizing effect in icings. Gelatin has been used for various food coatings. Coatings that included gelatin and antioxidants slowed the deterioration of fat components in meat products. Gelatins obtained from the skins of tuna and halibut, and from the tunics of jumbo flying squid, were hydrolyzed by the enzyme alcalase to produce antioxidant peptides. Hydrolysis yielded a ca. twofold increase in the antioxidant capacity of the gelatins. Gelatin has been used in sausage coatings as the first layer adhering to the meat. Dried fruits and nuts have been coated with gelatin. Those coated products were mixed with granulated sugar prior to drying. The treatment prevented loss of flavor and the products could be used in cake mixes. Flavor oils were emulsified with gelatin–water solutions and, after jellification and drying, the product was ground to a powder containing the bulk of the oil entrapped in the produced particles. Lyophilized, absorptive gelatin sponges can also be prepared. The lyophilized sponge has a unique pore size, bulk density, degree of cross-linking, mechanical stability, water-absorption capacity, water-intake rate, and thermal stability. The sponge can be utilized as a promising matrix for both food and medicinal applications.

Gelatin has been used for the macroencapsulation of flavors such as citrus essential oils. Gelatin capsules have been used to encapsulate seasonings for rice products or for inclusion in dehydrated soups. Gelatin–gum arabic coacervates have been used as flavor fixatives to be included in cake mixes, chewing gum, confections, and instant coffee, among many other products. Gelatin is also used to improve the appearance of certain food products. Texture is a significant parameter in terms of both food palatability and safety of consumption. In recent years, the significance of texture has been emphasized for the development of nursing-care foods. The textural design of these food products is now one of the most important tasks in the food industry. With regard to gelatin, the high flavor intensity of gelatin gel products is due to a lower melting temperature, which enhances flavor release. Due to their beneficial health effects, mainly cardioprotective and anticarcinogenic, flavonoids are regarded as functional dietary constituents. However, most of these compounds are bitter and/or astringent. Enhancing their contents in foods may thus result in off-tastes and low consumer acceptance. Among the proteins tested for their affinity toward epigallocatechin gallate, casein (**Supplemental G**), and gelatin hydrolysates, in particular fish gelatin, were some of the most promising carriers with an affinity on the order of 10^4 M^{-1}. A flexible open structure, as present in random coil proteins, was found to be significant. Porous gelatin carriers can be prepared by insufflating an inert gas (argon) into a concentrated solution of gelatin in the presence of a suitable polymeric surfactant in association with sodium dodecyl sulfate. This technique produces a morphology characterized by spherically symmetrical pores that are highly interconnected and, as a consequence, are mainly used for different purposes.

Another novel use of gelatin is in double emulsions. Water-in-oil-in-water double emulsions are composed of a water-in-oil emulsion dispersed as droplets within an aqueous phase. As a result of the presence of two aqueous domains divided by an oil layer, the inner aqueous compartment presents vast potential for the encapsulation and controlled release of hydrophilic bioactive ingredients. Water-in-oil-in-water double emulsions containing gelatin and sodium chloride in the inner aqueous phase were developed for controlled-release applications. The presence of gelatin and salt played a key role in the double emulsion's stability and release. In double emulsions with gelatin and no salt, gelation of the inner aqueous phase enhanced double emulsion stability and increased NaCl encapsulation efficiency. The amount of gelatin present also altered double-emulsion release behavior and stability. The encapsulation efficiency of the double emulsions with 10% gelatin was slightly higher than with 3% gelatin.

Regulations

According to EU and US legislation, gelatin is categorized as a food, rather than a food additive. Its use with other food ingredients is consequently commonly permitted where good manufacturing practices are followed. In particular cases, its utilization can be limited, to stay away from misleading situations such as injection of gelatin into ham to enhance water binding, which is forbidden in Germany. The advent of bovine spongiform encephalopathy (BSE) led to directives disqualifying the UK from selling gelatin produced from UK-bred animals over 30 months of age. At present, Northern Ireland is exempt from this directive, but circumstances are under frequent review.

Recipes with Gelatin

Fruit Jelly (Figure 9.1)

Ingredients

(Serves 5)
9 g (1 Tbsp) gelatin powder
30 mL (2 Tbsp) water
250 mL (1 cup) water
50 g (a little under ⅓ cup) white sugar
85 g (⅓ cup) canned fruit syrup
100 g canned fruit (**Hint 1**)

Preparation

1. Cut canned fruit into 1 cm cubes.
2. Put 30 mL (2 Tbsp) water in a bowl, add gelatin, and soak for 20 minutes (**Figure 9.1A**).

FIGURE 9.1 Fruit jelly. (**A**) Soak gelatin in water. (**B**) Add swollen gelatin to the sugar solution. (**C**) Pour fruit syrup–gelatin solution mix into molds. (**D**) Serve.

3. Heat water (250 mL) in a pan, add white sugar, and dissolve it fully at 80°C.
4. Turn off the heat, add the swollen gelatin (**2**) (**Figure 9.1B**) (**Hint 2**), and mix until dissolved.
5. Add canned fruit syrup to the gelatin solution (**4**).
6. Put the cut fruit (**1**) in molds, and pour in the gelatin solution (**5**) (**Figure 9.1C**). Cool to less than 10°C in the refrigerator or in an ice-water bath until set, and serve (**Figure 9.1D**).

$$\text{pH} = 4.42$$

Hints

1. Some fruits contain protease, so if you use fresh fruit, the gelatin will be hydrolyzed by the protease and the jelly will not set.
2. Heating gelatin at higher temperatures increases its smell, which is not advisable.

This recipe is for a typical jelly. Even beginner cooks can make it easily.

Fruit Gummy Candy (Figure 9.2)

Ingredients

(Serves ~10)
15 g (1½ Tbsp) gelatin powder
60 g (4 Tbsp) fruit juice†
15 mL (1 Tbsp) lemon juice
21 g (1 Tbsp) corn syrup
36 g (3 Tbsp) granulated sugar

†USE YOUR FAVORITE FRUIT JUICE.

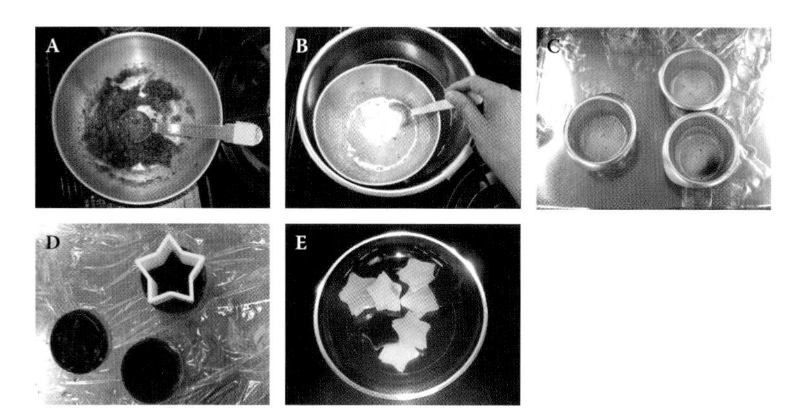

FIGURE 9.2 Fruit gummy candy. (**A**) Mix fruit juice, lemon juice, and gelatin. (**B**) Add corn syrup and granulated sugar to the gelatin dispersion and warm until dissolution. (**C**) Pour the dispersion into molds and cool. (**D**) and (**E**) Remove gummy candy from molds, and create your favorite shapes.

Preparation

1. Pour fruit juice and lemon juice in a bowl, add gelatin, and mix well (**Figure 9.2A**).
2. Add corn syrup and granulated sugar, and warm the bowl by placing in a hot-water bath until dissolved (contents should be at around 55°C) (**Figure 9.2B**).
3. Pour the gelatin dispersion into molds with a spoon, and cool to less than 10°C in the refrigerator or an ice-water bath (**Figure 9.2C**).
4. Remove gummy candy from the molds, and create your favorite shapes (**Figure 9.2D,E**).

pH of peach = 3.90

pH of orange = 3.99

pH of grape = 4.03

Gummy candy is very chewy, which makes it effective for strengthening children's teeth.

Aspic Jelly Salad (Figure 9.3)

Ingredients

(Serves ~10)
50 g (½) onion
50 g (½) carrot
70 g good quality processed ham (lump)
40 g (5) mushrooms
70 g chicken breast
600 mL (2⅖ cups) soup stock
100 g (3) asparagus
5 quail eggs
Parsley
10 g (a little over 1 Tbsp) gelatin powder
50 g (¼ head) lettuce†
10 cherry tomatoes†
50 g (½) cucumber†
Salt, pepper

†OR YOU CAN USE ANY OF YOUR FAVORITE VEGETABLES.

FIGURE 9.3 Aspic jelly salad. (**A**) Dice onion, carrot, and ham. Cut mushrooms into pieces. (**B**) Heat diced onion, carrot, and chicken breast in soup stock. (**C**) Add swollen gelatin to soup stock, season, and cool. (**D**) Place half a boiled quail egg in the thickened soup stock. (**E**) Mix all ingredients (except quail's eggs) with the thickened soup stock. (**F**) Place in molds and cool to set. (**G**) Garnish and serve.

Preparation

1. Dice onion, carrot, and ham into 8-mm cubes, cut mushrooms into 4 pieces (**Figure 9.3A**), and finely chop parsley.
2. Put the diced onion and carrot (**1**), chicken breast and soup stock in a pan (**Figure 9.3B**), and bring to a boil. When the chicken breast is cooked through, remove it from the stock, dice into 8-mm cubes, and season with salt and pepper.
3. Add the diced mushrooms (**1**) to the pan and heat for a few minutes, then remove all solid ingredients and boil soup stock down to 400 g (**Hint 1**).
4. Boil the diced ham for a few minutes. Boil asparagus with salt for 5 minutes, and cut them into 5-mm segments. Hard boil the quail eggs and slice in half.
5. Soak gelatin in 40 g water for 20 minutes (**Figure 9.1A**).
6. Add the swollen gelatin to the boiled down soup stock, season with salt and pepper, and cool to thicken (20°C) (**Figure 9.3C**).
7. Put ½ tsp of the thickened soup stock into each mold, and place half a boiled quail's egg on top of it (**Figure 9.3D**).
8. Mix all ingredients (except quail egg) and the rest of the thickened soup stock (**Figure 9.3E**), and pour into each mold (**Figure 9.3F**); cool to less than 10°C in the refrigerator or an ice-water bath until set.

9. Tear lettuce into bite-size pieces, and slice cherry tomatoes in two. Cut cucumber into 5-mm thick slices.
10. Put the aspic jelly (**8**) in each dish, and garnish with the torn lettuce, cut tomato, and cucumber (**Figure 9.3G**).

Hint

1. Before you start cooking, weigh the pan.

This recipe is good as an *hors d'oeuvre* or salad. Aspic jelly originally referred to jelly set by gelatin with meat or fish soup stock. You can add ketchup sauce, mustard sauce, sour cream sauce, and so forth, to aspic jelly salad.

Bavaroise (Bavarian Cream) (Figure 9.4)

Ingredients

(Serves ~5)
10 g (a little over 1 Tbsp) gelatin powder
45 mL (3 Tbsp) water
30 g (2) egg yolks
60 g (a little over ¼ cup) granulated sugar
200 mL (⅘ cup) milk
100 g (⅖ cup) fresh cream
Vanilla flavoring

FIGURE 9.4 *Bavaroise* (Bavarian cream). (**A**) Beat egg yolks and granulated sugar until mixture turns whitish. (**B**) Add milk to the yolk mixture and heat until fully incorporated. (**C**) Add the cooled gelatin yolk dispersion to whipped cream and stir. (**D**) Set in molds, remove, and serve.

Preparation

1. Soak gelatin in water for 20 minutes (**Figure 9.1A**).
2. Beat yolks in a bowl, then beat in granulated sugar until the mixture turns whitish (**Figure 9.4A**).
3. Add milk to the yolks and heat until yolks are completely incorporated over a hot-water bath at 60°C (**Figure 9.4B**) (**Hint 1**).
4. Add the swollen gelatin (**1**) to the egg yolk mixture and remove from heat; strain after gelatin dissolves.
5. Add vanilla flavoring and cool dispersion until it thickens (30°C).
6. Whip fresh cream until stiff peaks have formed.
7. Add the cooled gelatin dispersion (**5**) to the whipped cream and stir (**Figure 9.4C**) (**Hint 2**), and pour into molds. Then, cool them down to less than 10°C in the refrigerator or an ice-water bath until set.
8. Remove the *Bavaroise* from the molds (**Figure 9.4D**).

pH = 6.30

Hints

1. When egg yolk is heated at higher temperature, it coagulates. Therefore, use a hot-water bath to heat the egg yolk.
2. If you do not stir well, the whipped cream will separate from the gelatin–egg yolk dispersion and set into visible white spots.

This classic dessert is known for its smooth texture.

Marshmallows (*Guimauve*) (Figure 9.5)

Ingredients

(Serves ~10)
150 g (⅚ cup) granulated sugar
20 g (1 Tbsp) corn syrup
60 mL (4 Tbsp) water
70 g (2) egg whites
15 g (1⅔ Tbsp) plain gelatin powder
45 mL (3 Tbsp) water
30 g (a little under ⅓ cup) cornstarch
30 g (a little under ⅓ cup) powdered sugar
Vanilla flavoring

FIGURE 9.5 Marshmallows (*guimauve*). (**A**) Sprinkle cornstarch and powdered sugar on a baking tray. (**B**) Whip egg whites. (**C**) Bring water, granulated sugar, and corn syrup to a boil. (**D**) Add swollen gelatin to whipped egg whites and heated syrup. (**E**) Pour the marshmallow dispersion into the prepared baking tray. (**F**) Sprinkle cornstarch and powdered sugar on both sides of the marshmallow mix. (**G**) and (**H**) Cut into your favorite shapes.

Preparation

1. Soak gelatin powder in 45 g water for 20 minutes (**Figure 9.1A**).
2. Spread baking paper on a tray (18 × 25 cm), and sprinkle cornstarch and powdered sugar over it (**Figure 9.5A**).
3. Whip egg whites in a bowl (**Figure 9.5B**).
4. Heat granulated sugar and corn syrup in 60 g water to 120°C (**Figure 9.5C**).
5. Add the heated syrup (**4**) to the whipped egg whites (**3**) in several steps with stirring (**Hint 1**), and add the swollen gelatin (**Figure 9.5D**) (**Hint 2**) and vanilla flavoring.
6. Pour the marshmallow dispersion (**5**) into the baking tray (**2**) (**Figure 9.5E**) and cool to 5°C.
7. Remove marshmallow (**6**) with baking paper from the tray, peel away the baking paper, and sprinkle both sides of the marshmallow with cornstarch and powdered sugar (**Figure 9.5F**), and create your favorite shapes (**Figure 9.5G,H**) (**Hint 3**).

pH = 7.04

Hints

1. This meringue is called "Italian meringue."
2. Add swollen gelatin to the meringue while the meringue is warm. If the meringue cools down, preheat the swollen gelatin in a microwave oven.
3. You can cut the marshmallows into cubes with a knife.

This recipe is softer than commercial marshmallows. Marshmallows can be used as a garnish for coffee or hot chocolate.

Tips for the Amateur Cook and Professional Chef

- Gel strengths of every type of gelatin decrease below pH 5 and above pH 9.
- In the range pH 5 to 9, gel strength remains approximately invariable.
- In thin solutions, gelatin offers gels with "melt in the mouth" texture.
- At higher concentrations, gelatin provides elastic gum-like textures, which gradually dissolve in the mouth.
- A 250 Bloom gelatin at 0.25% inhibits crystallization of ice and sugar in frozen cream products.
- 2% gelatin gel contains less than 8 kcal per 100 g and consequently it is appropriate for low-calorie sweets and spreads.
- By controlling the drying process, an exceptionally finely pulverized gelatin can be manufactured with no crystalline character.
- To avoid clumping, cold water-soluble gelatin should be blended with all other fine-particle ingredients in the formulation.
- Gelatin hydrolysates cannot gel but they are used as dietary supplements, binding agents, foaming and emulsifying agents and carriers.

REFERENCES AND FURTHER READING

Alemán, A., Giménez, B., Montero, P., and M. C. Gómez-Guillén. 2011. Antioxidant activity of several marine skin gelatins. *LWT - Food Sci. Technol.* 44:407–13.

Barbetta, A., Rizzitelli, G., Bedini, R., Pecci, R., and M. Dentini. 2010. Porous gelatin hydrogels by gas-in-liquid foam templating. *Soft Matter* 6:1785–92.

Bettman, C. H. 1953. Preserving butter without refrigeration. U.S. Patent 2,612,454.

Block, H. W. 1961. Chocolate chiffon. U.S. Patent 2,983,617.

Bloom, O. T. 1925. Machine for testing jelly strength of glues, gelatins, and the like. U.S. Patent 1,540,979.

Bogue, R. H. 1922. *The chemistry and technology of gelatin and glue*. New York: McGraw-Hill.

Bohin, M. C., Vincken, J.-P., van der Hijden, H. T. W. M., and H. Gruppen. 2012. Efficacy of food proteins as carriers for flavonoids. *J. Agric. Food Chem.* 60:4136–43.

Boland, A. B., Delahunty, C. M., and S. M. van Ruth. 2006. Influence of the texture of gelatin gels and pectin gels on strawberry flavor release and perception. *Food Chem.* 96:452–60.

Borker, E., Stefanucci, A., and A. Lewis. 1966. Gelatin and gelatin products. Dessert gel strength testing. *J. Assoc. Offic. Agric. Chemists* 49:528–33.

Bronson, W. F. 1951. Technology and utilization of gelatin. *Food Technol.* 5:55–8.
Childs, W. H. 1957. Coated sausage. U.S. Patent 2,811,453.
Corben, L. D. and W. H. Hatch. 1962. Gelatin dessert and method of preparing the same. U.S. Patent 3,018,181.
Damodaran, S. 2007. Inhibition of ice crystal growth in ice cream mix by gelatin hydrolysate. *J. Agric. Food Chem.,* 55:10918–23.
Fiszman, S. M., Lluch, M. A., and A. Salvador. 1999. Effect of addition of gelatin on microstructure of acidic milk gels and yoghurt and on their rheological properties. *Int. Dairy J.* 9:895–901.
Funami, T. 2011. Next target for food hydrocolloid studies: Texture design of foods using hydrocolloid technology. *Food Hydrocolloids* 25:1904–14.
Glabau, C. A. 1955. The function of gelatin as a setting agent in bakery products. *Bakers Weekly* 166:38–43.
Glicksman, M. 1969. *Gum technology in the food industry*, 360–93. New York and London: Academic Press.
Green, B. K. 1957. Oil-containing microscopic capsules and method of making them. U.S. Patent 2,800,458.
Hamill, J. 1947. Food-coating formulation. U.S. Patent 2,427,857.
Haynes, L. C., Patel, P. N., Slade, L., and H. Levine. 2008. Process for manufacture of aerated confections with dry blend of sugar and gelatin. U.S. Patent 7,329,428.
Hernandez-Briones, A., Velazquez, G., Vazquez, M., and J. A. Ramırez. 2009. Effects of adding fish gelatin on Alaska pollock surimi gels. *Food Hydrocolloids* 23:2446–9.
Howard, R. D., Mulvaney, F., and G. Narsimham. 2003. Protein enhanced gelatin-like dessert. U.S. Patent 6,607,776.
Johnston-Banks, F. A. 1990. Gelatin. In *Food gels*, ed. P. Harris, 233–89. London: Elsevier Applied Science.
Ledward, D. A. 1986. Gelation of gelatin. In *Functional properties of food macromolecules*, ed. J. R. Mitchell and D. A. Ledward, 171–201. London: Elsevier Applied Science.
Ledward, D. A. 2000. Gelatin. In *Handbook of hydrocolloids*, ed. G. O. Phillips and P. A. Williams, 67–86. Boca Raton: CRC Press & Cambridge: Woodhead Publishing Limited.
Montero, P. and J. Borderías. 1991. Emulsifying capacity of collagenous material from the muscle and skin of hake (*Merluccius merluccius* L.) and trout (Salmo irideus Gibb): Effect of pH and NaCl concentration. *Food Chem.* 41:251–67.
Montero, P., Borderías, J., Turnay, J., and M. A. Leyzarbe. 1990. Characterization of hake (*Merluccius merluccius* L.) and trout (Salmo irideus Gibb) collagen. *J. Agric. Food Chem.* 38:604–9.
Muller, G. 1960. The manufacture of gelatin capsules. *Fette Seifen Austrichmittel* 62: 395–9.
Norland, R. E. 1990. Fish gelatin. In *Advances in fisheries technology and biotechnology for increased profitability*, ed. M. N. Voight and J. K. Botta, 325–33. Lancaster: Technomic Publishing Co.
Novak, L. J. 1957. Gelatin-fixed citrus oils. U.S. Patent 2,786,767.
Nussinovitch, A. 1997. *Hydrocolloid applications: Gum technology in the food and other industries*. London: Blackie Academic & Professional.
Nussinovitch, A. 2003. *Water soluble polymer applications in foods*. Oxford, UK: Blackwell Publishing.
Plowright, D. G. 1963. How edible gelatin is produced. *Food Manuf.* 38:182–7.
Sapei, L., Naqvi, M. A., and D. Rousseau. 2012. Stability and release properties of double emulsions for food applications. *Food Hydrocolloids* 27:316–23.
Selby, J. W. 1951. Uses of gelatin in food. *Food* 20:284–6.

Shyamkuwar, L., Chokashi, K. P., Waje, S. S., and B. N. Thorat. 2010. Synthesis, characterization, and drying of absorbable gelatin foam. *Drying Technol.* 28:659–68.

Stanley, J. P. and C. W. Bradley. 1959. Gelatin composition for capsules. U.S. Patent 2,870,062.

Steiner, A. B. and L. B. Rothe. 1949. Stabilizer for icings. U.S. Patent 3,474,019.

Supavititpatana, P., Wirjantoro, T. I., Apichartsrangkoon, A., and P. Raviyan. 2008. Addition of gelatin enhanced gelation of corn–milk yogurt. *Food Chem.* 106:211–6.

Tan, J. M. and M. H. Lim. 2008. Effects of gelatine type and concentration on the shelf-life stability and quality of marshmallows. *Int. J. Food Sci. Technol.* 43:1699–1704.

Tiemstra, P. J. 1964. Marshmallows. IV. Set and syneresis. *Food Technol.* 18:147–52.

Tobins, J. and G. Edman. 1951. A study of bloom gelatins. *Dairy Ind.* 16:633.

Williams, T. C. 1958. Function of colloids in jelly confectionery. *Confectionery Manuf.* 3:482–3.

10

Gellan Gum

Historical Background

Bacteria are a significant source of polysaccharides. Examples include dextran (**Supplemental I**) and xanthan gum (**Chapter 15**). However, of the numerous gums produced by bacteria, only a small number have commercial value, because only a few present major improvements over known and well-used hydrocolloids. An example of such a novel polysaccharide is gellan gum with its exceptional gelling abilities. Gellan gum is an extracellular polysaccharide secreted by the microorganism *Sphingomonas elodea* (ATCC 31461), previously referred to as *Pseudomonas elodea*. It is a hydrocolloid that can be produced on demand. Its quality is not strongly influenced by the different qualities of the raw materials. Gellan gum can be produced in two forms, that is, substituted or nonsubstituted, and it is advantageous for industries that need very low levels of it in a large variety of applications. An assortment of gel textures can be produced, for example, soft elastic gels from the substituted gum, and hard brittle gels from the nonsubstituted gum.

Structure and Chemical Composition

Gellan gum is a linear anionic heteropolysaccharide of ~ 0.5×10^6 Daltons. It is composed of the monosaccharide building units glucose, glucuronic acid, and rhamnose in a molar ratio of 2:1:1. Gellan gum has a tetrasaccharide repeating unit. In the polymer secreted by the bacteria (i.e., in its native form), there are about 1.5 O-acyl groups (acyl substituents) per recurring tetrasaccharide unit. The O-acyl substituent was first considered to be O-acetal, until it was found that gellan gum contains both O-acetyl and O-L-glyceryl substituents on the 3-linked glucose unit, the former tentatively assigned to the 6-position and the latter to the 2-position. Further information on the shape and structure of the gellan gum molecule can be located elsewhere.

Source, Production Supply, and Regulatory Status

Manufacture

A pure culture of the bacteria *S. elodea* is used to manufacture gellan gum. The fermentation media for gum production contain glucose as a carbon source, a nitrogen source, and inorganic salts. Sterile conditions, aeration, agitation, controlled pH and temperature need to be maintained. As the glucose can be metabolized by the bacteria, the gum secreted into the fermentation broth enhances its viscosity. Upon completion of the fermentation, the viable bacteria are killed by heat treatment, and then the broth is processed to produce the gum. The pasteurized broth can be treated with alkali to remove the acyl substituents on the gellan-gum backbone. Then the cellular debris is removed and the gum is recovered by precipitation with alcohol. As a result, a nonsubstituted form with a high degree of gellan gum purity is achieved. Two types of gellan gum are manufactured: the completely acylated native form (high-acyl gellan gum) and the deacylated form (low-acyl gellan gum). Two types of gels with dissimilar textures can be fabricated from these raw substances: cohesive elastic gels from the native material, and strong brittle gels from the deacylated material. Because the degree of acylation can be controlled by the deacylation step, numerous compositions with intermediate acyl contents can be produced, enabling the manufacture of many different gel textures and items for consumption.

Nutritional Aspects

Low-acyl (LA) gellan gum is sold as a free-flowing powder. In an oral LD50 test, gellan gum is regarded as safe at >5000 mg/kg. A 3-month diet showed no indications of toxicity when supplemented at up to 6% (per weight) of the diet. In a two-generation reproduction study, no unfavorable consequences were noted at up to 6% of the diet. Teratology studies demonstrated no dose-related consequences at up to 5% of the diet. In chronic dietary tests, male and female dogs exposed for 52 weeks to daily ingestion of ~1.0, 1.5, or 2.0 g/kg showed no toxicological outcomes. No clinical signs or changes in blood chemistry of monkeys subjected to 28 days of oral testing at doses of up to 3 g/kg were noted. Neither skin nor eye irritation was noted in rabbits. In humans, 23-day dietary tests yielded no unfavorable consequences on plasma biochemistry, hematology, or urinalysis at a daily dose of 200 mg/kg. Such results indicate that there is no reason for concern over the use of gellan gum as a food product.

Regulatory Status

Since 1998, gellan gum has been considered a natural food additive in Japan. It is also approved for food use in the United States, European Union, Canada, South Africa, Australia, most of Southeast Asia, and Latin America. Gellan gum is denoted as E418 in European Community Directive EC/95/2. The Joint FAO/WHO Expert Committee on Food Additives and the European Community Scientific Committee for food give gellan gum a nonspecified Acceptable Daily Intake. Combinations of LA and high-acyl (HA) gellan gum are generally referred to as *gellan gum*.

Functional Properties

Hydration

Low-Acyl (LA) Gellan Gum

The hydration temperature of LA gellan gum depends on the type and concentration of ions in the solution. The presence of sodium and for the most part calcium inhibits the hydration of LA gellan gum. Therefore, a sequestering agent such as sodium citrate is used to bind the soluble calcium to assist with hydration. Incomplete hydration occurs if the sodium chloride concentration exceeds 1.3%. After hydration of the gum, additional ions can be added to the hot solution as long as its temperature is kept above the setting temperature. At pH 4 and 6, the sequestering agent sodium hexametaphosphate is more efficient than tetrasodium diphosphate, disodium orthophosphate, or trisodium citrate dehydrate. LA gellan gum can be hydrated in sugar solutions of up to 80% total soluble solids by heating to boiling. Below pH 3.9, LA gellan gum will not be fully hydrated. Under such conditions, a concentrated acid solution should be added to the hot gum solution, and prolonged heating avoided to minimize hydrolytic degradation. At pH 3.5, LA gellan gum should be held for up to 1 hour at 80°C, thus minimally affecting the quality of the formed gel. If milk is used instead of water as the medium, then LA is readily dispersible and hydration at ~80°C is possible without using a sequestering agent.

High-Acyl (HA) Gellan Gum

Hydration of HA gellan gum is much less dependent than that of LA gellan gum on the concentration of ions in solution. In general, heating to ~85 to 95°C is adequate to hydrate the gum in both water and milk media. At 40–50°C, dispersion of HA gellan gum swells, producing a viscous suspension. If the heating is continued there is a sudden, major reduction in viscosity at 80–90°C, when complete hydration is achieved. The swelling stage can be avoided if the gum is added directly to hot water (>80°C), when the gum has been previously dispersed in other powders of the recipe or in oil or glycerol if included in the formulation. Hydration of HA gellan gum is inhibited by the presence of sugars, and therefore the gum is first hydrated in less than 40% soluble solids and then additional sugar is added to the hot hydrocolloid solution. In contrast to its LA counterpart, HA gellan gum hydrates well below pH 4.0.

Mechanism of Gelation and Gellan-Gum Gel Properties

Numerous studies have suggested that gellan gum gelation takes place via the creation of double helices, followed by their ion-induced association. This particular gelation mechanism proposes that heating and cooling in the absence of gel-promoting cations support the construction of fibrils via double-helix formation between the ends of neighboring molecules. In the presence of gel-promoting cations, these fibrils associate, resulting in gel formation. Divalent ions are valuable in LA gellan gel manufacture, and 3–7% of the recommended level of monovalent cations is sufficient for such uses. The controlled release of divalent ions into the gellan gum solution, as has been done throughout alginate gelation (**Chapter 3**), can be employed to form uniform gellan-gum gels, with the limitation that these gels show signs of syneresis. Gel strength increases with increasing ion concentration until a maximum is reached. Further addition of ions results in a decrease in gel strength. In various applications, gelling ions do not need to be added since there are sufficient ions in the medium or in the other formulation ingredients. When ion addition is needed, gel-promoting ions are added to the hot gum solution, with the temperature being kept higher than that required for setting. Significant thermal hysteresis between the setting and melting temperatures of LA gellan-gum gel has been reported. Under nearly all conditions, LA gellan-gum gels are not reversible below 100°C. Exceptions are gels formulated with a low level of monovalent ions (predominantly potassium) and milk gels. HA gellan gum gels are easily formed by cooling hot solutions. In this case, cations do not need to be added for gel formation and the formed gels' properties are much less dependent on the concentration of the ions in the medium. HA gellan gum gels show no thermal hysteresis; in other words, they set and melt at the same temperature. HA gellan gum gels form supporting gels at concentrations above ~0.2% gum in comparison to 0.05% gum in LA gellan gum gels. Moreover, HA gellan gum gels do not undergo syneresis.

Comparison to Other Hydrocolloids

Native gellan gum dispersed in cold water yields extremely high viscosities. Nevertheless, it is exceptionally susceptible to salt concentration. In solutions of xanthan (**Chapter 15**) or gellan gum at identical concentrations with increasing concentrations of sodium chloride, the viscosity of the native gellan gum is largely dependent upon salt concentration, whereas that of xanthan gum is not. Native gellan gum solutions appear to be highly thixotropic and it seems that high viscosities are the outcome of gel-like network formation. Comparable thixotropic behavior is observed with solutions containing a blend of xanthan and locust bean gum (LBG) (**Chapter 8**). Once the gellan gum has been dissolved in cold water with the aid of a sequestering agent, extremely viscous solutions can be obtained. At a concentration of 1%, gellan gum is less pseudoplastic (or shear-thinning) than xanthan gum, but more pseudoplastic than high-molecular-weight sodium alginate (**Chapter 3**). Gellan gum was compared to κ-carrageenan (**Chapter 4**), and both were found to be ion-dependent. In certain applications, it is also possible to produce gellan gum gels via methods, which are appropriate for alginates. Alginate precipitation can be induced by the addition of acid, and this is another effective means of isolating gellan gum. Gellan gum gels produced as a result of hydrogen-ion addition are extremely strong. Blends with other hydrocolloids are sometimes necessary in the manufacture of different food products. The benefits of blending hydrocolloids can be synergistic. In the field of hydrocolloids, nongelling agents (e.g., xanthan and guar gum), or gelling and nongelling agents (e.g., carrageenan and LBG) are frequently combined to obtain amplified viscosity or better-quality properties of the formed gels, such as elevated elasticity. Viscosity-forming agents such as guar gum, LBG, xanthan gum, carboxymethyl cellulose (**Chapter 5**), or tamarind gum (**Supplemental L**) were added to gellan gum; their concentration was then increased while holding the total gum concentration constant, and a progressive decrease in hardness and modulus was noted, whereas brittleness remained essentially constant and elasticity increased slightly.

The addition of thickeners to gelling agents also assists in decreasing syneresis, improving freeze-thaw stability and in some cases, reducing undesirable interactions between the components. When blends of LA gellan gum and agar (0.50% and 0.25% total gum concentration, respectively, in 4 mM Ca^{2+}) were tested, hardness and modulus decreased as the mix became richer in the agar component. The higher the gum concentration, the more pronounced the decrease. In this case, gel brittleness and elasticity remained almost constant. When blends of κ-carrageenan and LA gellan gum (in 0.16 M K^+) were prepared, a rapid drop in hardness was observed (from 4.5 to 2 N·cm^{-2}) in going from a 0.5% LA gellan gum to a 0.5% 80:20 blend containing 80% gellan gum. When LA gellan gum and xanthan–LBG were combined, the gels became less brittle, less hard and stiff, and elasticity increased in parallel to an increase in the xanthan–LBG fraction. Similar textural changes are induced when LBG is replaced with other gums, such as cassia gum and konjac mannan (**Chapter 12**). It is important to note that even without the addition of other gums, a variety of textures can be achieved by means of different concentrations of LA gellan gum by itself or in a blend with different proportions of the HA form. Starches (**Chapter 14**) provide a thick, pasty consistency to foods. The addition of gellan gum does not noticeably influence starch viscosity. Upon cooling of the latter, addition of >0.1% gellan gum produces a firmer texture. This property is significant when incorporating gellan gum into starch-based products such as puddings and pie fillings, where improved flavor is of enormous significance.

Adding gellan gum and gelatin (**Chapter 9**) together results in a range of textures, dependent upon the relative proportions of the gums in the blend. The properties of the gellan gum–gelatin mix are dependent upon pH, temperature, ionic strength, time, total and relative hydrocolloid concentrations, and gelatin type. The addition of gellan gum to gelatin elevates the latter's setting temperatures, a fact which is of significance to the manufacture of multilayered desserts. The mixed gel is built as a gelatin gel entrapped within a gellan gum gel structure. Manipulation of polymer concentration and ionic strength promotes phase inversion from a gellan gum continuous phase to a system in which gelatin gum forms the supporting matrix. Attention to these interactions might help in developing food products with novel textures.

Food and Other Applications

Gellan gum has been checked for use in numerous applications, for example, in baked goods and cereals, films and coatings, and generally in foods and pet foods. Gellan gum can be used as a fining agent for alcoholic beverages such as wines and beers. In general, a 0.1% gellan gum is prepared in a sodium-citrate medium and added prior to fermentation or for fining mature beer or wine during the final packaging stage at 1 to 100 ppm. LA gellan gum can replace carrageenan–LBG (**Chapters 4 and 8**) at approximately 33 to 50% of the concurrent level. Because such gels break easily, the addition of xanthan–LBG (**Chapters 15 and 8**) helps simulate the consistency of the substituted gels. The inclusion of propylene glycol should be considered if freeze-thaw properties are essential.

Pet food might contain large meat-like pieces, which are manufactured by alginate–calcium reaction. Alginate (**Chapter 3**) can also be utilized to fabricate a block enclosing ground meat. Alginate continues to be the vehicle for chunk production, while low grades of κ-carrageenan are used for binding in a gelled block. The carrageenan is gelled via potassium ions and LBG is included to manage the gel's elastic properties. LA gellan gum has been suggested as a constituent of pet-food formulations. LA gellan gum can substitute the carrageenan at approx. half of the latter's concentration. Gellan gum can be used on its own, or in a blend with supplementary hydrocolloids in low-fat (up to 20%) ground meat, poultry, or seafood products, where the micelle structure of the gellan gum dispersion makes it a suitable fat substitute. Mixed gels of konjac mannan (**Chapter 12**) and gellan gum were incorporated into reduced-fat frankfurters and compared with reduced-fat and high-fat controls for physicochemical, textural and sensory properties, and storage stability. It appears feasible to incorporate a mixed konjac mannan/gellan gum gel in reduced-fat frankfurters for acceptable sensory merits with reasonable shelf life. Additional applications consist of brine for cured meat and seafood products, where gellan gum binds water and then gels during cooking. Benefits of gellan gum include its ability to function at low concentrations, 0.01–3% of the total system's weight, and its good pH and thermal stability. Gellan gum can be utilized for breadings and batters, and with cheese, chicken, fish, potatoes, and vegetables, with the aim of reducing fat. Low-fat, reduced-sodium meat batters formulated with KCl and $MgCl_2$ or $CaCl_2$ with gellan gum were also evaluated. Fat and sodium reduction through incorporation of gellan gum and either of the dicationic salts produced less rigid, more ductile structures. Related to reduced-calorie foods, the influence of gellan gum on flow behavior, particle-size distribution, microstructure, surface tension, and phase separation in a fat-free fermented dairy drink, as a suspension of protein particles, was also studied. Proteins were present as large aggregated particles produced by mixing, diluting, and homogenizing a milk protein gel. In the absence of gellan gum, sedimentation occurred. Gellan caused a significant decrease in sedimentation but the phase behavior was altered and syneresis was observed. Flow behavior measurements indicated that a weak gel structure was created in the solution due to the presence of gellan gum.

Foods coated with gellan gum batter coatings contain low levels of fat but retain the pleasing qualities of fried foods, such as crispiness and juiciness. Alginate- or gellan-gum-based coating formulations on fresh-cut papaya pieces improved water-vapor resistance, affected gas exchange, and carried agents that helped maintain the overall quality of the minimally processed fruit. Formulations containing glycerol and ascorbic acid exhibited slightly improved water-barrier properties as compared to uncoated samples. The incorporation of sunflower oil into the alginate- or gellan-gum-based formulations resulted in an increase in water-vapor resistance of the coated samples. In general, coatings improved firmness of the fresh-cut produce during the studied period. Furthermore, the addition of ascorbic acid as an antioxidant in the coatings helped to preserve the natural ascorbic acid content of the fresh-cut papaya, thereby retaining its nutritional quality throughout storage.

Gellan gums are employed for the most part in confectioneries, jams and jellies, fabricated foods, water-based gels, pie fillings and puddings, pet foods, icings and frostings, and other dairy products. LA gellan gum is of commercial value at very low levels of usage. An additional decrease of ~33% in its concentration can be attained by means of clarified gellan gum, as compared to its nonclarified counterparts. Gellan gum and polydextrose were utilized to manufacture low-calorie or sugar-free jelly sweets, where the final total solids content exceeded 80%. This is not considered to be achievable with other confectionery systems, for example, gelatin (**Chapter 9**), pectin (**Chapter 13**), or agar (**Chapter 2**). In comparison with other gums, gellan gum does not noticeably add to the viscosity of the depositing mix and is hence a good gelling agent for these low-calorie systems. Gellan gum is used in cocoa and chocolate items for consumption and in the development of alcoholic and nonalcoholic beverages.

Emulsions can also be stabilized by different hydrocolloids, including gellan. Characteristics of caseinate (**Supplemental G**), gellan gum, and caseinate/gellan mixture oil-in-water emulsions were studied. Addition of $CaCl_2$ to the gellan solutions induced shear-thinning behavior, as well as the development of viscoelasticity. Both the viscosity and the elastic modulus of polysaccharide solutions were attenuated by the presence of protein. Emulsions without gellan gum in the aqueous phase were almost Newtonian, with relatively low viscosity values. The emulsion was stabilized against both flocculation and creaming, while the rheological behavior strongly depended on the structural state of the polysaccharide matrix, which was influenced by the presence of oil droplets and casein aggregates acting as filler particles.

In selected Japanese foods, such as *mitsu-mame* jelly cubes (**Chapter 2**), *yokan*, *mizu yokan* (**Chapter 2**), and *tokoroten* noodles, gellan gum can replace agar to yield comparable or improved consistencies at a concentration of ~33–50% of the conventionally used agar concentration. These products are firmer than their agar counterparts. Gellan gum can be used in bakery fillings. Its inclusion in fruit fillings results in a smooth texture and improved mouthfeel relative to customary fillings made only with modified starch (**Chapter 14**). Gellan gum's exceptional water-binding mechanisms enable clean, fast release of the fruit aroma. Its use in bakery fillings also results in improved storage and heat stability, and reduced boil-out and moisture-loss rates. Orange gels including 15% fruit pulp with low sugar content can be produced by using gellan gum or a mixture of gellan, xanthan, and LBGs (3:1:1). Use of the gum mixture produced a low-sugar orange product having mechanical characteristics similar to those of the reference gel, though some differences in texture were perceived. The low-sugar gels were slightly lighter in color and slightly more bitter and refreshing than the reference sample. Production of reduced-calorie grape juice jellies with a mixture of gellan gum, xanthan gum, and LBG is also possible. A jelly produced with white grape juice, total sugar content of 39.3°Brix, and 0.54% of the total gum added in the proportions of 1:1:1.7 (gellan:xanthan:LBG) received a maximum overall evaluation. The textural characteristics were found to be similar to those of a reference product, a previously developed reduced-calorie grape juice jelly with low-methoxy pectin (**Chapter 13**). Fruit-based gels can be manufactured with gellan gum as the gelling agent. Textural attributes of the gellan gum gels, formed with different concentrations of the gum (0.5–3.0%) and sugar and/or pineapple juice, were determined. The fracture stress/energy markedly increased with an increase in the concentration of gellan gum, while fracture strain exhibited a marginal effect. The change in these compressive textural parameters was more pronounced for samples with added sugar compared with gels without sugar. The use of gellan gum provides an innovative method for developing fruit juice-based gels as a convenience food because of attractive transparent appearance and textural attributes. The lower viscosity of a starch–gellan gum combination at elevated temperatures improves heat transfer and allows the product to cook more rapidly: processing times are decreased, flavor distortion is reduced and gelation temperatures are more rapidly decreased. Furthermore, gellan gum sustains its gel strength over a broad pH range, 3.5 to 8.0. Gellan gum and additional gelling agents have been shown to stabilize water-in-oil dispersions. Gellan gum can be utilized to prepare gels that can be dried and later rehydrated to their original shape. Possible uses for gellan gum pieces manufactured by such a method include dried rehydratable particulates for instant foods. Additional hydrocolloids can be mixed with gellan gum for improved performance, in the sense that a larger variety of rheological properties become accessible. Moreover, the transparency of gellan gum gels and their slow-release properties can be used to improve commonly used products, in parallel to developing exceptional new ones.

Recipes with Gellan Gum

Fruit-Juice Jelly (Figure 10.1)

Ingredients

(Serves 5)
0.5 g (a little over ¼ tsp) low-acyl (LA) gellan gum powder
50 g (a little under ⅓ cup) white sugar
1.5 g (a little under ⅓ tsp) citric acid, anhydrous
1 g (a little over ⅕ tsp) trisodium citrate dehydrate
150 mL (⅗ cup) water
150 g (⅗ cup) fruit juice†

†YOU CAN USE YOUR FAVORITE FRUIT JUICE.

Preparation

1. Put LA gellan gum, white sugar, citric acid, and trisodium citrate dehydrate in a pan (**Figure 10.1A**), and mix well.
2. Add water to the powder mixture, heat to a boil (at 100°C) and boil until the powder dissolves fully.
3. Remove the gellan gum solution from the heat, add fruit juice (**Figure 10.1B**), and mix well.
4. Pour the gellan gum solution (**3**) into molds moistened with water (**Hint 1**), and cool them down to less than 40°C in the refrigerator or an ice-water bath (**Figure 10.8C**) until the solution sets.

pH of pineapple jelly = 3.72

FIGURE 10.1 Fruit-juice jelly. (**A**) Thoroughly mix LA gellan gum, white sugar, citric acid, and trisodium citrate dehydrate. (**B**) Add fruit juice to gellan gum solution. (**C**) Cool until set.

Hint

1. LA gellan gum jelly is very sticky, and it is hard to remove from the molds if they have not been premoistened.

This jelly is resistant to freezing, and has the same texture as that of unfrozen jelly.

Low-Solids Jam (Figure 10.2)

Ingredients

(Makes ~350 g)
1.0 g (a little over ½ tsp) LA gellan gum powder
100 g (a little under ⅗ cup) white sugar
0.2 g trisodium citrate dehydrate
0.4 g (⅕ tsp) potassium sorbate
100 mL (⅖ cup) water
180 g frozen fruit†
1.0 g (⅕ tsp) citric acid solution (50 wt%)

†YOU CAN USE YOUR FAVORITE FROZEN FRUIT.

Preparation

1. Mix citric acid and water, to make a 50 wt% citric acid solution.
2. Put LA gellan gum, white sugar, trisodium citrate dehydrate, and potassium sorbate in a pan, and mix well.
3. Add water to the powder mixture (**Figure 10.2A**) and then add frozen fruit (**Hint 1**), and heat to boiling for 2 minutes (**Figure 10.2B**).
4. Remove the jam from the heat, add citric acid solution (**1**), and pour into glass bottles (**Figure 10.2C**) (and see **Chapter 13**).

pH of blueberry jam = 3.51

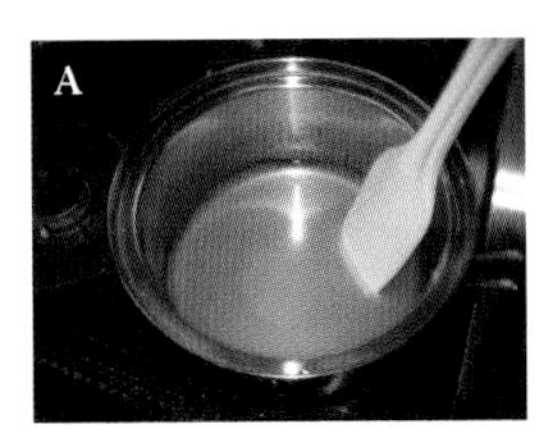

FIGURE 10.2 Low-solids jam. (**A**) Mix LA gellan gum, white sugar, trisodium citrate dehydrate, and potassium sorbate with water. (**B**) Add frozen fruit and heat until boiling. (**C**) Pour into containers.

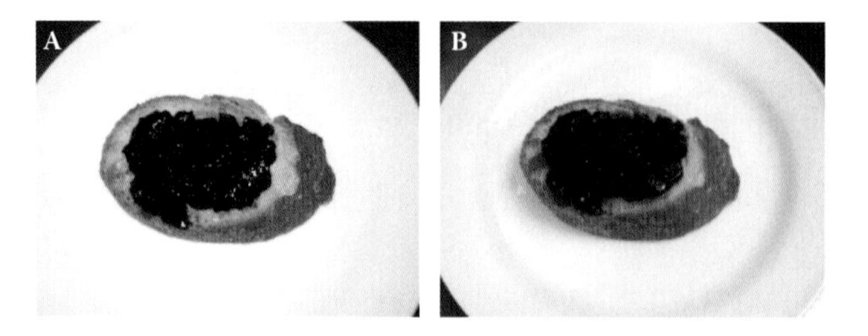

FIGURE 10.3 Serving suggestions for gellan gum–fruit-based jam. (**A**) Untoasted jam. (**B**) Toasted jam.

Hint

1. If you like jam which spreads easily, chop frozen fruit finely.

This jam recipe is more resistant to heat. You can eat this jam on bread (**Figure10.3A**), or toast it (**Figure 10.3B**). You can also bake pound cake or cookies with it.

Liqueur Jelly (Figure 10.4)

Ingredients

(Serves 5)
1.5 g (⅚ tsp) LA gellan gum powder
80 mL (⅓ cup) water
50 g chocolate liqueur (17% alcohol)
100 mL (⅖ cup) milk
20 g (a little over 2 Tbsp) white sugar

Preparation

1. Soak LA gellan gum in water (**Figure 10.4A**).
2. Mix chocolate liqueur, milk, and white sugar in a pan, and heat it (**Figure 10.4B**).
3. Add the gellan gum dispersion (**1**) to the chocolate liqueur mixture with heating, and continue to heat until gellan gum dissolves (**Figure 10.4C**).
4. Remove the gellan gum and liqueur dispersion from the heat, and pour it into a square mold.
5. Cool the solution down to less than 40°C in the refrigerator or an ice-water bath (**Figure 10.4D**) until it sets.
6. Cut liqueur jelly into cubes (**Figure 10.4E**).

$$pH = 6.74$$

This recipe is for adults because alcohol remains in the jelly. Gellan gum can set jelly even if the alcohol concentration in the liqueur or the liqueur concentration is

FIGURE 10.4 Liqueur jelly. (**A**) Soak LA gellan gum in water. (**B**) Mix in chocolate liqueur, milk, and white sugar while heating. (**C**) Add gellan gum dispersion to chocolate liqueur mixture, and heat until gum dissolves. (**D**) Cool the gellan gum–liqueur dispersion in a mold until set. (**E**) Cut liqueur jelly into cubes and serve.

higher. In this case, since chocolate liqueur is used, milk is added for taste. However, berry, fruit, or unflavored liqueur can be used without milk. You can add these liqueurs instead of the chocolate liqueur and milk (150 g, i.e., ⅗ cup).

Pulp-Suspension Fluid Gel (Figure 10.5)

Ingredients

(Serves 5)
0.2 g high-acyl (HA) gellan gum powder
0.2 g trisodium citrate dehydrate
270 mL (a little over 1 cup) water
50 g (a little under ⅓ cup) white sugar
0.7 g citric acid, anhydrous
0.4 g potassium citrate
80 g (⅓ cup) fruit juice†
50 g orange

†YOU CAN USE THE SAME QUANTITY OF YOUR FAVORITE FRUIT JUICE.

Preparation

1. Separate out pulp from orange sections (**Figure 10.5A**).
2. Put HA gellan gum and trisodium citrate dehydrate in a pan and mix well.

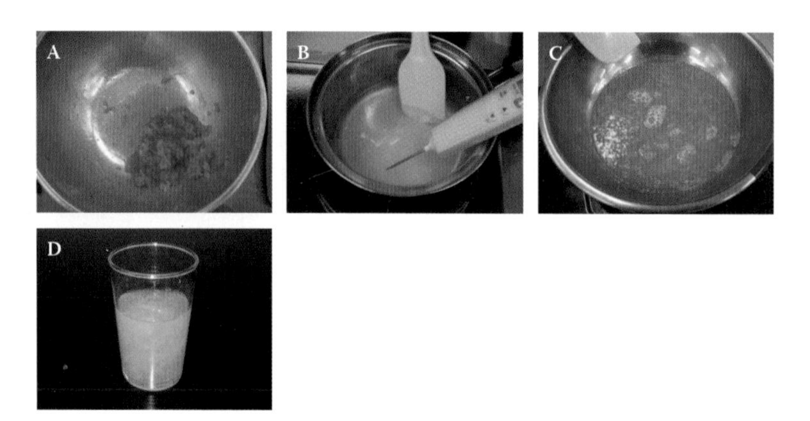

FIGURE 10.5 Pulp-suspension fluid gel. (**A**) Separate pulp from orange sections. (**B**) Mix HA gellan gum, trisodium citrate dehydrate, and water, and heat until the powders dissolves. (**C**) Add orange pulp to gellan gum solution. (**D**) Serve.

3. Add water to the powder mixture (**2**) and heat to 90°C until the powder dissolves fully (**Figure 10.5B**).
4. Add white sugar, citric acid, potassium citrate, and fruit juice to the gellan gum solution (**3**) with heating.
5. Remove the gellan gum solution from the heat, add orange pulp (**1**) (**Figure 10.5C**), cool to less than 40°C at room temperature and serve (**Figure 10.5D**).

pH of pineapple gel with orange pulp = 3.52

Without the orange pulp, this recipe produces a smooth drink.

Herb-Suspension Oil-Free Dressing (Figure 10.6)

Ingredients

(Serves 5)

0.2 g HA gellan gum powder

0.1 g trisodium citrate dehydrate

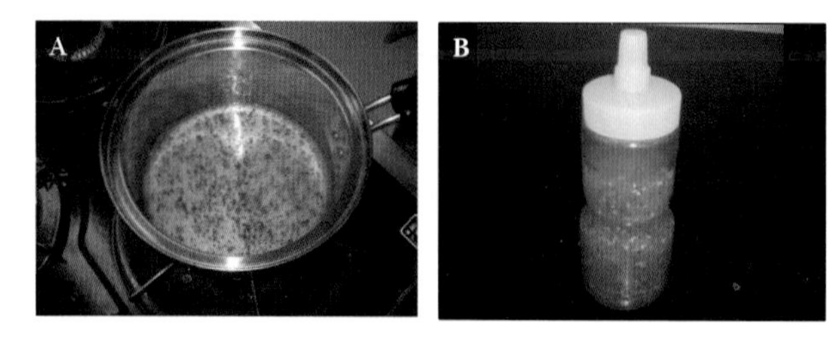

FIGURE 10.6 Herb-suspension oil-free dressing. (**A**) Heat all ingredients. (**B**) Pour into a bottle for easy use.

100 mL (⅖ cup) water
20 mL (1⅓ Tbsp) vinegar
5 g (1⅔ tsp) white sugar
4 g (⅔ tsp) salt
1 g dried parsley or basil
Sodium monoglutamate, black pepper

Preparation

1. Put HA gellan gum and trisodium citrate dehydrate in a pan and mix well.
2. Add water to the powder mixture and heat to 80°C until the powder dissolves fully.
3. Add vinegar, white sugar, salt, sodium monoglutamate, black pepper, and dried parsley to the gellan gum solution with stirring at 25°C (**Hint 1**).
4. Heat the dressing to 85°C (**Figure 10.6A**), and pour into a bottle (**Figure 10.6B**).

pH = 3.67

Hint

1. Lowering the temperature to below the gelling temperature of gellan gum enables uniform dispersion of the solids.

This is a sour type of dressing. It will suit any vegetable.

Tips for the Amateur Cook and Professional Chef

- The firm and brittle texture of LA gellan gum complements the flavor of fruit-juice jellies.
- Combinations of HA and LA gellan gum can be used to produce jellies with a variety of textures.
- The presence of sodium or calcium ions inhibits proper hydration.
- Addition of sequestrating agent promotes hydration.
- The presence of sugars has two major effects on the properties of LA gellan gum gels: first, the ion requirements for optimum gel properties are reduced and second, above ~40% sugar, gels become less firm and less brittle.
- Addition of sugar to HA gellan gum gels generally results in increasing the force required to break the gel.

REFERENCES AND FURTHER READING

Alfonso, T. and M. L. Perez-Chabela. 2009. Textural properties and microstructure of low-fat and sodium-reduced meat batters formulated with gellan gum and dicationic salts. *LWT - Food Sci. Technol.* 42:563–9.

Anon. 1995. A user's guide to Kelcogel gellan gum. *Confectionery Prod.* 61:750–1.

Baird, J. K. and W. W. Smith. 1989. An analytical procedure for gellan gum in food gels. *Food Hydrocolloids* 3:407–11.

Carroll, V., Miles, M. J., and V. J. Morris. 1982. Fibre-diffraction studies of the extracellular polysaccharide from *Pseudomonas elodea*. *Int. J. Biol. Macromol.* 4:432–3.

Chalupa, W. F. and G. R. Sanderson. 1994. Process for preparing low-fat fried foods. U.S. Patent 6027 (930115).

Chilvers, G. R. and V. J. Morris. 1987. Coacervation of gelatin-gellan gum mixtures and their use in micro-encapsulation. *Carbohydr. Polym.* 7:111–20.

Clark, R. C. 1990. Flavour and texture factors in model gel systems. In *Food technology international, Europe*, ed. A. Turner, 272–7. London: Sterling Publications International.

Clark, R. C. and D. R. Burgum. 1989. Blends of acyl gellan gum with starch. U.S. Patent 4,869,916.

Damasio, M. H., Costell, E., and L. Duran. 1997. Sensory quality of low-sugar orange gels with gellan, xanthan and locust bean gums. *Z Lebensm. Unters Forsch A* 204:183–8.

Dartey, C. K. 1993. Applications of gellan gum as a fining agent in alcoholic beverages. *Res. Disclosure* 348:256.

Doner, L. W. and D. D. Douds. 1995. Purification of commercial gellan to monovalent cation salts results in acute modification of solution and gel-forming properties. *Carbohydr. Res.* 273:225–33.

Duxbury, D. D. 1993. Fat reduction without adding fat replacers. *Food Processing - USA* 54:68,70.

Gao, Y. C., Lelievre, J., and J. Tang. 1993. AQ constitutive relationship for gels under large compressive deformation. *J. Texture Studies* 24:239–51.

Gaspar, C., Laureano, O., and I. Sousa. 1998. Production of reduced-calorie grape juice jelly with gellan, xanthan and locust bean gums: Sensory and objective analysis of texture. *Z. Lebensm. Unters Forsch A* 206:169–74.

Giese, J. 1995. Developments in beverage additives. *Food Technol.* 49:63–5, 68–70,72.

Graham, H. D. 1991. Isolation of gellan gum from foods by use of monovalent cations. *J. Food Sci.* 56:1342–6.

Gunning, A. P. and V. J. Morris. 1990. Light-scattering studies of tetramethyl ammonium gellan. *Int. J. Biol. Macromol.* 12:338–41.

Hershko, V. and A. Nussinovitch. 1995. An empirical model for the stress-strain relationships of hydrocolloid gels in tension mode. *J. Texture Studies* 26:675–84.

Jansson, P. E., Lindberg, B., and P. A. Sandford. 1983. Structural studies of gellan gum, an extracellular polysaccharide elaborated by *Pseudomonas elodea*. *Carbohydr. Res.* 124:135–9.

Juming, T., Tung, M. A., and Z. Yanyin. 1995. Mechanical properties of gellan gels in relation to divalent cations. *J. Food Sci.* 60:748–52.

Kang, K. S., Colegrov, G. T., and G. T. Veeder. 1980. Heteropolysaccharide produced by bacteria and derived products. European Patent 0 012 552.

Kang, K. S., Colegrov, G. T., and G. T. Veeder. 1982. Deacetylated polysaccharide S-60. U.S. Patent 4,326,052.

Kang, K. S., Colegrov, G. T., and G. T. Veeder. 1982. Polysaccharide S-60 and bacterial fermentation process for its preparation. U.S. Patent 4,326,053.

Kang, K. S. and G. T. Veeder. 1983. Fermentation process for preparation of polysaccharide S-60. U.S. Patent 4,377,636.

Kiani, H., Ebrahimzadeh-Mousavi, M.-A., Emam-Djomeh, Z., and M. S. Yarmand. 2008. Effect of gellan gum on the stability and physical properties of acidified milk protein solutions. *Aust. J. Dairy Technol.* 63:87–92.

Kuo, M. S., Dell, A., and A. J. Mort. 1986. Identification and location of L-glycerate, an unusual acyl substitution in gellan gum. *Carbohydr. Res.* 156:173–87.

Laaman, T. R. and R. J. Tye. 1991. Application of gellan gum to meat systems. *Res. Disclosure* 323:212.

Lelievre, J., Mirza, I. A., and M. A. Tung. 1992. Failure testing of gellan gels. *J. Food Eng.* 16:25–37.

Lin, K.-W. and H.-Y. Huang. 2003. Konjac/gellan gum mixed gels improve the quality of reduced-fat frankfurters. *Meat Sci.* 65:749–55.

Maga, J. A., Kim, C. H., and C. L. Wolf. 1991. The effect of gellan gum addition on corn grits extrusion. *Food Hydrocolloids* 5:435–41.

Miyoshi, E., Takaya, T., and K. Nishinari. 1994. Gel-sol transition in gellan gum solutions. I. Rheological studies on the effects of salts. *Food Hydrocolloids* 8:505–27.

Morita, Y. and A. Takaragawa. 1996. Manufacturing of sol-like food with particles. Japan Patent 8–23893.

Nakamura, K., Harada, K., and Y. Tanaka. 1993. Viscoelastic properties of aqueous gellan solutions: The effects of concentration on gelation. *Food Hydrocolloids* 7:435–47.

Norton, I. T. 1992. Water in oil dispersion. European Patent Application 0,473,854 A1.

Nussinovitch, A. 1997. *Hydrocolloid applications: Gum technology in the food and other industries*. London: Blackie Academic & Professional.

Nussinovitch, A. 2003. *Water soluble polymer applications in foods.* Oxford, UK: Blackwell Publishing.

Nussinovitch, A. 2010. *Polymer macro- and micro-gel beads: Fundamentals and applications*. New York: Springer.

Nussinovitch, A., Ak, M. M., Normand, M. D., and M. Peleg. 1990. Characterization of gellan gels by uniaxial compression, stress relaxation and creep. *J. Texture Studies* 21:37–49.

Nussinovitch, A., Kaletunc, G., Normand, M. D., and M. Peleg. 1990. Recoverable work versus asymptotic relexation modulus in agar, carrageenan and gellan gels. *J. Texture Studies* 21:427–38.

Nussinovitch, A., Peleg, N., and E. Mey-Tal. 1995. Continuous monitoring of changes in shrinking gels. *Lebensmittel Wissenschaft und Technologie* 28:347–9.

Ogawa, E. 1993. Osmotic pressure measurements for gellan gum aqueous solutions. *Food Hydrocolloids* 7:397–405.

O'Neill, M. A., Selvendran, R. R., and V. J. Morris. 1983. Structure of the acidic extracellular gelling polysaccharide produced by *Pseudomonas elodea*. *Carbohydr. Res.* 124:123–33.

Owen, G. 1989. Gellan gum-quick setting gelling systems. In *Gums and stabilisers for the food industry*, vol. 5, ed. G. O. Phillips, D. J. Wedlock, and P. A. Williams, 345–9. Oxford: IRL Press.

Papageorgiou, M., Kasapis, S., and R. K. Richardson. 1994. Steric exclusion phenomena in gellan/gelatin systems. I. Physical properties of single and binary gels. *Food Hydrocolloids* 8:97–112.

Rinaudo, M. 1988. Gelation of ionic polysaccharides. In *Gums and stabilisers for the food industry,* vol. 4, ed. G. O. Phillips, D. J. Wedlock, and P. A. Williams, 119. Oxford: IRL Press.

Saha, D. and S. Bhattacharya. 2010. Characteristics of gellan gum based food gels. *J. Texture Studies* 41:459–71.

Sanderson, G. R. 1989. The functional properties and applications of microbial polysaccharides—a supplier's view. In *Gums and stabilisers for the food industry,* vol 5, ed. G. O. Phillips, D. J. Wedlock, and P. A. Williams, 333–44. Oxford: IRL Press.

Sanderson, G. R. 1990. Gellan gum. In *Food gels*, ed. P. Harris, 210–32. London and NY: Elsevier Applied Science.
Sanderson, G. R., Bell, V. L., Clark, R. C., and D. Ortega. 1987. The texture of gellan gum gels. In *Gums and stabilisers for the food industry,* vol. 4, ed. G. O. Phillips, D. J. Wedlock, and P. A. Williams, 219–29. Oxford: IRL Press at Oxford University Press.
Sosa-Herrera, M. G., Berli, C. L. A., and L. P. Martinez-Padilla. 2008. Physicochemical and rheological properties of oil-in-water emulsions prepared with sodium caseinate/gellan gum mixtures. *Food Hydrocolloids* 22:934–42.
Stokke, B. T., Elgsaeter, A., and S. Kitamura. 1993. Macrocyclization of polysaccharides visualized by electron microscopy. *Int. J. Biol. Macromolecules* 15:63–8.
Sutherland, I. W. 1992. The role of acylation in exopolysaccharides including those for food use. *Food Biotechnol.* 6:75–86.
Sworn, G. 2009. Gellan gum. In *Handbook of hydrocolloids,* 2nd ed., ed. G. O. Phillips and P. A. Williams, 204–28. Cambridge: Woodhead Publishing Limited.Tanaka, Y., Sakurai, M., and K. Nakamura. 1993. Ultrasonic velocities in aqueous gellan solutions. *Food Hydrocolloids* 7:407–15.
Tapia, M. S., Rojas-Grau, M. A., Carmona, A., Rodriguez, F. J., Soliva-Fortuny, R., and O. Martin-Belloso. 2008. Use of alginate- and gellan-based coatings for improving barrier, texture and nutritional properties of fresh-cut papaya. *Food Hydrocolloids* 22:1493–503.
Upstill, C., Atkins, E. D. T., and P. T. Atwool. 1986. Helical conformations of gellan gum. *Int. J. Biol. Macromolecules* 8:275–88.
Watase, M. and K. Nishinari. 1993. Effect of potassium ions on the rheological and thermal properties of gellan gum gels. *Food Hydrocolloids* 7:449–56.
Willoughby, L. and S. Kasapis. 1994. The influence of sucrose upon gelation of gellan gum in large deformation compression analysis. *Food Sci. Technol. Today* 8:227–33.
Wolf, C. L., LaVelle, W. M., and R. C. Clark. 1989. Gellan gum/gelatin blends. U.S. Patent 4,876,105.

11

Gum Arabic

Introduction

Gums exude from trees and shrubs in tear-like, striated nodules or amorphous lumps, and then dry in the sun to form hard, glassy, different-colored exudates. Exudate gums have been utilized in food applications for years, for emulsification, thickening, and stabilization. The oldest and best-known of all natural gums is gum arabic. This gum is safe for human consumption based on a long and harmless history of use, as well as recent toxicological studies.

Common Names, Economic Importance, and Distributional Range

Gum arabic is also known as gum acacia, Turkey gum, gum Senegal, and many other descriptive and colorful local names such as: kher, Sudanese gum arabic, three-thorn acacia, *acacia à gomme* (French), *gommier blanc* (French), *gummiarabikumbaum* (German), *Senegal akazie* (German), and *acacia del Senegal* (Spanish). Gum arabic is also called *hashab* after the local name of the tree or *kordofan* after the main production area in the Sudan. The trees are of environmental significance when they are used as boundaries, barriers, or support. These trees are planted to reduce soil erosion and for ornamental reasons, as well as to improve soils. The tree is used as fuel (i.e., for charcoal production) and the gum has applications in both food and medicine. The native distributional range of the 500 species of *Acacia* covers tropical and subtropical areas of Africa, India, Australia, Central America, and southwest North America, but only a small number are of commercial importance. The significant production areas are Sudan, French West Africa, and several smaller neighboring African countries.

Gum Arabic Production

Approximately 90% of the world supply of gum arabic comes from Sudan. *Acacia senegal* is found in the drier parts of Sudan and the northern Sahara, and throughout the area from Senegal to the Red Sea and on to Eastern India. In Sudan, *A. senegal* gum is drawn in equal parts from natural stands and cultivated areas, and is gathered by tapping the trees. Following superficial injury of the bark, bulky nodules or tears of gum form alongside the strip or wound on the exposed surface and are left to dry and harden. After 5 weeks, the initial gum collection is complete, with supplementary collections from the same trees at approx. 15-day intervals until the end of February, that is, five or six collections overall. This tapping technique can only be applied to *Acacia* trees with thin, fibrous bark and not to the fleshy-bark species. Yield per tree does not exceed ~300 g. Most *Acacia* gums are water-soluble. Indian acacia gums are noticeably inferior to African *A. senegal* or *seyal* in adhesive properties. After collection, the gum is divided into two major grades: cleaned and hand-picked selected. Gum arabic can be sold in this condition, or post-manufacturing and handling, that is, ground, sieved, or granulated. Untreated gum is precleaned to get rid of the bark, sand, and fines, and foreign material that make up less than 0.5% of food-grade powdered gum. Processing, which involves sieving, decanting, centrifuging, concentrating, pasteurizing, and atomizing followed by spray-drying, yields a product with no insoluble matter, which hydrates more quickly than its unprocessed counterpart. The drying temperature influences the gum's functional behavior. Heat either from spray drying or roller drying causes the gum solution to be somewhat turbid or opalescent.

Gum Arabic Properties

Gum arabic contains no pathogens and no more than 10^3 microbes per gram. Owing to the elevated temperatures involved, spray-dried preparations contain no more than ~40% of the common count (~4×10^2 microorganisms per gram). A decrease in viable bacteria in the gum can also be achieved by ethylene oxide or propylene oxide treatments. Heating carried out throughout processing to decrease the microflora can lead to precipitation of the arabinogalactan–protein complex, which supports stabilization and emulsification in a variety of food products.

Gum Chemical Characteristics

Gum arabic from *A. senegal* is made up of ~3.8% ash, 0.34% nitrogen, 0.24% methoxyl, 17% uronic acid, and the following sugar constituents after hydrolysis: 45% galactose, 24% arabinose, 13% rhamnose, 16% glucuronic acid, and 15% 4-O-methyl glucuronic acid. The gum is a somewhat acidic complex polysaccharide manufactured as a combination of calcium, magnesium, and potassium salts. It has a molecular mass of ~580,000 Da. Gums from dissimilar sources display large differences in content, amino-acid composition, uronic-acid content, and molecular weight. The three most important fractions have been identified by hydrophobic affinity chromatography: a low-molecular-weight arabinogalactan (AG), a very high-molecular-weight arabinogalactan–protein complex, and a low-molecular-weight glycoprotein (GI). These constituents make up 88%, 10%, and 1% of the molecule, respectively, and they include 20%, 50%, and 30% polypeptides, in that order. The protein is positioned on the outside of the AG–protein unit. The overall conformation of the gum arabic molecule is explained by the "wattle blossom" model in which ~five bulky AG blocks, ~200,000 Da each, are positioned along the GI polypeptide chain, which may include up to 1,600 amino-acid residues.

Viscosity and Acid Stability

Even though gum arabic is highly branched, it has a dense structure. Solutions including less than 10% gum arabic have low viscosities and Newtonian characteristics. Above ~30% gum arabic, the hydrated molecules efficiently overlap and steric interactions result in higher solution viscosities and increasing pseudoplastic behavior. This gum is unique in that it can be used to prepare solutions at very high concentrations. Therefore, high levels of gum can be used in numerous food products. Its stability in acid solutions is useful for the stabilization of citrus oil emulsions. Solution viscosities can be altered by the addition of acid or alkali, because these change the electrostatic charges on the macromolecule. A lower pH (more compact polymer volume) leads to lower viscosity, while a higher pH (extension of the molecule) results in maximal viscosity in the region of pH 5.0–5.5. At still higher pH, the ionic strength of the solution increases until the repulsive electrostatic charges are masked, yielding a compact conformation with lower viscosity.

Applications of Gum Arabic

Gum arabic is used in five major food areas: confections, beverages, and emulsions, flavor encapsulation, baked goods, and brewing. Combinations of gum tragacanth (**Supplemental H**) and gum arabic at a 4:1 ratio create minimum viscosity, which chemically, commercially, and in practice produce a thin, pourable emulsion with high quality and shelf-life stability. Fish-oil emulsions are required as dietary supplements or in health foods, and are also emulsified with gum arabic and tragacanth. The vulnerability of lipids to oxidation is one of the primary problems in oil-in-water emulsions. The effects of key formula ingredients, including gum arabic and xanthan gum (**Chapter 15**), significantly reduce the oxidation of walnut oil in the prepared emulsions.

Numerous different emulsifiers are on hand to incorporate into emulsions. A few of them are exclusively emulsifiers, such as Spans and Tweens, and some have both emulsifying and stabilizing properties, such as gums, milk proteins (**Supplemental G**), and modified starches (**Chapter 14**). The performance of gum arabic is considered fairly unique, and very few other polysaccharide–protein systems have a similar stabilizing mechanism. A valuable emulsifier should sorb rapidly at the new interface created throughout emulsification, decrease interfacial tension substantially to facilitate droplet disruption, and prevent the flocculation of new droplets by providing a protective layer around them. The effects of adding gum arabic at 0–4% on the stability of oil-in-water emulsions stabilized with flaxseed protein concentrate and soybean protein concentrate (**Supplemental K**) were studied. Emulsions stabilized by both proteins in the presence of 2% gum arabic showed better stability than those without the gum arabic; stabilization occurred via an increase in the emulsion viscosity of the flaxseed protein concentrate-stabilized emulsion and via competitive adsorption between the gum arabic and the soybean protein concentrate layer to give steric repulsion in the soybean protein concentrate-stabilized emulsion, respectively.

As stated, gum arabic is becoming more and more recognized as a key ingredient in the industrial manufacture of beverages. It stabilizes emulsions and suspensions, harmonizes their texture and offers a good mouthfeel. Its high-fiber content and low caloric value, added to its high-quality intestinal tolerance, make gum arabic an efficient ingredient consistent with the current standards of food composition. The creaming stability of beverage emulsions is strongly influenced by the concentration of free biopolymer in the aqueous phase. When the biopolymer concentration exceeds a certain threshold, the droplets become flocculated through a depletion mechanism. In other words, depletion flocculation leads to an increase in creaming instability and apparent viscosity of the emulsions. The threshold concentration of gum arabic is significantly lower than that of modified starch. The threshold concentration decreases with increasing droplet size because the strength of the depletion attraction increases with droplet size. In general, these results show that depletion flocculation by gum arabic and modified starch can have an adverse effect on the stability of beverage emulsions. Glycerol may also take part in stabilizing beverages. Apart from being a plasticizer, the helpful outcome of glycerol on beverage-emulsion stability indicates that it can perform as a co-emulsifier for gum arabic and offer an additional decrease in interfacial tension between the oil and aqueous phases. An important positive effect of glycerol and vegetable oil on emulsion turbidity proves that these constituents can serve as suitable natural cloudifying agents to improve the turbidity of gum arabic-based beverage emulsions. This observation can be attributed to these factors' ability to reduce the refractive index of emulsion droplets, which scatter light.

Gum arabic is employed in the manufacture of a broad assortment of confections, from soft lozenges and pastilles to hard gums. Substituting gum arabic with different quantities of modified starches is another option. Low levels of gum arabic (up to 2.0%) are included in chewy sweets based on gelatin (**Chapter 9**) to improve product adhesion, reduce elasticity, and produce extra-fine sugar crystallization with a smooth texture. Additional information on the use of gums in confectioneries and on confections based on gum arabic can be located elsewhere. In icings with high-sugar content, the humidity of the manufactured goods is controlled in parallel to molding and rolling properties. In glazes that are applied warm to baked goods, the addition of gum arabic preserves adhesion between the two surfaces. Concentrated gum arabic solutions that are sprayed or brushed onto pastries or biscuits before baking yield an attractive glossy coating after the water evaporates. A novel edible gum arabic coating for enhancing shelf life and improving postharvest quality of fresh produce has also been reported. Coating tomato fruit with gum arabic has been found to enhance shelf life and postharvest quality. Gum arabic in aqueous solutions of up to 20% was applied as a novel edible coating to green-mature tomatoes which were stored at 20°C and 80–90% relative humidity for 20 days. The results suggested that by using 10% gum arabic as an edible coating, the ripening process could be delayed and the storage life of tomatoes stored at 20°C and at the breaker stage could be extended up to 20 days without any spoilage or off-flavors.

Gum arabic is also used as an encapsulating agent for flavors in dry foods, such as soups, beverages, and dessert mixes. A characteristic formulation includes 7% oil-based flavor and 28% gum arabic, and results in 20% flavor in the dried material. The oil droplets must be fully coated before spray drying, to eliminate the loss or oxidization of volatile oils. The performance of gum arabic and soy protein isolate in a paprika oleoresin microcapsule preparation and its storage was also evaluated. Paprika oleoresin emulsions with a ratio of paprika oleoresin to wall material of 1:4 were prepared using high-pressure homogenization, and then spray dried. In both treatments, carotenoid retention in the microcapsules increased as the inlet air temperature was increased from 160 to 200°C, and the yellow fraction was more stable than the red fraction at all temperatures tested. Microencapsulation of orange juice was reported by spray drying using different gums, including gum arabic and tricalcium phosphate as an anti-adherence agent. The encapsulant that provided major protection for sugar and vitamin C was composed of 2.5% gum arabic and 1.5% tricalcium phosphate.

Drying of fruit juices and other products with high sugar content presents technical difficulties because of their hygroscopicity and thermoplasticity at high temperatures and humidity. For this reason, addition of maltodextrin and gums, as well as other substances such as pectins, calcium silicate, and carboxymethylcellulose, has been used in the production of powdered juices. Moisture sorption isotherms describe the relationship between water activity and the equilibrium moisture content of a food product. Knowledge of water sorption isotherms and isosteric heat of adsorption is essential for various food processes, such as drying, to estimate shelf-life stability, which is very important, mainly for food powders. These properties give information on the sorption mechanism and interactions between food components and water. They also help to establish the final moisture content and permit an estimation of energy requirements of drying processes. Sorption isotherms of lemon juice powders with and without 18% maltodextrin or 18% gum arabic as additives were determined at 20–50°C. Addition of additives was shown to affect the isotherms such that, at the same water activity, lemon juice–gum arabic or lemon juice–maltodextrin samples presented lower equilibrium moisture content and were not as affected by varying temperatures.

Foamed sodium caseinate (**Supplemental G**) solutions were optimally stabilized with 0.3% sodium alginate (**Chapter 3**) and karaya gum (**Supplemental H**) at pH 7.0, 0.2% at pH 8.0. The addition of sodium alginate increased foam stability of the solution, but did not enhance foaming ability. Surface tensions and solution turbidities were related to foaming ability, and specific viscosities were related to foam stability. In beers and lagers, the interaction between charged uronic-acid residues and proteins assists in stabilizing the foam, facilitates adhesion to the glass while drinking, and diminishes cloudiness in the drink (~250 ppm of high-quality gum). In wines, low levels of gum react with inherent proteins to sediment them, after which the wine is decanted. Given that gum arabic is exceptional in its properties of emulsification and in the production of mouthfeel and low solution viscosities, it is difficult to replace, particularly in the area of food applications.

Recipes with Gum Arabic

Fruit Juice (Figure 11.1)

Ingredients

(Serves 5)

2.0 g (½ tsp) gum arabic powder

0.4 g gum tragacanth powder

2.0 g (⅖ tsp) citric acid, anhydrous

FIGURE 11.1 Fruit juice. (**A**) Dissolve a mixture of gum arabic, gum tragacanth, citric acid, sodium citrate, and white sugar in water. (**B**) Add fruit juice, fruit essence, and coloring to the solution. (**C**) Serve.

1.0 g (⅕ tsp) trisodium citrate dehydrate
110 g (a little over ⅗ cup) white sugar
800 mL (3⅕ cups) water
70 g (a little under ⅓ cup) fruit juice†
1 g fruit essence†
Food coloring

†**THE SAME FRUIT SHOULD BE USED.**

Preparation

1. Mix gum arabic, gum tragacanth, citric acid, sodium citrate, and white sugar in a pan.
2. Add water, heat until the powders dissolve fully (**Figure 11.1A**) (**Hint 1**), and continue heating for 5 min.
3. Add fruit juice, fruit essence, and food coloring to the gum arabic solution (**Figure 11.1B**), and cool (optional).
4. Pour the fruit juice into glasses (**Figure 11.1C**).

pH = 3.63

Hint

1. Gum arabic and gum tragacanth dissolve at over 75°C.

This recipe is for an imitation fruit juice. Gum arabic is used as the emulsifier, and it reinforces the taste of the fruit juice. In this recipe, orange juice, orange essence, and red and yellow food coloring were used.

Sugar-Free Candy (Figure 11.2)

Ingredients

(Makes 70 g)
1 g (¼ tsp) gum arabic powder
60 g (3⅛ Tbsp) sorbitol powder

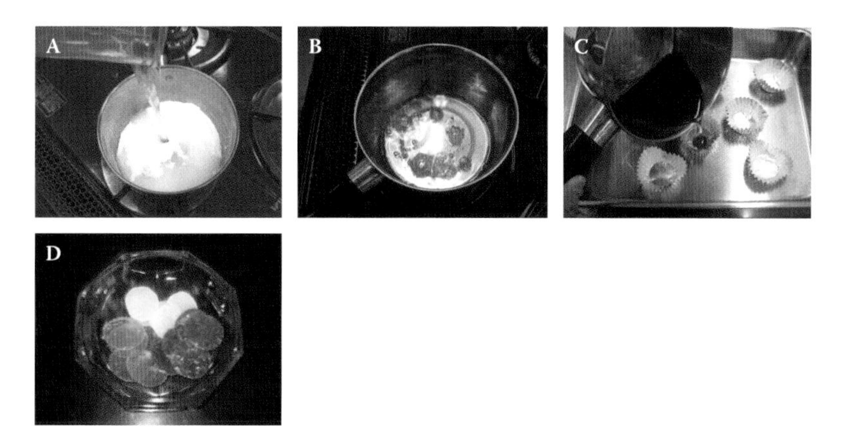

FIGURE 11.2 Sugar-free candy. (**A**) Mix sorbitol, mannitol, and water in a pan. (**B**) Reduce the volume of sorbitol and mannitol solution. (**C**) Pour the candy dispersion into receptacles. (**D**) Colorful sugar-free candy.

10 g (1⅓ tsp) mannitol powder
70 mL (a little under ⅓ cup) water
Flavoring, coloring, citric acid (optional)

Preparation

1. Mix sorbitol, mannitol, and water in a pan (**Figure 11.2A**), and heat until the total weight is 70–75 g (**Figure 11.2B**) (**Hint 1**).
2. Add gum arabic to the solution, add flavor, color, and citric acid, and mix well.
3. Drop the candy dispersion on a baking paper or pour it into round aluminum cups (**Figure 11.2C**), and cool at room temperature.

pH of candy without citric acid = 6.29

Hint

1. Sorbitol and mannitol dissolve at 75–80°C, and the temperature of the solution at the end of heating is 130°C. When the solution is heated to more than 130°C, it starts to smell bad.

This recipe has a wide variety of applications. You can use your favorite colors and flavoring (**Figure 11.2D**). You can make sugar-free candy without gum arabic, but adding gum arabic gives the candy a smooth surface.

Brown Gravy (Figure 11.3)

Ingredients

(Makes 400 g)
36 g (3 Tbsp) butter†
27 g (3 Tbsp) all-purpose flour
0.5 g gum arabic powder
450 mL (1⅘ cups) chicken or beef soup stock
Salt, black pepper
Garlic, ketchup, Dijon mustard, Worcestershire sauce, red wine, and so forth (optional)

†WHEN YOU USE THIS GRAVY FOR POULTRY OR BEEF, YOU CAN USE THE DRIPPINGS FROM THE MEAT.

Preparation

1. Put gum arabic and soup stock in a bowl, and whip with a mixer (**Figure 11.3A**).
2. Put butter in a saucepan, heat on a low flame, add flour with stirring, and continue to heat for 5 minutes until the flour becomes golden brown (roux) (**Figure 11.3B**).
3. Gradually add the gum arabic dispersion (**1**) to the roux (**Figure 11.3C**) (**Hint 1**), heat to boiling with stirring, and simmer for 20 minutes, stirring occasionally (**Figure 11.3D**).
4. Season with salt and pepper.

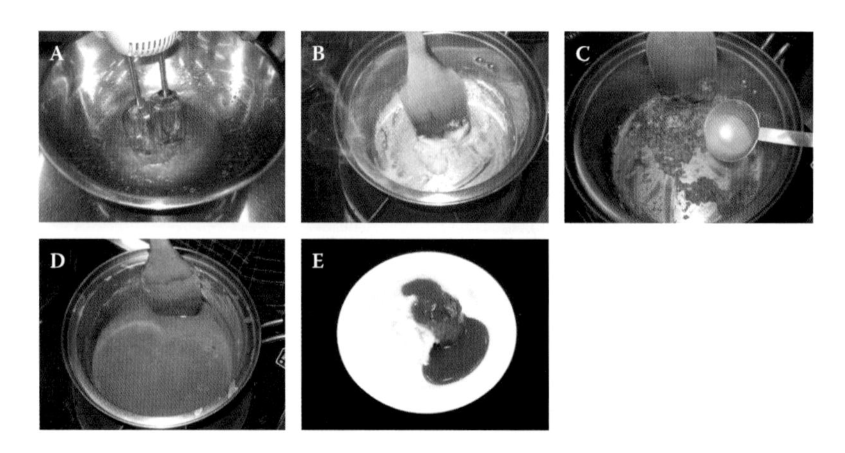

FIGURE 11.3 Brown gravy. (**A**) Whip together gum arabic and soup stock. (**B**) Heat butter, gradually add flour and cook until browned. (**C**) Add the gum arabic dispersion to the roux gradually. (**D**) Simmer and stir occasionally. (**E**) Pour the gravy over mashed potatoes.

5. Pour the gravy over mashed potatoes (**Figure 11.3E**), roast beef, roast chicken or turkey, and so forth.

$$pH = 4.53$$

Hint

1. If you add the gum arabic dispersion all at once, lumps of flour will form, and the gravy will not thicken.

This recipe is often made with beef broth, and served with roasts, meatloaf, mashed potatoes, and so forth. Gum arabic is used as an emulsifier in this recipe, giving the gravy a smooth texture.

Tips for the Amateur Cook and Professional Chef

- Gum arabic is unique in that it is extremely soluble and is not very viscous at low concentrations.
- To achieve high viscosities, ~40–50% gum arabic should be added.
- Solutions of gum arabic are slightly acidic, having a pH of ~4.5–5.5, and hence are in the area of maximum viscosity.
- The addition of electrolytes to a gum arabic solution lowers its viscosity.
- Gum arabic solutions undergo a decrease in viscosity with age.
- Baking properties of rye and wheat flour can be improved by adding small amounts of gum arabic, that is, 0.08–0.20 parts by weight of gum arabic per 1,000 parts flour.

REFERENCES AND FURTHER READING

Ali, A., Maqbool, M., Ramachandran, S., and P. G. Alderson. 2010. Gum arabic as a novel edible coating for enhancing shelf-life and improving postharvest quality of tomato (*Solanum lycopersicum* L.) fruit. *Postharvest Biol. Technol.* 58:42–7.

Anderson, D. M. W. 1993. Some factors influencing the demand for gum arabic (*Acacia senegal* (L.) Willd.) and other water-soluble tree exudates. *Forest Ecol. Manage.* 58:1–18.

Anderson, D. M. W., Brown Douglas, D. M., Morrison, N. A., and W. Weiping. 1990. Specifications for gum arabic (*Acacia senegal*): Analytical data for samples collected between 1904 and 1989. *Food Addit. Contam.* 7:303–21.

Anderson, D. M. W. and F. J. McDougall. 1987. Degradative studies of gum arabic (*Acacia senegal* (L.) Willd) with special reference to the fate of the amino acids present. *Food Addit. Contam.* 4:247–55.

Anderson, D. M. W., Millar, J. R. A., and W. Weiping. 1991. Gum arabic (*Acacia senegal*): Unambiguous identification by ^{13}C-NMR spectroscopy as an adjunct to the revised JECFA specification, and the application of ^{13}C-NMR spectra for regulatory/legislative purposes. *Food Addit. Contam.* 8:405–21.

Benech, A. 2008. Gum Arabic—A functional hydrocolloid for beverages. *Agro Food Industry Hi-Tech* 19:58–9.

Blake, S. M., Deeble, D. J., Phillips, G. O., and A. D. Plessis. 1988. The effect of sterilizing doses of γ-irradiation on the molecular weight and emulsifying properties of gum arabic. *Food Hydrocolloids* 2:407–15.

Caius, J. F. and K. S. Radha. 1939. The gum arabic of the bazaars and shops of Bombay. *J. Bombay Nat. Hist. Soc.* 41:261–71.

Chanamai, A. and D. J. McClements. 2001. Depletion flocculation of beverage emulsions by gum arabic and modified starch. *J. Food Sci.* 66:457–63.

Gharibzahedi, S. M. T., Mousavi, S. M., Hamedi, M., Khodaiyan, F., and S. H. Razavi. 2012. Development of an optimal formulation for oxidative stability of walnut-beverage emulsions based on gum arabic and xanthan gum using response surface methodology. *Carbohydr. Polym.* 87:1611–9.

Imeson, A. P. 1992. Exudate gums. In *Thickening and gelling agents for food*, ed. A. P. Imeson, 66–97. Glasgow: Blackie Academic & Professional, an imprint of Chapman & Hall, Bishop Briggs.

Jafari, S. M., Beheshti, P., and E. Assadpoor. 2012. Rheological behavior and stability of D-limonene emulsions made by a novel hydrocolloid (Angum gum) compared with Arabic gum. *J. Food Eng.* 109:1–8.

Martinelli, L., Gabas, A. L., and J. Telis-Romero. 2007. Thermodynamic and quality properties of lemon juice powder as affected by maltodextrin and arabic gum. *Drying Technol.* 25:2035–45.

Mirhosseini, H., Ping Tan, C. P., and A. R. Taherian. 2008. Effect of glycerol and vegetable oil on physicochemical properties of Arabic gum-based beverage emulsion. *Eur. Food Res. Technol.* 228:19–28.

Mirhosseini, H., Tan, C. P., Hamid, N. S. A., and S. Yusof. 2008. Optimization of the contents of arabic gum, xanthan gum and orange oil affecting turbidity, average particle size, polydispersity index and density in orange beverage emulsion. *Food Hydrocolloids* 22:1212–23.

Mothe, C. G. and M. A. Rao. 1999. Rheological behavior of aqueous dispersions of cashew gum and gum arabic: Effect of concentration and blending. *Food Hydrocolloids* 13:501–6.

Naddaf, L., Avalo, B., and M. Oliveros. 2012. Spray-dried natural orange juice encapsulants using maltodextrin and gum arabic. *Revista Tecnica de La Facultad de Ingenierla Universida del Zulia* 35:20–7.

Nussinovitch, A. 1997. *Hydrocolloid applications: Gum technology in the food and other industries*. London: Blackie Academic & Professional.

Nussinovitch, A. 2003. *Water soluble polymer applications in foods*. Oxford, UK: Blackwell Publishing.

Nussinovitch, A. 2010. *Plant gum exudates of the world sources, distribution, properties and applications*. Boca Raton, FL: CRC Press, Taylor & Francis Group.

Pitalua, E., Jimenez, M., Vernon-Carter, E. J., and C. I. Beristain. 2010. Antioxidative activity of microcapsules with beetroot juice using gum Arabic as wall material. *Food Bioprod. Process.* 88:253–8.

Randall, R. C., Phillips, G. O., and P. A. Williams. 1989. Effect of heat on the emulsifying properties of gum arabic. In *Food colloids*, ed. R. D. Bee, P. J. Richmond and J. Mingins, 386–90. Cambridge: Royal Society of Chemistry.

Rascón, M. P., Beristain, C. I., García, H. S., and M. A. Salgado. 2011. Carotenoid retention and storage stability of spray-dried encapsulated paprika oleoresin using gum arabic and soy protein isolate as wall materials. *LWT – Food Sci. Technol.* 44:549–57.

Reidel, H. 1986. Confections based on gum arabic. *Confect. Prod.* 52:433–4, 437.

Wang, B., Wang, L.-J., Li, D., Adhikari, B., and J. Shi. 2011. Effect of gum Arabic on stability of oil-in-water emulsion stabilized by flaxseed and soybean protein. *Carbohydr. Polym.* 86:343–51.

Williams, P. A., Phillips, G. O., and R. C. Randall. 1990. Structure-function relationships of gum arabic. In *Gums and stabilisers for the food industry 5*, ed. G. O. Phillips, D. J. Wedlock and P.A. Williams, 25–36. Oxford: IRL Press at the Oxford University Press.

Yang, S. T., Kim, M. S., and C. O. Park. 1993. Effects of sodium alginate, gum karaya and gum arabic on the foaming properties of sodium caseinate. *Korean J. Food Sci. Technol.* 25:109–17.

12

Konjac Mannan

Historical Background

The first description of konjac gel (*Kon-nyaku* in Japanese) and its preparation are found in an old Chinese poem by Zuo Shi and its annotation written in the third century. The Japanese note that the manufacture of konjac gel was initiated in Korea along with Buddhism in the 6th century as a medicine. Nevertheless, it took quite a long time before it became an accepted foodstuff due to two main developments. The first was a manufacturing technique developed by T. Nakajima (1745–1826) for konjac flour by pulverizing dried fragments of konjac tuber. The second was a method by K. Mashiko (1745–1854) that improved this practice, resulting in a cleaner konjac flour. Today the flour is produced in modern factories; however, the principles behind its production follow previously accumulated knowledge and experience.

The Plant and the Tuber

Konjac (*Lasioideae amorphophallus*) is a perennial plant and a member of the family Araceae. The geographical origin of the konjac plant is not certain, but is thought to be Southeast Asia. There are numerous species of konjac plants in the Far East and Southeast Asia that belong to the *Amorphophallus*. Only *Amorphophallus konjac* K. Koch is cultivated in Japan (**Figure 12.1**). These plants contain konjac mannan (KM) in their tubers. The main component of the konjac tuber is KM, making up 8 to 10% of the raw tuber, along with starch, lipid, and minerals. KM accumulates in egg-shaped cells covered with scale-like cell walls. These cells are located in the tuber parenchyma and their size and number increase with distance from the epidermis. The cultivation process includes the planting of seed tubers (*Kigo*) and/or one-year-old tubers in the spring. From the tubers emerge new shoots consuming it completely. The plants develop throughout the summer, and put out new tubers. In the late autumn, the plants die and the new tubers are dug up from the soil. The new tuber has seed tubers on top of its suckers. Two-year-old tubers are used to manufacture konjac flour. One-year-old tubers and seed tubers are kept under heated conditions throughout the winter to avoid cold damage and to facilitate repeating the same process the following spring. In countries other than Japan, other kinds of konjac plants are cultivated.

FIGURE 12.1 *Amorphophallus konjac* K. Koch cultivated in Japan.

Manufacture

Two-year-old tubers are conveyed to a warehouse in containers. They are washed with water and brushed. Later, they are sliced into thin chips, which are then dried in a hot-air drier. The dried chips are called *Arako* in Japanese. They are pulverized and konjac flour is obtained. The flour is composed of tough KM particles that are polished to remove the impurities surrounding the KM cells. Then separation is effected by wind shifting. The polished flour is called *Seiko*. In this process, a microfine powder called *Tobiko* (a dust byproduct) is also collected using a dust collector. The main ingredients of *Tobiko* are starch, fine KM powder, and ash. *Seiko*, the commercial konjac flour, has a light color, a fish-like smell, and a somewhat harsh taste. In many companies, konjac flour is washed with an aqueous ethanol solution to get rid of the microfine powder settled on the surface and the impurities trapped inside the konjac particles. Washing serves to whiten the flour. After purification, the scale-like pattern on the surface of the flour is more obvious. Analytically, konjac flour before purification consists of 7.2 g water, 2.2 g protein, 2.3 g lipid, 82.6 g carbohydrate, 0.5 g fiber and 5.2 g ash per 100 g of sample, whereas results for the purified konjac flour are 7.5 g water, 0.8 g protein, 0.9 g lipid, 88.6 g carbohydrate, 0.5 g fiber, and 1.7 g ash per 100 g flour. The values for protein include all nitrogen-containing substances. After washing, the carbohydrate content increases and the concentration of the other components, as well as the fish-like smell, decrease.

Structure

KM is the main component of konjac flour. This glucomannan's main chain consists of D-glucose and D-mannose linked by β-D-1,4 bonds. The glucose-to-mannose ratio has been reported to be 1 to 1.6. The interested reader can refer to other literature sources which deal with the structural units of the main chain, the crystalline form of

KM, the fiber pattern of the annealed KM, and other related data. KM contains acetyl groups in the main chain. The acetyl content has been estimated at 1 per 19 sugar residues. Konjac flour forms very viscous solutions. The weight average molecular weight (M_w) is ~1 x 10^6, but depends on the species of the konjac plant.

Technical Data

Many factors determine the quality of konjac flour. These include the size of the KM particles, viscosity, whiteness, moisture, and presence and type of impurities. Different bacteria within the flour might purify the konjac gel and decrease the molecular weight of the KM. Typical analytical results for components of ordinary and purified Japanese konjac tuber flour are, respectively: viscosity for an aqueous solution of 1% konjac at 35°C after 4 hours of stirring at 90 rpm, 15.0–15.2 and 17.0–18.0 mPa·s; whiteness, 66–68 and 73; water, 6.5 and 6.6 g/100 g of konjac flour; protein, 2.1 and 1.1 g/100 g; lipid, 1.3 and 0.3 g/100 g; carbohydrate, 84.6 and 89.2 g/100 g; fiber 0.5 and 0.6 g/100 g, ash 5.0 and 2.2 g/100 g; sulfur dioxide, 0.65 and 0.17 g/kg; arsenic and lead not detected; trimethylamine as nitrogen-containing substances, 490 and 85 ppm; number of microbes < 300 per gram for both ordinary and purified tuber flour and absence of coliform bacteria for both flours. The viscosity of aqueous KM solutions increases with stirring time, reaching a constant value after 2 hours. Viscosity of a 2% aqueous solution is 12 times higher than that of a 1% solution. Viscosity of KM solutions decreases in parallel to decreasing pH, but not decreasing salt concentration. KM interacts synergistically with additional polysaccharides such as xanthan gum (**Chapter 15**), carrageenan (**Chapter 4**), and agar (**Chapter 2**), and forms thermoreversible gels. On the one hand, addition of sugar increases to some extent the strength of mixed KM and κ-carrageenan gels at higher KM composition; on the other, the strength is reduced if the gel contains less KM.

Food Applications

Konjac flour has been used as a food ingredient for more than a thousand years. Addition of calcium hydroxide (mild alkali) changes an aqueous solution of konjac flour into a strong, elastic, and irreversible gel. Such a preparation serves as a popular traditional Japanese food called *Kon-nyaku*. There are numerous applications and functional uses of KM. In confectionery, KM is used as a viscosity former, texture improver, and moisture enhancer. In jellies, KM contributes to gel strength and texture improvement (**Figure 12.2** and **Figure 12.3**). In yoghurt, the hydrocolloid helps suspend fruit, increase viscosity, and induce gelation. In puddings, KM is responsible for thickening or improving mouthfeel. In pasta, the gum improves water-holding capacity (**Figure 12.4** and **Figure 12.5**). In beverages, it elevates the fiber content and improves mouthfeel. In meat products, the gum serves as a bulking agent, fat replacer, and moisture enhancer. Konjac flour has been reported to act synergistically with certain gums in dissimilar types of meat products. Acceptable sensory traits of reduced-fat frankfurters formulated with mixed konjac flour–gellan gum (**Chapter 10**) gels have been reported. In addition, partial substitution of fat with mixed konjac

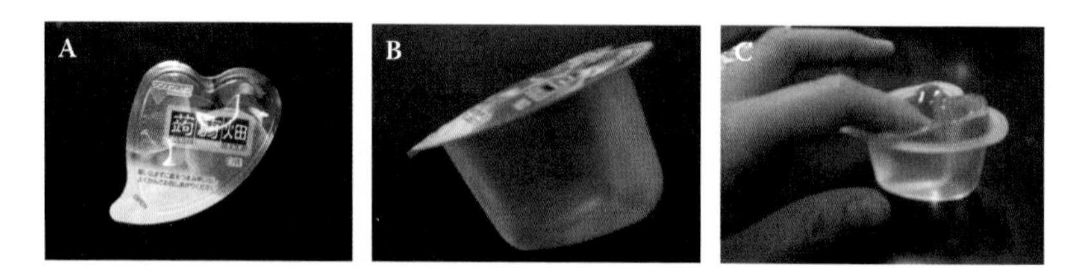

FIGURE 12.2 The very elastic Japanese konjac jelly. (**A**) and (**B**) Packages of konjac jelly. (**C**) Elastic konjac jelly.

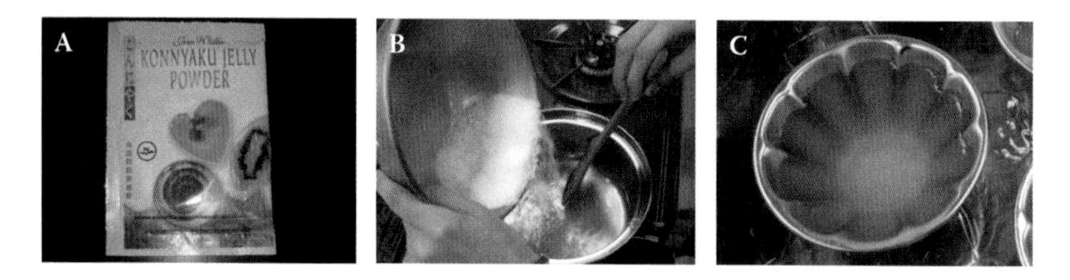

FIGURE 12.3 Konjac jelly powder mix (product of Singapore). (**A**) Packet of konjac jelly powder mix. (**B**) Making konjac jelly. (**C**) Set konjac jelly.

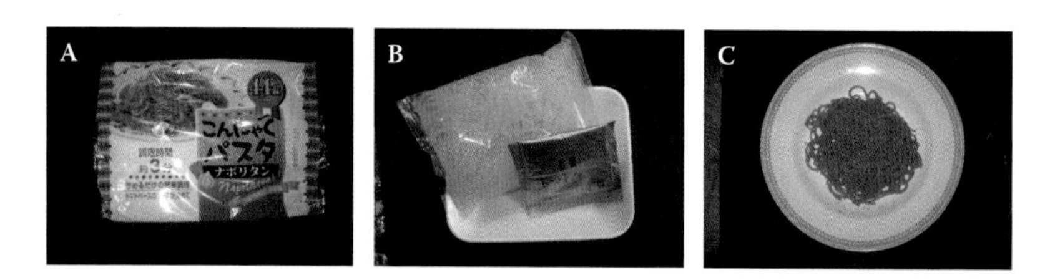

FIGURE 12.4 Japanese konjac pasta. (**A**) Packet of konjac pasta. (**B**) Contents of konjac pasta packet. (**C**) Prepared Konjac pasta.

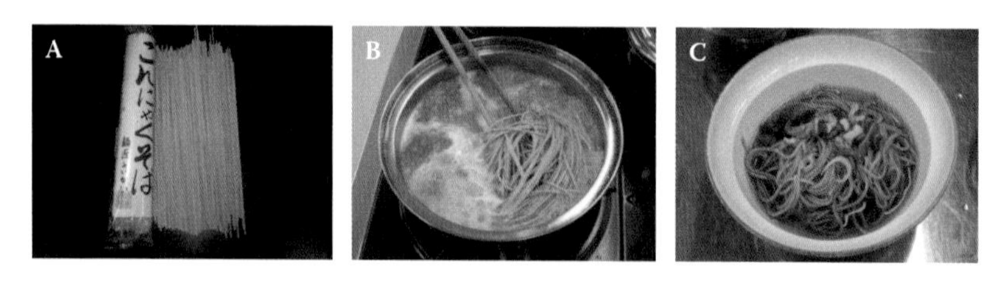

FIGURE 12.5 Japanese konjac buckwheat noodles. (**A**) Packet of konjac buckwheat noodles. (**B**) Boiling konjac buckwheat noodles. (**C**) Prepared Konjac buckwheat noodles with green onion.

flour–potato starch (**Chapter 14**) gels produces reduced-fat frankfurters with textural characteristics and an overall sensory quality comparable to reduced-fat or high-fat controls. Adding mixed konjac–potato starch gels to reduced-fat frankfurter can create a product with the same textural and sensory attributes as regular high-fat frankfurters. There has been a recent tendency to prepare gels that include both KM and other hydrocolloids, with related health claims. KM solution reduces both serum cholesterol and serum triglycerides. KM influences glucose tolerance and absorption. Nevertheless, the alkali-treated gel foods do not produce such an effect.

Recipes with Konjac Mannan

Kon-Nyaku (Konjac) (Figure 12.6)

Ingredients

(Makes 750 g)

30 g (2 Tbsp) konjac flour (*seiko*)

1 L (4 cups) hot water (50–70°C)

1.5 g calcium hydroxide

100 mL (⅖ cup) water

Preparation

1. Put konjac flour in a bowl, and pour in hot water (**Figure 12.6A**). Stir well with a whisk or hand mixer for 5–10 minutes, and then let stand for 1–2 hours until it is swollen (**Figure 12.6B**) (**Hint 1**).
2. Mix calcium hydroxide and 100 mL water in another cup (**Figure 12.6C**), pour into the swollen konjac dispersion (**1**) (**Figure 12.6D**), and mix well for several minutes (**Figure 12.6E**) (**Hint 2**).
3. Pour dispersion (**2**) into square molds (**Figure 12.6F**) (**Hint 3**), level the surface, and let sit for about 10 minutes until the surface becomes smooth.
4. Boil konjac in molds for 1 hour (at 80–90°C) (**Figure 12.6G**).
5. Heat water in another pan, remove the konjac gels from the molds, and boil them for 5 more minutes to remove the bitter taste (**Figure 12.6H**).

pH = 11.0

Hints

1. Swelling konjac flour will lead to an elastic konjac gel.
2. If the calcium hydroxide dispersion and the konjac dispersion are not well mixed, the konjac gel will not set completely.
3. Using a mold with straight sides is better for removal of konjac gel from the mold. Perform this process quickly so that the konjac will not solidify or dry out.

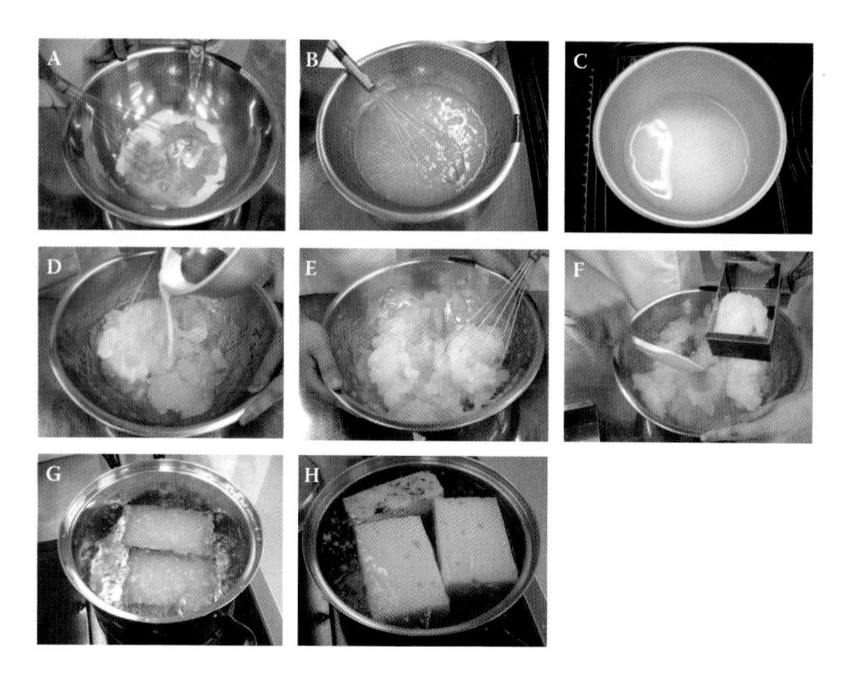

FIGURE 12.6 *Kon-nyaku* (konjac). (**A**) Mix konjac flour and hot water. (**B**) Let konjac soak up water until swollen. (**C**) Mix calcium hydroxide and water. (**D**) Pour calcium hydroxide dispersion into the swollen konjac dispersion. (**E**) Whisk for a few minutes. (**F**) Pour konjac dispersion into molds. (**G**) Boil konjac in molds for 1 hour. (**H**) Remove the konjac gels from molds and boil again.

This recipe is a typical Japanese *Kon-nyaku*, which is called *Ita kon-nyaku* (board konjac), and it is usually used for simmered or stir-fried dishes. There are many kinds of konjac, which have various shapes or tastes (**Figure 12.7**) in Japan. *Ito kon-nyaku* (string konjac) is the second most popular konjac product. **Figure 12.8** shows how to make *Ito kon-nyaku*.

FIGURE 12.7 The many kinds of konjac. (**A**) Upper left: string konjac. Upper right: round konjac. Lower left: board konjac. Lower right: ball konjac. (**B**) Left: konjac with green laver. Center: konjac with sesame. Right: regular konjac.

FIGURE 12.8 Preparation of *Ito kon-nyaku* (string konjac). (**A**) An apparatus for manufacturing *Ito kon-nyaku*. (**B**) *Ito kon-nyaku*.

Konjac Jelly (Figure 12.9)

Ingredients

(Serves 5)
4 g (⅘ tsp) konjac flour (*seiko*)
6 g (1 Tbsp) κ-carrageenan powder
10 g (a little over 1 Tbsp) white sugar
400 g (1⅗ cups) fresh fruit juice

Preparation

1. Mix konjac flour, carrageenan, and white sugar well.
2. Put fruit juice in a pan, and whisk in the blended powders (**Figure 12.9A**) (**Hint 1**).
3. Heat the konjac dispersion on medium heat until it boils at more than 80°C (**Figure 12.9B**), and remove from heat.
4. Pour the konjac dispersion into molds (**Figure 12.9C**), and cool in a refrigerator or ice-water bath until the konjac jelly has set.
5. Remove konjac jelly from the mold (**Figure 12.9D**).

$$pH = 3.81$$

Hint

1. Stir well to prevent lumps.

This recipe is not usually made at home, but it is a popular jelly product in Japan. You can use agar–agar, gelatin, or gellan gum instead of carrageenan.

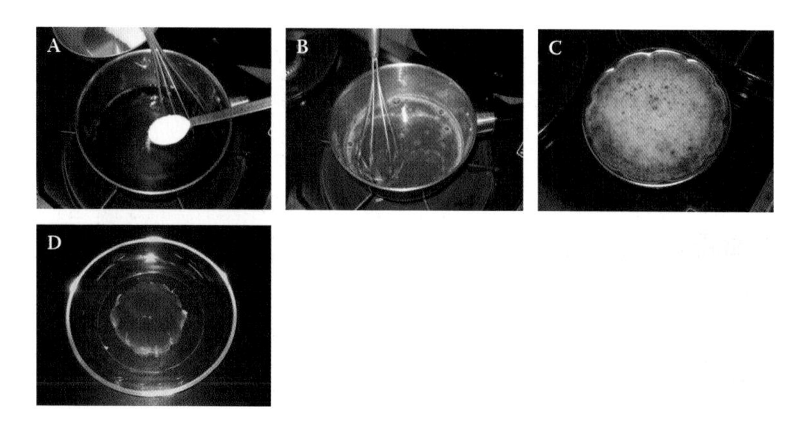

FIGURE 12.9 Konjac jelly. (**A**) Mix konjac flour, carrageenan, and white sugar and add to fruit juice in a pan with stirring. (**B**) Heat the konjac dispersion slowly until it boils. (**C**) Pour the konjac dispersion into molds and cool until set. (**D**) Remove the prepared gel from the mold.

Okara (Soybean Fiber) Konjac (Figure 12.10)

Ingredients

(Makes 1.2 kg)

40 g (2⅔ Tbsp) konjac powder (*seiko*)

800 mL (3⅕ cups) hot water at 50°C

3 g (1½ tsp) calcium hydroxide powder

20 mL (1⅓ Tbsp) water

1 egg

300 g *okara*

Preparation

1. Gradually add calcium hydroxide to 20 mL water in a cup (**Figure 12.10A**), then add egg (**Figure 12.10B**) and mix well.
2. Put *okara* in a bigger bowl, and break up it (**Figure 12.10C**).
3. Put 800 mL hot water in another bowl, and add konjac powder to it, stirring slowly (**Figure 12.10D**) until it thickens (about 5 minutes) (**Figure 12.10E**).
4. Pour the konjac dispersion (**3**) into the *okara* (**2**) and stir with a spatula (**Figure 12.10F**).
5. Add the calcium hydroxide–egg mixture (**1**) to the konjac and *okara* mixture (**4**), and knead by hand (**Figure 12.10G**); push into square molds (**Figure 12.10H**) or form into favorite shapes (**Hint 1**).
6. Boil *okara* konjac (**5**) for 30 minutes (**Figure 12.10I**), chill in water until cold; drain.

pH = 9.72

FIGURE 12.10 *Okara* (soybean fiber) konjac. (**A**) Gradually add calcium hydroxide to water. (**B**) Add egg to calcium hydroxide dispersion. (**C**) Break up the *okara*. (**D**) Stir konjac powder into hot water. (**E**) Continue to stir until the solution thickens. (**F**) Pour the konjac dispersion into the *okara* and stir with a spatula. (**G**) Add the calcium hydroxide and egg mixture to the konjac and *okara* mixture and knead by hand. (**H**) Push into molds or form into favorite shapes. (**I**) Boil *okara* konjac in water.

Hint

1. Fill molds or form quickly, as the dough hardens quite fast.

This recipe is used as a substitute for meat. It is high in fiber and has less fat and protein. Some dishes using this recipe are introduced below.

Okara Konjac Recipes

BBQ *Okara* Konjac (Figure 12.11)

Ingredients

(Serves 5)
300 g *okara* konjac
18 mL (1 Tbsp) soy sauce
9 g (1 Tbsp) white sugar
15 mL (1 Tbsp) *sake*
27 g (3 Tbsp) potato starch
BBQ sauce, vegetable oil for deep-frying

FIGURE 12.11 BBQ *okara* konjac. (**A**) Slice *okara* konjac. (**B**) Mix soy sauce, white sugar, and *sake* and marinate the sliced *okara* konjac. (**C**) Sprinkle potato starch onto it. (**D**) Deep-fry the *okara* konjac. (**E**) Heat the deep-fried *okara* konjac and pour BBQ sauce over it. (**F**) Serve.

Preparation

1. Cut *okara* konjac into 2–3 mm thick slices (**Figure 12.11A**).
2. Mix soy sauce, white sugar, and *sake* in a bowl, marinate the *okara* konjac slices in it for 20 min. (**Figure 12.11B**), sprinkle potato starch over them (**Figure 12.11C**) **Hint 1**, and let stand for 5 min.
3. Heat vegetable oil in a frying pan and deep-fry the *okara* konjac at 170°C (**Figure 12.11D**).
4. Heat the deep-fried *okara* konjac in a frying pan or grill it, and pour BBQ sauce over it (**Figure 12.11E**) and serve.

Hint

1. Put potato starch and okara konjac in a plastic bag. Close bag and shake to coat *okara* konjac.

Fried *Okara* Konjac (Figure 12.12)

Ingredients

(Serves 5)

300 g *okara* konjac

500 mL (2 cups) soup stock

1 bag (about 80 g) fried chicken seasoning

Vegetable oil for deep-frying

Preparation

1. Cut *okara* konjac into big bite-size pieces with a spoon (**Figure 12.12A,B**) (**Hint 1**).
2. Heat soup stock in a pan, add the cut *okara* konjac (**1**) (**Figure 12.12C**), and heat for 10 min.

FIGURE 12.12 Fried *okara* konjac. (**A**) and (**B**) Cut *okara* konjac into big bite-size pieces. (**C**) Put the cut *okara* konjac in heated soup stock. (**D**) Deep-fry the seasoned *okara* konjac. (**E**) Drain off the oil.

3. Sprinkle fried chicken seasoning over the boiled *okara* konjac.
4. Heat vegetable oil in a frying pan, and deep-fry the *okara* konjac at 170°C (**Figure 12.12D**).
5. Drain off the oil (**Figure 12.12E**).

Hint

1. Using a spoon to cut *okara* konjac increases the surface of the cut konjac, making it easier to season.

Regulatory Status

KM is classified as generally recognized as safe (GRAS) by the Food and Drug Administration (FDA). In Sweden, KM is regarded as a functional food due to its ability to reduce serum cholesterol. KM has a provisional European classification number as a food additive (E425). In Japan, KM is regarded as both a food ingredient and a food additive for purposes such as thickening and stabilization. It is important to note that for regulatory purposes, KM flour and the hydrocolloid KM are different entities. The Food Chemical Codex lists the current uses of konjac flour in the United States as gelling agent, thickener, film former, emulsifier, and stabilizer. Konjac flour is used as a binder in the meat industry.

Tips for the Amateur Cook and Professional Chef

- Use konjac flour as you would cornstarch. The viscosity achieved with KM is much higher than that achieved with cornstarch.
- Konjac flour is an ideal thickener for low-carbohydrate or grain-free diets. It provides a rich, viscous texture without added calories, fat, or carbohydrates.
- Konjac flour can be included in baked goods such as cookies, cakes, bread, and biscuits.

- For bread baking, 1 tsp of konjac flour is added to 2 cups of wheat flour.
- KM gels at pH higher than 10.5.
- Adding calcium hydroxide to a konjac dispersion makes the gel set quickly and become hard. Adding sodium carbonate makes the konjac gel black and rough. Adding potassium carbonate makes the konjac gel light pink and smooth.

REFERENCES AND FURTHER READING

Chen, H. L, Cheng, H. C., Liu, Y. J., Liu, S. Y., and W. T. Wu. 2006. Konjac acts as a natural laxative by increasing stool bulk and improving colonic ecology in healthy adults. *Nutrition* 22:1112–9.

Doi, K. 1995. Effect of konjac fiber (glucomannan) on glucose and lipids. *Eur. J. Clin. Nutr.* 49:S190–7.

Gao S. J. and K. Nishinari. 2004. Effect of deacetylation rate on gelation kinetics of konjac glucomannan. *Colloids and Surfaces B – Biointerfaces* 38:241–9.

Gao, S. J. and K. Nishinari. 2004. Effect of degree of acetylation on gelation of konjac glucomannan. *Biomacromolecules* 5:175–85.

Huang, H. Y. and K. W. Lin. 2004. Influence of pH and added gums on the properties of konjac flour gels. *Int. J. Food Sci. Technol.* 39:1009–16.

Huang, L., Takahashi, R., Kobayashi, S., Kawase, T., and K. Nishinari. 2002. Gelation behavior of native and acetylated konjac glucomannan. *Biomacromolecules* 3:1296–303.

Kato, K. and K. Matsuda. 1969. Studies on chemical structure of konjac mannan. I. Isolation and characterization of oligosaccharides from partial acid hydrolyzate of mannan. *Agric. Biol. Chem.* 33:1446–53.

Kato, K., Watanabe, T., and K. Matsuda. 1970. Studies on chemical structure of konjac mannan. 2. Isolation and characterization of oligosaccharides from enzymatic hydrolyzate of mannan. *Agric. Biol. Chem.* 34:532–9.

Kohyama, K. and K. Nishinari. 1997. New application of Konjac glucomannan as a texture modifier. *JARQ – Jap. Agric. Res. Quart.* 31:301–6.

Kritchevsky, D., Tepper, S. A., Davidson, L. M., and D. M. Klurfeld. 1992. Effect of konjac mannan preparations on lipids in cholesterol-fed rats. *FASEB J.* 6:A1654.

Li, B., Xia, J., Wang, Y., and B. J. Xie. 2005. Structure characterization and its antiobesity of ball-milled konjac flour. *Eur. Food Res. Technol.* 221:814–20.

Lin, K. W. and H. Y. Huang. 2003. Konjac/gellan gum mixed gels improve the quality of reduced-fat frankfurters. *Meat Sci.* 65:749–55.

Lin, K. W. and C. Y. Huang. 2008. Physicochemical and textural properties of ultrasound-degraded konjac flour and their influences on the quality of low-fat Chinese-style sausage. *Meat Sci.* 79:615–22.

Maekaji, K. 1974. Mechanism of gelation of konjac mannan. *Agric. Biol. Chem.* 38:315–21.

Millane, R. P. and T. L. Hendrixason. 1994. Crystal-structures of mannan and glucomannans. *Carbohydr. Polym.* 25:245–51.

Nguyen, T. A., Do, T. T., Nguyen, T. D., Pham, L. D., and V. D. Nguyen. 2011. Isolation and characteristics of polysaccharide from *Amorphophallus corrugatus* in Vietnam. *Carbohydr. Polym.* 84:64–8.

Ohta, Y. and K. Maekaji. 1980. Preparation of konjac mannan gel. *J. Agric. Chem. Soc. Jap.* 54:741–6.

Ohta, Y. and K. Maekaji. 1981. Rheological properties of konjac mannan gel. *J. Agric. Chem. Soc. Jap.* 55:415–9.

Sugiyama, N., Shimahar, H., Andoh, T., and M. Takemoto. 1973. Studies on mannan and related compounds. 3. Konjac–mannanase from tubers of *Amorphophallus konjac* C. Koch. *Agric. Biol. Chem.* 37:9–17.

Suto, S., Aita, K., Ohkubo, M., and H. Nagasawa. 1996. Preparation, compressive properties and liquid crystal-formability of konjac gel. *Sen-I Gakkaishi* 52:137–42.

Vuksan, V. 2008. Konjac-mannan and ginseng compositions and methods and uses thereof. U.S. Patent number 7,326,404.

Williams, P. A., Clegg, S. M., Langdon, M. J., Nishinari, K., and L. Piculell. 1993. Investigation of the gelation mechanism in κ-carrageenan/konjac mannan mixtures using differential scanning calorimetry and electron spin resonance spectroscopy. *Macromolecules* 26:5441–6.

Williams, P. A., Day, D. H., Langdon, M. L., Phillips, G. O., and K. Nishinari. 1991. Synergistic interaction of xanthan gum with glucomannans and galactomannans. *Food Hydrocolloids* 4:489–93.

13

Pectin

Introduction

Pectins are natural polymers that are found in all land plants. Pectin is a structural carbohydrate that was discovered by Vauquelin in the 18th century. Braconnot was the first to typify it as a fruit constituent, responsible for gel formation. He suggested the word *pectin*, from the Greek word meaning "to congeal or solidify." Citrus peel or apple pomace was used for commercial extraction of pectin. Isolation of pectins from suitable plant material started toward the beginning of the 20th century and has been on the rise ever since. Pectic substances are fundamental structural components of the cell, serving as cementing material in the middle lamellae of primary cell walls. Acidic extraction of pectin followed by its isolation by precipitation and drying yields a powder with typical properties. The water content of the powder is less than 10% by weight, and the product must be kept in vapor-tight packaging under cool, dry conditions. Marketable pectins generally have a particle size of ~0.25 mm and a density of ~0.7 g/cm^3. They include for the most part polymerized galacturonic acid, which has been partly esterified with methanol. The percentage of the partially esterified portion of polymerized galacturonic acid greatly influences the practical properties of the pectin, and both low- and high-methoxy pectins (LMP and HMP, respectively) are sold. At low pH, HMPs with the addition of enough sugar create fruit-system gels.

Pectin is regularly employed as a gelling agent in jams and jellies, which are customarily manufactured fruit-based products. Its heat stability under acidic conditions makes it a perfect candidate for the conditions occurring when texturization or stabilization are required in acidic foodstuff systems. In homemade jams, the natural pectin content in the fruit pulp is responsible for the gelation. More uniform preparations can be obtained in the food industry when already-produced pulp is used. HMPs can form gels at low pH when adequate quantities of sugar are included. LMPs produce gels in the presence of bivalent ions by means of a different mechanism. Today, the increasing amounts of pectin produced are utilized outside their customary industry, as components of the confectionery industry, as stabilizers in milk manufacturing, and for pharmaceutical purposes.

Nomenclature

Pectin and pectic substances are heteropolysaccharides composed for the most part of galacturonic acid and galacturonic-acid methyl-ester residues. The amended classification for pectic substances, approved by the American Chemical Society, defines *pectic substances* as complex colloidal carbohydrate derivatives that can be located or are manufactured from vegetation. They contain a large amount of anhydrogalacturonic acid units, considered to be present in a chain-like combination. Polygalacturonic acid carboxyl groups can be esterified by methyl groups to a certain extent and partially or completely neutralized by one or more bases. *Protopectin* is considered a water-insoluble parent pectin material that is located within plants and which, with limited hydrolysis, yields pectin or pectinic acid. *Pectinic acids* are the colloidal polygalacturonic acids containing a minor proportion of the methyl-ester groups. Under appropriate conditions, pectinic acids can produce gels in water with the inclusion of sugars and acid or, if suitably low in methoxyl content, with certain ions. Salts of pectinic acids are either normal or acid pectinates. *Pectins* are those water-soluble pectinic acids of changeable methyl-ester content and degree of neutralization, which are able to create gels with sugar and acid under appropriate conditions. *Pectic acid* is a term related to pectic materials composed of colloidal polygalacturonic acids which are in essence free of methyl-ester groups.

Structure

Marketable pectins are made up for the most part of polymerized, somewhat methanol-esterified- (1,4) linked α-D-galacturonic acid. The pectin molecule can have 200–1,000 linked galacturonic-acid units. In a number of pectins, methyl-ester groups are to some extent substituted by amide groups, to an upper limit of 80%. Throughout extraction, only a fraction of the pectin molecules can be extracted by nondegradative means, while dilute acids are usually employed. Consequently, the composition of the resulting pectin varies to a great extent. Neutral sugars, such as arabinose, galactose, glucose, rhamnose, and xylose comprise ~5–10% of the galacturonic acids. They are bound to the main galacturonate chain, are inserted into the main chain (rhamnose), or are part of the contaminating polysaccharides such as glucans and xyloglucans (**Supplemental L**). Pectins from apple, beet, cabbage, carrot, cherry, citrus, onion, pumpkin, potato, and strawberry have the same neutral-sugar composition, quite the opposite of pectins from mountain-pine pollen, Japanese kidney beans, and duckweed, which include large quantities of xylose or apiose. Pectin can be extracted from pumpkin and sugarbeet by bacterial enzymes. Classification of pectic substances from avocado, banana, lime, mango, pawpaw, star apple, and tomato, to mention only a few, in terms of their gelation properties can be found elsewhere.

Sources and Properties

Citrus peel and apple pomace are the most important sources of commercial pectin. Peels are provided for pectin manufacture after the juice has been squeezed and the essential oils extracted. The peels are conveyed to the extraction location, and then a water wash is employed to eliminate as much of the water-soluble material as possible, except for pectin. Then extraction is begun. Alternatively, the peel is dried for subsequent processing. Pectin factories are sometimes located near plants that supply them directly with the raw material, such as manufacturers of apple or citrus juice, or cider. Apple pomace, which in the past was the most important raw material for pectin, has been almost entirely replaced by citrus peel because the latter contains 15–20% extra pectin on a dry-weight basis. During the Second World War, sugarbeet waste served as a resource for pectin manufacture. In view of the fact that this pectin included acetyl ester, today's improved sources are preferred and, in fact, additional raw materials exist.

Pectins can vary as a consequence of ripening and these dissimilarities can control the effectiveness of the extraction method. It is important to note that pectin extracted from the primary cell wall may have extra branches of neutral sugars in comparison with pectin extracted from the middle lamella. Neutral-sugar side chains are distributed nonuniformly alongside the main chain. Consequently, models illustrating smooth and hairy sections inside pectin which has been extracted by a mild process can be assumed for pectic substances from citrus, cherry, carrot, and sugarbeet. A report of pectic fractions from dissimilar sources can be located elsewhere. Substituents such as acetyl groups (in potato and sugarbeet pectins) can prevent gelation. Apple and citrus exhibit only a very low degree of acetylation, and the acetyl groups may be positioned in the hairy regions. Active pectin oligomers have also been identified in ripening tomato fruits. Furthermore, associations of pectin with boron in the cell walls of squash and tobacco have been described. Since pectin can come from diverse agricultural sources, it is not surprising that dissimilar pectins have dissimilar substituents positioned in different positions.

The methanol ester of galacturonate residues is for the most part quite abundant. A degree of esterification (DE) of ~70% was observed for apple or citrus pectins that were not subjected to de-esterification. This value is high compared to the low DE values for pectins extracted from pear, sunflower heads, potato, and tobacco. DE is classified as the ratio of esterified galacturonic-acid units to the overall number of galacturonic-acid units in the molecule. These values can be further influenced by the ripeness of the raw material and alterations in the extraction process. Ester-group distribution depends on the source. There is confirmation of random intramolecular distribution in mildly extracted apple pectins, contradicting a study which found some regularity. Nonrandom distribution has been found in commercial pectins and data on de-esterification by fungal enzymes, pectin structure, conformation in solution and gels, and other properties can be found elsewhere.

Pectin Manufacture

Numerous regions worldwide produce pectin. A small number of examples are Hercules (factories in Denmark, Germany, and Florida, US), Unipectine (France), Pektin-Fabrik (Germany), General Foods Corp. (US), and Pectina de Mexico. Less important pectin manufacturers are located in Switzerland, Brazil, Israel, Argentina, and some other European countries.

Although industrialized pectin-manufacturing processes are generally known, the differences or minor changes in the processes, that is to say the detailed conditions employed, are kept classified by the producers—they are regarded as trade secrets. As a general rule, the fresh or dried raw material (apple pomace, citrus peel, or a number of other substances such as sunflower bottoms and sugarbeet waste) is extracted in demineralized water that has been acidified to provide a pH of 1.5 to 3.0, at 70°C for ~3 hours. For citrus peels, pretreatment by blanching and washing to eliminate pectinesterase (PE) activity and remove glucosides, sugars, and citric acid is widespread. PE, or pectinmethylesterase, is the enzyme that catalyzes the hydrolysis of the ester bonds of pectic substances to yield methanol and pectic acid. Dried peel is stable throughout storage, making its shipping over large distances practical. Dried citrus peel includes ~20–30% pectin. Dried apple pomace yields 10–15% pectin.

During the extraction, a specific degree of pectin de-esterification occurs. Therefore, conditions to fit the required manufactured good need to be selected. pH, temperature, and time need to be carefully controlled for an efficient extraction. Rapid-setting high-ester pectins are usually extracted at temperatures close to boiling. At these temperatures, hydrolysis of the parent pectic materials proceeds more rapidly, viscosity is decreased and diffusion becomes possible. Such a procedure may take under 1 hour with just a little de-esterification, while lesser extraction temperatures and extended extraction times support de-esterification to produce slow-setting HMP or even LMP. Once the extract has been separated out, the peels can be utilized for farm animals; at the same time, the viscous liquid containing 0.3–1.5% dissolved pectin is clarified by filtration and centrifugation. At this point, the clear pectin extract can be additionally de-esterified by maintaining a controlled pH and temperature. The extract can be concentrated and, subsequent to preservation with sulfur dioxide, sold as "liquid pectin."

Isolation of pectin by alcohol precipitation or by precipitation as an insoluble salt can be easily performed. By means of the previous course of action, the pectin precipitates out, and the alcohol is recovered by distillation. Combining the extract with methanol, ethanol, or 2-propanol leads to alcohol precipitation. In a number of procedures, the extract is concentrated by evaporation prior to precipitation to minimize distillation expenditures. Alternative procedures involve separating pectin with Al^{3+} or Cu^{2+} as an insoluble salt. These metal ions can then be removed with acidified alcohol washes, followed by a wash in alkaline alcohol to neutralize the product. The resultant alcohol-wetted pectin is pressed, dried, and milled, or is de-esterified in the alcohol suspension.

De-esterification can be achieved with an acid or base. If ammonia is employed, followed by replacement of a number of the methyl-ester groups by amide groups, the product is termed *amidated pectin*. For pectin production, as for different hydrocolloids, a combination and standardization phase is essential. Following the addition of such a phase, the traded blends display comparable performance with respect to firmness of the resultant gel and the time required to gel HMPs under predetermined constant conditions. In a comparable manner, LMPs are standardized in terms of their calcium reactivity.

Commercial Availability, Specifications, and Regulatory Status

Marketable pectin is classified as partial methyl esters of polygalacturonic acids and their salts. Pectin is extracted from plant organs that are fit for human consumption and no organic precipitants, excluding methanol, ethanol, or isopropanol, are utilized. Amidated pectins can be manufactured by ammonia treatment. Standardization can be accomplished by dilution with sugars. Buffer salts can yield the desired setting conditions. Saleable pectins are divided into HMPs and LMPs in conformity to their DE, that is, the percentage of galacturonic acid subunits that are methyl-esterified. If the DE is > 50% it is a HMP, if 50% > DE > negligible amounts, it is labeled a LMP.

Pectate is a polymerized galacturonic acid with insignificant or no esterification. The degree of amidation (DA) is the percentage of galacturonic-acid subunits that are amidated. HMPs employed for gel-making can be separated into rapid-set, medium-set, and slow-set, depending on the time required for solidification: as a general rule, the higher the DE (in HMPs), the shorter the product setting time. HMPs are standardized on a regular basis to 150 grade of USA-SAG, meaning that 1 part pectin can solidify 150 parts sucrose into a jelly with the standard properties of 65°Brix, pH 2.2–2.4, and 23.5% SAG (indication of gel strength). Purity of marketable pectins is also characterized by a number of requirements: galacturonic acid content > 65%, DA < 25%, loss on drying not more than 12%, no more than 1% acid-insoluble ash, alcohol residues of all kinds not more than 1%, nitrogen not more than 2.5%, and no more than 50 mg/kg sulfur dioxide. HMP loses about 5% of its USA-SAG grading when stored at 20°C in a dry atmosphere, while LMP is more stable and under favorable conditions, loss is undetectable. The microbiological purity of pectins is specified in numerous cases by the manufacturer, given that it is used for the most part in acidic media and thus yeast and mold counts are relevant. Characteristic specifications may comprise a total plate count at 37°C of less than 500/g, a yeast and mold count at 25°C of less than 10/g, and negative test results for *Escherichia coli*, salmonella, and staphylococcus.

Pectin is a key ingredient in land plants and can therefore be eaten in large amounts. Pectin passes to the large intestine without enzymatic degradation; somewhere microorganisms make use of it as a carbon source. Nevertheless its hydrolyzation in the intestinal tract generates almost no calories. From a toxicological viewpoint, there are no restrictions on its use. However, although pectins are in general identified as not dangerous for use in human foods, the Food and Drug Administration (FDA) has not issued any definite restrictions or guidelines. The possible dietary advantages of citrus pectin and fiber have been assessed. Additional health features of pectin are significant and have been studied by many researchers. Examples include the function of pectin in cholesterol regulation, and citrus pectin and cholesterol interactions in the regulation of hepatic cholesterol homeostasis and lipoprotein metabolism in the guinea pig, among other studies. Recently, a pectin-supplemented enteral diet was reported to lessen the severity of methotrexate-induced enterocolitis in rats. The physiological effects of low-molecular-weight pectin were discussed. Such pectins, which can retain their activities, are important because high viscosity reduces their usability. This preparation exhibited high solubility and a repressive effect on lipid accumulation in the liver. Pectin formulations have also been utilized for colonic drug delivery.

Solution Properties

In general, pectin is not soluble under the conditions in which it forms a gel. The powder should be dispersed in water at over 60°C, at decreased mixing rates and then at full speed. Ignoring the manufacturer's advice could result in the creation of lumps, which are not easy to dissolve. Good-quality dissolution is attained by combining pectin with five times its own weight of sugar. Different blending media, such as a 65% sugar solution or alcohol to wet the pectin for small-scale laboratory use, are advised. If a high-shear mixer is not employed, boiling for 1 minute is essential to ensure complete dissolution.

Viscosity

Among many parameters, the viscosity of pectin solutions relies on their concentration, presence of calcium or comparable nonalkali metals, pH, the chemical properties of the pectin, DE, and average molecular weight. Up to approximately 0.5% (i.e., dilute) pectin solutions are Newtonian, and only to some extent influenced by calcium ions. Salts of monovalent cations decrease pectin solution viscosity, due to reductions at high ionic strength. In parallel to an increase in average molecular weight, an increase in solution viscosity is detected. The molecular weight of pectin can be approximated by means of methods that determine intrinsic viscosity. Pseudoplastic solutions can be achieved with concentrations higher than 1%. In contrast to dilute solutions with no calcium, such solutions increase in viscosity if the pH is decreased within the typical application range of 2.5–5.5. Pectins in the presence of calcium form thixotropic solutions whose viscosity increases with increasing pH within the abovementioned range. In fact, dissimilar textures can be easily achieved by combining pectin types and concentrations, ion concentrations, and pH. Based on viscometry measurements, the average molecular weight of commercial pectin usually falls between 50 and 150×10^3. It is vital to note that by means of other techniques, for instance light-scattering, other results (~10^6 Da or higher) have been found owing to intermolecular associations and aggregation of pectin molecules.

Pectin Gel Types and Properties

HMP gels can be effectively prepared following fine dissolution. Jam manufacturing methods can be located elsewhere. Briefly, they comprise heating the sugar and fruit fraction in quantities which will yield 65% soluble solids in the final batch. The pectin is mixed in solution form and heat-treated under vacuum to obtain the requested soluble-solids content. The vacuum must be broken prior to heating to pasteurization. After that the citric acid is adjoined to decrease the pH to 3.0–3.1. The mixture is cooled to filling temperatures and gelation occurs in the container itself. For HMP gelation, low pH, a high soluble-solids concentration, and suitable temperatures are required to accomplish the required demands. A HMP gel cannot be melted following solidification. Stirring during gelation results in lower-strength gels, or no gelation at all. For gel structuring, a 3D network is essential for holding water, sugar, and other solutes. Junction zones in the HMP gel network have been previously described. In line with this model, 3 to 10 polymer-chain segments with a helical structure create aggregates of parallel chains, which are limited in size by steric barriers, entropic factors, and probably rhamnose insertions. Local crystallization is maintained by intermolecular hydrogen bonds, and is most likely reinforced by hydrogen-bonding with water molecules in one set of triangular channels, and hydrophobic attractions between methyl groups forming columns in a second set of triangular channels. In molecular gel networks, no less than two types of bonding are engaged. One is strong and accounts for the elastic properties of the gel and the other is weaker, and capable of reforming following disruption. Sugars play an active role in the creation of the pectin-gel network by associating with pectin molecules by means of hydrogen bonding to form secondary links, which reinforce the molecular network structure. The ageing process of aqueous HMP–sucrose gels can be tracked by low-amplitude oscillation. Dynamic mechanical measurements facilitate determination of the point at which the system experiences the sol–gel transition. The HMP–sucrose system is very responsive to temperature differences throughout ageing, in particular in the lower temperature range. The gel's viscoelastic behavior indicates changes with ageing temperature, most likely because of differences in the mobility of the pectin chains, and as a result, in the lifetime of the junction zones. HMP–sugar gels are produced by a combination of hydrogen bonding and hydrophobic effects. Extents of the latter are affected by the solute used and temperature, gel strength, and rate of structure development. Industrially significant results that occur throughout the sol–gel transition have been thought about, together with a profile of complex viscosity for the period of gelation, and the effects of rate of cooling on pectin concentration. Structural developments in HMP–fructose gels have also been described. Weaker pectin networks are shaped under thermal circumstances that are unfavorable to the development of hydrophobic interactions. Gelling time and elastic modulus have a complex reliance on temperature, which can be attributed to the different thermal behaviors of the intermolecular interactions that stabilize the nonpermanent crosslinks of these physical networks.

LMP gels do not need a high solids content or low pH, but they do require the presence of calcium, which can be supplied by the fruit pulp if a fruit product is being made. Calcium binding to LMP cannot be described as a straightforward electrostatic interaction: it engages intermolecular chelate binding of the cation directed to the structuring of macromolecular aggregates. An "egg-box" model has been proposed for primary junction zones in the LMP molecular gel network. Chain segments with 14 or more residues having a ribbon-like symmetry are considered to form parallel-oriented aggregates. Chelate bonds with oxygen atoms from both galacturonan chains formed by calcium ions are created when calcium ions fit into "cavities" in the structure. Even though they vary from those involving HMP gels, the concepts of good manufacturing practice need to be maintained, and the constituents suitably chosen. Major differences between high- and low-ester systems are the ability to melt a LMP gel and the propensity with which solidification takes place in the LMP system, relative to the slow rate of the HMP gel. Amidated LMPs are typically able to jellify preserves, jams, and jellies with calcium ions originating from fruit and water. Nonamidated LMPs normally necessitate a higher calcium level and the addition of extra calcium is frequently required to obtain suitable gel formation. DA and DE control the readiness of LMP reactions with calcium to encourage gel formation. LMPs with a DE of 25–35% (nonamidated), and pectins with 20–30% DE and 18–25% DA are highly reactive with calcium, and are consequently used in systems with low calcium and low soluble-solids content. Pectins with a low ester content of 35 to 45% (nonamidated) and those with 30 to 40% DE and 10–18% amidation, due to their lower calcium reactivity, can serve in high-calcium or high-soluble-solids content systems.

As already noted, pectin gel properties can be characterized by the conventional SAG method, based on Cox and Higby's work and adopted by the Institute of Food Technologists (IFT). Additional methods involve the use of a Leatherhead Food Research Association (LFRA) texture analyzer, a Boucher Electronic Jelly Tester, a Food Industrial Research Association (FIRA) tester, universal testing machines from various manufacturers, a Herbstreith pectinometer, a spreadometer, or a Bostwick Consistometer. In addition to the diversity and plurality of gel-consistency measurements, additional measurements of time–temperature relationships have also been expanded in the industry. Gelation is dependent on the degree of methyl esterification, amidation, pectin concentration, water activity, the presence of calcium ions, and pH.

Applications

Pectins are widely used in the manufacture of jams, jellies, or analogous gels. In general, regular jam is prepared from HMP while the LMPs are employed when a yielding and more spreadable texture is requested. If fruit pulp is to be included in the jam, an elevated gelation temperature is utilized and, consequently, solidification starts more or less without delay when the containers are filled, with little or no floatation of the particles. If very big containers are used for jam-filling, pectins with lower filling temperatures should be considered, to reduce flavor and color destruction, particularly in the interior of the container. A number of hydrocolloids, particularly pectin, have been recently shown to improve color stability in a model gel system, whereas other hydrocolloids had no or even unfavorable effects. It was proposed that pectin plays a role in the color degradation of jam products. By increasing LMP concentration in the formulation from 0.3 to 1%, there was a 13% increase in the content of total monomeric anthocyanins. Since pectin is a polyuronic acid, its color-stabilizing effect may be based on electrostatic interactions between the anthocyanin flavilium cation and the dissociated carboxylic groups of the pectin.

LMPs mixed with calcium in a quantity related to the gelation temperature and the quality of the structured texture are used to produce low-sugar (less sweet) gums. When manufacturing jellies with no particles, slow-setting pectins that coagulate for an extended time following filling, letting air bubbles float up to the surface and escape from the product, are preferred. Slow-setting HMP is employed for confections. The solids content (~78%) of such preparations is high, in contrast to jams at ~65% or low-sugar jams at ~30–55%. For baking purposes, a heat-resistant gel is typically manufactured with a soluble-solids content of 45–75%. Thermostable gels are usually manufactured from HMPs but can be produced from LMPs if calcium citrate is employed in the formulation to raise the gelation temperature once the system has set. Fruit preparations for dairy products are frequently sold as semigel/thixotropic items for consumption with characteristic soluble-solids contents of 30–65% and a pH of 3.6–4.0, habitually prepared with 0.3–0.6% LMP.

Pectins can also be employed in bakery fillings and glazes. Heat-resistant high-sugar jams are manufactured at a solids content of ~70% with rapid-set pectin. An additional requirement for such consumables is mechanical stability: the less the gel is ruptured, the lower the syneresis at elevated heating temperatures. Nonamidated LMPs are recommended for the manufacture of bakery jams with reasonable consistency. LMP gels are fabricated with ~65% soluble solids and a quite high quantity of calcium-reactive, LMP. Prior to being applied to the baked goods, water is added, then the gel is heated to ~85°C to encourage melting and hot coating of the product. Upon cooling, a shiny coating is formed.

Pasteurized or sterilized acidified milk products (pH ~ 3.5–4.2) can be obtained using high-ester pectins with DE > ~70. Acidification can be generated by either fermentation or addition of fruit juice. If the casein is not stabilized, an unattractive grain-like consistency is acquired. The pectin, inserted before homogenization, sorbs onto the casein particles, which have a positive charge in the nonstabilized milk. If the quantity of added pectin is minute, the charge will be neutralized and the system will likely collapse as a result of the removal of repulsive forces. If pectin addition is continued, however, a new repulsive force builds up, resulting in stabilization of the acidified milk system. HMP was used to stabilize acidified milk drinks against flocculation of milk protein (**Supplemental G**). Upon dilution of the stable samples with water, increased sorption of pectin onto the casein (**Supplemental G**) aggregates was detected and the stability decreased. In both water and serum-simulating buffer, a certain amount of pectin was firmly—possibly irreversibly—bound to the casein aggregates, and it could be inferred that the sorbed pectin formed a multilayer on the casein aggregates. It is possible that the stability in acidified milk drinks is the result of a mechanism resembling steric stabilization combined with secondary adsorption. Soluble soybean polysaccharide (**Supplemental J**) and HMP can both stabilize casein particles in low-pH milk. Soluble soybean polysaccharide functionality seems to complement pectin, as at pH 4.6, pectin stabilizes acid dispersions, whereas under acidic conditions (pH 3.2–4.0), soluble soybean polysaccharide sorbs to the surface of casein micelles and prevents aggregation. The dissimilarities in functionality are caused by the different molecular structures of the two polysaccharides.

A combination of alginate (**Chapter 3**) and pectin polymers resulted in capsules with high encapsulation efficiency, notable stability in a milk system, and significantly improved stress tolerance as seen by high folic acid retention during cheese pressing and even distribution in a cheese matrix. Addition of encapsulated folic acid showed excellent stability during cheese ripening relative to that of free folic acid. These results suggest that cheddar cheese may be an effective medium for folic acid delivery, particularly if alginate–pectin capsules are used.

Cloud stabilization in beverages is also achieved with pectin. Such stability relies on the nature and quantity of the pectin being used. Natural clouding agents can be manufactured from orange and lemon peels by means of enzyme preparations to hydrolyze the pectin in the peel. The chemical and physical properties of the clouds were assessed separately from the drink's properties, taste, and stability. The cloudiness of the fabricated drinks stabilized after 42 days of storage at 25°C. Additional reports on the physicochemical nature of pectin related to marketable orange juice clouding can be found elsewhere.

Confectionery manufacturers employ slow-set HMPs to produce fruit jellies and jelly centers. LMPs are also employed to confer thixotropic behavior at low concentrations, or to obtain a cold-set type of gelation if diffusion of calcium ions occurs. Use of pectin in confections enables the production of foods with tailor-made textural properties, the liberation of fine flavors, and compatibility with continuous processing. HMP was also used for fabrication of coatings that slow down lipid migration in a confectionery product.

LMPs are employed as gelling agents and texture formers in numerous highly dissimilar food items for consumption, such as imitation caviar, dessert jellies, and meat products. Pectin and carrageenan (**Chapter 4**) are extensively used as gelling agents in dairy products. When low-ester amidated pectins with varying calcium reactivity were added to dairy desserts, they affected the microstructure, rheology, and sensory characteristics of the dairy desserts containing carrageenan and starch (**Chapter 14**). Increased calcium reactivity of pectin resulted in a microstructure with increased colocalization of pectin and protein and reduced phase separation. The results can be explained by a combination of factors related to the gel strength of the pectin, convergence of pectin, and carrageenan gelation temperatures, and hydrogen bonding between amide groups on pectin and proteins. Successful production of low-fat frankfurters by inclusion of hydrocolloids in the formulation without extra fat addition markedly reduced cholesterol content in the final low-fat product. Replacing fat with a carrageenan or carrageenan–pectin gel in low-fat frankfurter formulations provided improved functional characteristics as compared to low-fat controls. Carrageenan was more effective when it was used at higher concentrations, improving the functionality of the pectin gel in low-fat frankfurters. Alginate–pectin mixtures have a synergistic consequence in terms of gel-formation properties. Xanthan gum (**Chapter 15**) and pectin can be used jointly as a suitable stabilizer for salad dressings. The integration of pectin in sherbet and water ice gave better product satisfaction by decreasing the enlargement of ice crystals. Galactomannans (**Chapter 8**) in a mixture with pectin serve to stabilize ice cream. Pectins are employed to stabilize emulsions. Modified pectins in whey-protein (**Supplemental G**) emulsions were seen to stabilize the whey protein at sufficiently high concentrations. The use of citrus pectin as an ingredient in a brand of instant noodles for weight reduction was also studied. Citrus LMP was appropriate for the production of a brand of instant noodles, with a concentration of 12% by weight. The developed instant noodles with citrus pectin were 31.4 kcal lower than the standard formula instant noodles without pectin. Physical properties and satisfactory evaluations from 30 volunteer consumers of developed instant noodles with citrus pectin were moderately and very satisfied, similar to results with the standard-formula instant noodles without pectin. The efficacy on weight and body shape reduction (waist, buttock, and top of the arm) of the developed instant noodles with citrus pectin was better than that of the standard formula instant noodles without pectin. For several particular utilizations, the most suitable pectin must be selected. Frozen fruit preparations are improved by including pectin in the manufactured good. Pectins can be employed for coating, in formulas of spray-dried instant tea and for numerous additional products.

Recipes with Pectin

Apple Jam (Figure 13.1)

Ingredients

(Makes 750 g)

1 kg (3–4) apples (**Hint 1**)

400 mL (1⅗ cups) water

45–70 mL (3–4 and ⅔ Tbsp) lemon juice

300–450 g (1¾–2⅔ cups) white sugar (40–60% sarcocarp and boiled juice)

Salt

Preparation

1. Wash, core, and cut the apples (**Figure 13.1A**), and soak them in 1% salt water (1 tsp salt per 2 cups water) (**Hint 2**).
2. Put the cut apples and water in a pan, and simmer for about 20 minutes, until the apples soften (**Figure 13.1B**) (**Hint 3**).
3. Separate the fruit from the peel in a strainer (**Figure 13.1C**).
4. Put the fruit in a pan, add lemon juice and white sugar (**Figure 13.1D**), and boil down to 70% weight before heating at 100°C (**Figure 13.1E**).
5. Pour the apple jam into a glass bottle, cap it, and boil it for sterilization (**Hint 4**).

pH = 3.13–3.22

FIGURE 13.1 Apple jam. (**A**) Wash, core, and cut apples. (**B**) Simmer apples until they soften. (**C**) Separate the apple flesh from its peel. (**D**) Heat apple flesh, lemon juice, and white sugar. (**E**) Continue until a ~30% reduction in weight is achieved.

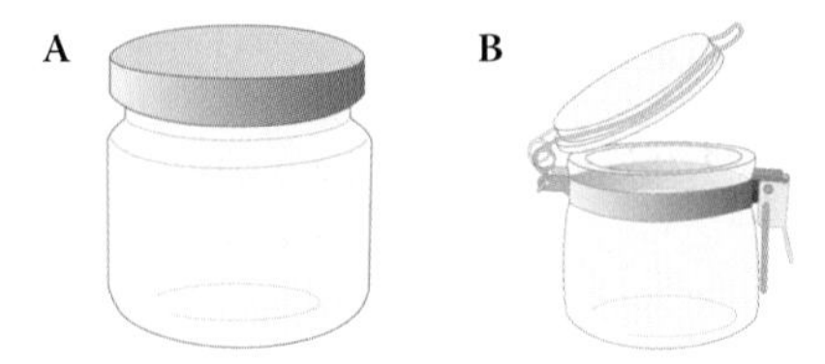

FIGURE 13.2 Jams and preserves should be put in glass bottles (**A**) or sealed containers (**B**).

Hints

1. Choose redder and sour apples so that the jam will have a beautiful red color.
2. Soaking the cut apples in salt water prevents them from browning.
3. Simmering the apple flesh and skin together enhances the apple jam's red color.
4. Sterilize jams and preserves to store them for long periods. Jams and preserves should be put in glass bottles (**Figure 13.2A**) or sealed containers (**Figure 13.2B**). Sterilization should proceed as outlined below.

This recipe is made without adding pectin powder. The apples themselves contain a high concentration of high-methoxy pectin (HMP). If HMP is added to the recipe, the jam will become harder; on the other hand, if low-methoxy pectin (LMP) is added, the jam will become resistant to heat and the amount of sugar can be decreased.

How to Sterilize Jams and Preserves

1. Wash the bottles and boil them for 15 minutes before using (**Figure 13.3A**). A dish towel or stainless-steel netting should be placed on the bottom of a pan while the bottles are boiling. After boiling, drain water from the bottles and place them on a clean dish towel.
2. Pour jams and preserves into the bottles (**Figure 13.3B**).

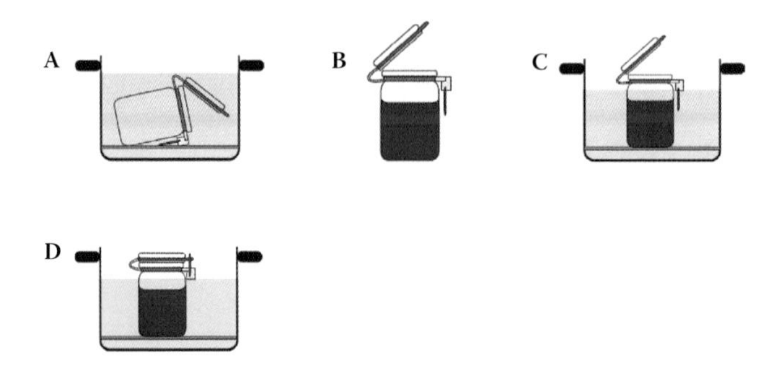

FIGURE 13.3 Steps in the sterilization of jams and preserves. (**A**) Wash and boil the container before use. (**B**) Pour jam or preserves into the container. (**C**) Boil the container with an open lid for deaeration. (**D**) Boil the container again, with the lid closed, for sterilization.

3. When sealed containers are being used, boil them for 10 minutes with the lids open (**Figure 13.3C**) to deaerate, and then close the lids. When glass bottles are being used, this process can be omitted.
4. Boil the bottles with lid closed for 20 minutes for sterilization (**Figure 13.3D**).

Orange Marmalade (Figure 13.4)

Ingredients

(Makes 800 g)

1 kg (4–5) oranges

75 mL (5 Tbsp) lemon juice

400–500 g white sugar (40–50% orange)

Preparation

1. Wash oranges with their peels well, then separate the peel from the pulp and remove the white part of the rind (**Figure 13.4A**).
2. Julienne the peels and put them in water for 1 hour (**Figure 13.4B**), replacing with fresh water a few times (**Hint 1**).
3. Put the orange pulp (**1**), julienned peels (**2**), lemon juice, and white sugar in a pan (**Figure 13.4C**) and simmer at 90°C until thickened (about 30 minutes) (**Figure 13.4D**).
4. Pour the marmalade into glass bottles or sealed containers and boil them for sterilization (see how to sterilize jams and preserves above) (**Figure 13.4E**).

pH = 3.38

FIGURE 13.4 Orange marmalade. (**A**) Peel the fruit and remove the albedo (the white part of the rind). (**B**) Soak julienned orange peels in water. (**C**) Combine the pulp, soaked peels, lemon juice, and white sugar. (**D**) Let thicken. (**E**) Prepared orange marmalade.

Hint

1. If you do not like bitter taste, boil the cut orange peel in 2% salt water (2 tsp salt per 2 cups water) for 10 minutes, then wash thoroughly with tap water.

This recipe is also made without adding pectin powder. Orange peels, both outside and inside, contain a high concentration of HMP.

Milk Jam (Low Sugar) (Figure 13.5)

Ingredients

(Makes 300 g)
4 g (1¼ tsp) LMP powder
30 g (a little over 3 Tbsp) white sugar (**Hint 1**)
30 g (2½ Tbsp) granulated sugar
60 g (a little under 3 Tbsp) corn syrup
200 mL (⅘ cup) milk

Preparation

1. Mix LMP, white sugar, and granulated sugar in a bowl, and add to milk in a pan with stirring with a whisk at room temperature (**Figure 13.5A**) (**Hint 2**).
2. Heat the pectin dispersion with stirring until it reaches 50°C, and add corn syrup to it (**Figure 13.5B**).
3. Continue to heat with stirring until it reaches 90°C (**Figure 13.5C**) (**Hint 3**).

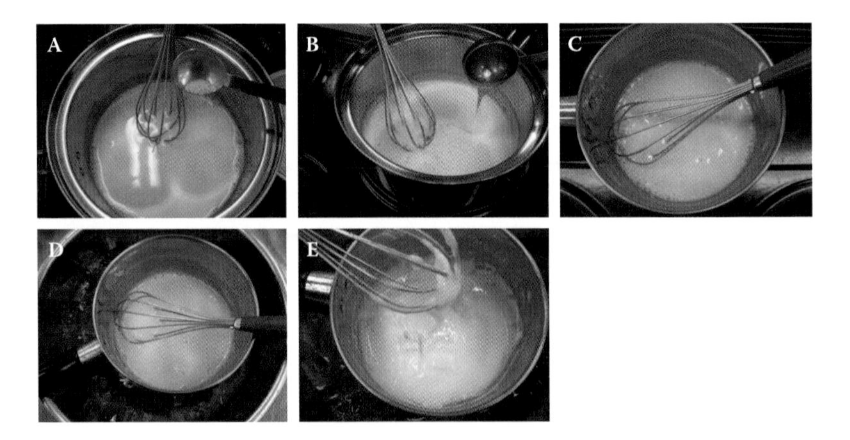

FIGURE 13.5 Milk jam (low sugar). (**A**) Stir in a mixture of LMP, white sugar, and granulated sugar to room temperature milk. (**B**) Heat the mixture and add corn syrup while stirring. (**C**) Continue to heat while stirring. (**D**) and (**E**) Cool the milk jam in an ice-water bath. Stir occasionally.

4. Cool the milk jam in an ice-water bath with occasional stirring (**Figure 13.5D,E**).

pH = 6.39

Hints

1. White sugar prevents milk fat from solidifying.
2. Using lukewarm milk (20–30°C) is better for LMP dispersion in milk. Using hot milk (more than 50°C) causes the milk protein to separate.
3. Heating milk jam to more than 90°C will cause the milk fat to solidify and the milk jam to toughen.

This recipe cannot be sterilized because milk fat solidifies at higher temperatures. The milk jam should be stored in the refrigerator.

Nappage Neutre (Clear Glaze) (Figure 13.6)

Ingredients

(Makes 200 g)
60 mL (4 Tbsp) water
140 g (a little over ⅗ cup) granulated sugar
10 g (1½ tsp) corn syrup
8 g (2½ tsp) LMP powder
15 g (1¼ Tbsp) granulated sugar
120 mL (a little under ½ cup) water
6 mL (a little over 1 tsp) lemon juice

Preparation

1. Put 60 g water, 140 g granulated sugar, and corn syrup in a pan, and heat it to boiling (at 100°C) (**Figure 13.6A**).
2. Mix LMP and 15 g granulated sugar in another pan (**Hint 1**), pour 120 mL water into it (**Figure 13.6B**), and heat with stirring until pectin and sugar dissolve (at 100°C).
3. Add the heated sugar solution (**1**) to the pectin solution (**2**) and mix well; continue to heat it until it boils (**Figure 13.6C**).
4. Remove the pectin dispersion from the heat, and add lemon juice.
5. Pour the *nappage neutre* into a glass bottle (**Figure 13.6D**), and cool it at room temperature.

pH = 2.58

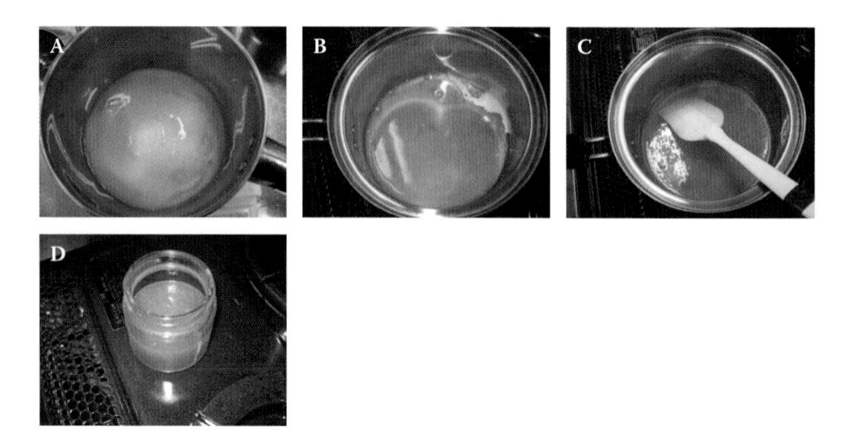

FIGURE 13.6 *Nappage neutre* (clear glaze). (**A**) Boil a mixture of water, granulated sugar, and corn syrup. (**B**) Add water to a mixture of LMP and granulated sugar. (**C**) Add the heated sugar solution to the pectin solution, mix well, and then bring to a boil. (**D**) Pour the *nappage neutre* into a glass bottle.

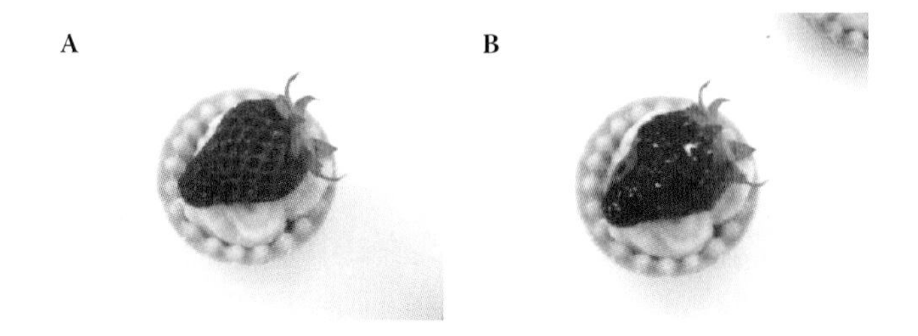

FIGURE 13.7 Strawberry tarts. (**A**) Strawberry tarts without *Nappage neutre*. (**B**) Strawberry tarts brushed with a diluted solution of *Nappage neutre*.

Hint

1. Pectin and sugar need to be mixed to disperse pectin in water.

This recipe is used over fruits on tarts. The *nappage neutre* is diluted in an equal volume of water, warmed in a microwave oven, and brushed over the fruits. **Figure 13.7** shows the difference between strawberry with (**B**) and without (**A**) the glaze.

Dessert Base (Figure 13.8)

Ingredients

(Serves 5)
6.5 g (a little over 2 tsp) LMP powder
65 g (a little over ¼ cup) granulated sugar
170 mL (⅔ cup) water

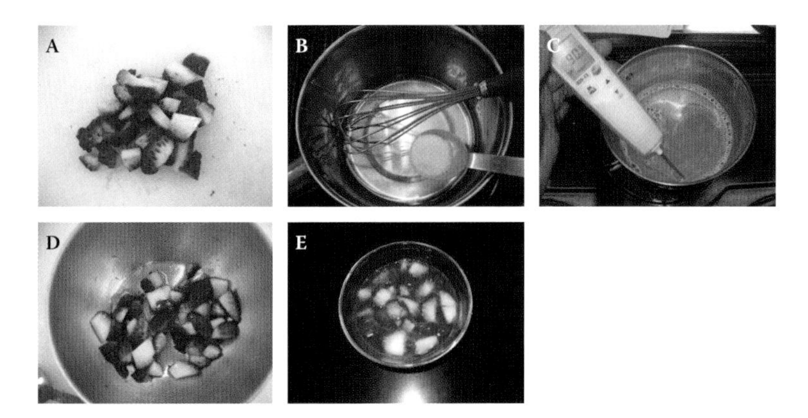

FIGURE 13.8 Dessert base. (**A**) Cut strawberries into large chunks. (**B**) Add a mixture of LMP and granulated sugar to water in a pan and stir. (**C**) Heat pectin dispersion to 90°C while stirring. (**D**) Add strawberry pieces to the citric acid and sodium citrate solution and mix well. (**E**) Prepared and cooled dessert base.

80 g (a little under ⅓ cup) fruit flesh
3 g (⅗ tsp) citric acid, anhydrous
2 g (⅓ tsp) sodium citrate dehydrate
Hot water, flavoring (optional)

Preparation

1. Cut fruit flesh into large chunks (**Figure 13.8A**).
2. Mix LMP and granulated sugar in a bowl, and add to water with stirring in a pan (**Figure 13.8B**).
3. Heat the pectin dispersion (**2**) with stirring until it reaches 90°C (**Figure 13.8C**).
4. Mix citric acid and sodium citrate in a bowl, add a little hot water and dissolve.
5. Add the cut fruit flesh (**1**) to the citric acid and sodium citrate solution (**4**) and mix well (**Figure 13.8D**).
6. Mix pectin dispersion (**3**) and fruit mixture (**5**), and add flavoring.
7. Pour the dessert base (**6**) into glass bottles, and boil at 90°C for 20 minutes to sterilize (**Hint 1**).
8. Cool at room temperature or in water (**Figure 13.8E**).

$$\mathrm{pH} = 3.30$$

Hint

1. If the dessert base is used soon after its preparation, it does not need sterilizing.

This recipe is used for milk desserts. Add equal amounts of milk to dessert base (**Figure 13.9A**) and mix well (**Figure 13.9B**). The LMP in the dessert base reacts with the calcium in the milk.

FIGURE 13.9 Milk dessert. (**A**) Add equal amounts of milk to dessert base. (**B**) Mix well.

Tips for the Amateur Cook and Professional Chef

- Even when sufficient pectin and sugar are present in the system, no gel will form until the pH is reduced to below the critical value of about 3.6, known as the limiting pH value.
- LMP does not require sugar for gel formation.
- LMP is sensitive to divalent metal cations.
- LMP is ordinarily used at 0.5–1.5% in the preparation of gels with calcium salts.
- In confectionery products, the main relevant characteristic of pectin is its ability to gel quickly when the pH is lowered to a certain critical level by the addition of acid.
- In comparative studies of the emulsifying efficiencies of pectin, tragacanth, karaya, and gum arabic was comparable or superior to the other gums in the preparation of stable oil-in-water emulsions of olive, cottonseed, or mineral oil.

REFERENCES AND FURTHER READING

Amado, R. and H. Neukom. 1984. Isolation and partial degradation of pectic substances of potato cell walls in phosphate buffer. *Abstracts: 9th Triennial Conf. Eur. Assoc. Potato Res.*, 103.

Anonymous 1978. E440(a)-Pectin, E440(b)-Amidated pectin. In *Council Directive of July 1978 laying down specific criteria of purity for emulsifiers, stabilizers, thickeners and gelling agents for use in food stuffs. Official J. Eur. Comm.* L 223, 16.

Anonymous 1981. Amidated pectin, pectins, in specifications for identity and purity of carrier solvents, emulsifiers and stabilizers, enzyme preparation, flavoring agents, food colors, sweetening agents and other food additives. *FAO Food and Nutrition Paper 19,* 10–4, 152–5. Rome: Food and Agriculture Organization of the United Nations.

Anonymous 1981. Evaluation of certain food additives. *25th report of the Joint FAO/WHO Expert Committee on Food Additives*. Geneva: World Health Organization.

Anonymous 1981. Pectin. In *Food chemicals codex*, 3rd ed., 215. Washington DC: National Academy Press.

Arltoft, D., Madsen, F., and R. Ipsen. 2008. Relating the microstructure of pectin and carrageenan in dairy desserts to rheological and sensory characteristics. *Food Hydrocolloids* 22:660–73.

Ashford, M., Fell, J., Attwood, D., Sharma, H., and. P. Woodhead. 1994. Studies on pectin formulations for colonic drug delivery. *J. Controlled Release* 30:225–32.

Baker, R. A. 1994. Potential dietary benefits of citrus pectin and fiber. *Food Technol.* 48:133–4.

Boothby, D. 1983. Pectic substances in developing and ripening plum fruit. *J. Sci. Food Agric.* 44:1117.

Braconnot, H. 1825. Recherches sur un nouvel acide universellement rependu dans tous les vegetaux. *Ann. Chim. Phys. Ser.* 2, 28:173.

Brake, N. C. and O. R. Fennema. 1993. Edible coatings to inhibit lipid migration in a confectionery product. *J. Food Sci.* 58:1422–5.

British Nutrition Foundation. 1990. *Complex carbohydrates in foods*. London: The Report of the British Nutrition Foundation's Task Force, BNF.

Broomfield, R. W. 1988. Preserves. In *Food industries manual*, ed. M. D. Ranken, 335–55. Glasgow: Blackie.

Buhl, S. 1990. Gelation of very low DE pectin. In *Gums and* Stabilisers for the Food Industry, vol. 5, ed. G. O. Phillips, D. J. Wedlock, and P. A. Williams, 233–41. Oxford: IRL Press.

Cambell, L. A. and G. H. Palmer. 1978. Pectin. In *Topics in dietary fiber research*, ed. G. A. Spiller, 105. New York: Plenum.

Candogan, K. and N. Kolsarici. 2003. The effects of carrageenan and pectin on some quality characteristics of low-fat beef frankfurters. *Meat Sci.* 64:199–206.

Cerda, J. J. 1994. The role of pectin in cholesterol regulation. *Abstracts of Papers of the American Chemical Society*, 208, 80–AGFD.

Christensen, O. and G. A. Towle. 1973. Pectin. In *Industrial Gums*, 2nd ed., ed. R. L. Whistler, 429–61. New York: Academic Press Inc.

Christensen, P. E. 1954. Methods of grading pectin in relation to the molecular weight (intrinsic viscosity) of pectin. *Food Res.* 19:163.

Christensen, S. H. 1986. Pectins. In *Food hydrocolloids*, vol 3, ed. M. Glicksman, 206–27. Boca Raton: CRC Press Inc.

Cox, R. E. and R. H. Higby. 1944. A better way to determine the jellying power of pectins. *Food Industry* 16:441.

Dasilva, J. A. L., Goncalves, M. P., and M. A. Rao. 1994. Influence of temperature on the dynamic and steady shear rheology of pectin dispersions. *Carbohydr. Polym.* 23:77–87.

Dasilva, J. A. L. and M. A. Rao. 1995. Rheology of structure developments in high-methoxyl pectin/sugar systems. *Food Technol.* 49:70, 72–3.

De Vries, J. A., Rombouts, F. M., Voragen, A. G. J., and W. Pilnik. 1982. Enzymatic degradation of apple pectins. *Carbohydr. Polym.* 2:25–33.

Doseburg, J. J. 1965. Pectic substances in fresh and preserved fruits and vegetables. *I.B.V.T. Communication No. 25.* Wageningen, The Netherlands: Institute for Research on Storage and Processing of Horticultural Produce.

Einhornstoll, U., Glasenapp, N., and H. Kunzek. 1996. Modified pectins in whey-protein emulsions. *Nahrung Food* 40:60–7.

Elshamei, Z. and M. Elzoghbi. 1994. Production of natural clouding agents from orange and lemon peels. *Nahrung Food* 38:158–66.

Fernandez, M. L., Sun, D. M., Tosca, M., and D. J. McNamara. 1994. Citrus pectin and cholesterol interact to regulate hepatic cholesterol homeostasis and lipoprotein metabolism: A dose-response study in guinea pigs. *Am. J. Clin. Nutr.* 59:869–78.

Food Chemicals Codex. 1972. 2nd en., 580–1. Washington DC: National Academy of Sciences.

Guillon, F., Thibault, J.-F., Rombouts, F. M., Voragen, A. G. J., and W. Pilnik. 1989. Enzymic hydrolysis of the "hairy" fragments of sugar-beet pectins. *Carbohydr. Res.* 190:97–108.

Hu, H. and P. H. Brown. 1994. Localization of boron in cell-walls of squash and tobacco and its association with pectin—Evidence for a structural role of boron in the cell-wall. *Plant Physiol.* 105:681–9.

Huber, D. J. 1984. Strawberry fruit softening: The potential roles of polyuronides and hemicellulose. *J. Food Sci.* 49:1310–5.

Jensen, S., Rolin, C., and R. Ipsen. 2010. Stabilization of acidified skimmed milk with HM pectin. *Food Hydrocolloids* 24:291–9.

Jitpukdeebodintra, S. and A. Jangwang. 2009. Instant noodles with pectin for weight reduction. *J. Food Agric. Env.* 7:126–9.

Karpovich, N. S., Telichuk, L. K., Donchenko, L. V., and M. A. Totkailo. 1981. Pectin and raw material resources. *Pishch Promst.* (Moscow) 3:36–9.

Kertesz, Z. I. 1951. *The pectic substances*. New York: Interscience Publishers Inc.

Klavons, J. A., Bennett, R. D., and S. H. Vannier. 1994. Physical/chemical nature of pectin associated with commercial orange juice cloud. *J. Food Sci.* 59:399–401.

Klurfeld, D. M., Weber, M. M., and D. Kritchevsky. 1994. Dose-response of colonic carcinogenesis to pectin and guar gum. *FASEB J.* 8:A152.

Kohn, R. and O. Luknar. 1977. Intermolecular calcium ion binding on polyuronates—Polygalacturonate and polyguluronate. *Collect. Czech. Chem. Commun.* 42:731–44.

Kopjar, M., Pilizota, V., Tiban, N. N., Subaric, D., Babic, J., and D. Ackar. 2007. Effect of different pectin addition and its concentration on color and textural properties of raspberry jam. *Deutsch Lebensm. Runds* 103:164–8.

Kopjar, M., Pilizota, V., Tiban, N. N., Subaric, D., Babic, J., Ackar, D., and M. Sajdl. 2009. Strawberry jams: Influence of different pectins on color and textural properties. *Czech. J. Food Sci.* 27:20–8.

Lewinska, D., Rosinski, S., and W. Piatkiewicz. 1994. A new pectin-based material for selective LDL-cholesterol removal. *Artificial Organs* 18:217–22.

Madziva, H., Kailasapathy, K., and M. Phillips. 2006. Evaluation of alginate–pectin capsules in Cheddar cheese as a food carrier for the delivery of folic acid. *LWT – Food Sci. Technol.* 39:146–51.

Maier, T., Fromm, M., Schieber, A., Kammerer, D. R., and R. Carle. 2009. Process and storage stability of anthocyanins and non-anthocyanin phenolics in pectin and gelatin gels enriched with grape pomace extracts. *Eur. Food Res. Technol.* 229:949–60.

Mao, Y., Kasravi, B., Nobeak, S., Wang, L. Q., Adawi, D., Roos, G., Stenram, U., Molin, G., Bengmark, S., and B. Jeppsson. 1996. Pectin supplemented enteral diet reduces the severity of methotrexate-induced *Enterocolitis* in rats. *Scand. J. Gastroenterol.* 31:558–67.

Mariana-Atena Poiana, M.-A., Alexa, E., and C. Mateescu. 2012. Tracking antioxidant properties and color changes in low-sugar bilberry jam as effect of processing, storage and pectin concentration. *Chem. Central J.* 6:4. http://journal.chemistrycentral.com/content/6/1/4. doi:10.1186/1752-153X-6-4.

Mascaro, L. J., and P. K. Kindell. 1977. Apiogalacturonan from *Lemna minor*. *Arch. Biochem. Biophys.* 183:139–48.

Matora, A. V., Korshunova, V. E., Shkodina, O. G., Zhemerichkin, D. A., Ptitchkina, N. M., and E. R. Morris. 1995. The application of bacterial enymes for extraction of pectin from pumpkin and sugar beet. *Food Hydrocolloids* 9:43–6.

Matsuura, Y. 1984. Chemical structure of polysaccharide of cotyledons of kidney beans. *J. Agric. Chem. Soc. Japan* 58:253–9.

May, C. D. 1992. Pectins. In *Thickening and gelling agents for food*, ed. A. Imeson, 124–52. Glasgow: Blackie Academic & Professional, an imprint of Chapman & Hall.

Melotto, E., Greve, L. C., and J. M. Labavitch. 1994. Cell-wall metabolism in ripening fruit. Biologically active pectin oligomers in ripening tomato (*Lycopersicon esculentum* Mill.) fruits. *Plant Physiol.* 106:575–81.

Michel, F., Doublier, J. L., and J. F. Thibault. 1982. Investigation on high methoxyl pectins by potentiometry and viscometry. *Prog. Food Nutr. Sci.* 6:367–72.

Nakamura, A., Yoshida, R., Maeda, H., and M. Corredig. 2006. The stabilizing behaviour of soybean soluble polysaccharide and pectin in acidified milk beverages. *Int. Dairy J.* 16:361–9.

Nwanekezi, E. C., Alawuba, O. C. G., and C. C. M. Mkpolulu. 1994. Characterization of pectic substances from selected tropical fruits. *J. Food Sci. Technol. India* 31:159–61.

Pedersen, J. K. 1980. Pectin. In *Handbook of water-soluble gums and resins*, ed. R. L. Davidson, 15–1 to 21. New York: McGraw Hill.

Pilnik, W. and P. Zwiker. 1970. Pektine. *Gordian* 70:202–4, 252–7, 302–5, 343–6.

Rao, M. A. and H. J. Cooley. 1993. Dynamic rheological measurements of structure developments in high-methoxyl pectin/fructose gels. *J. Food Sci.* 58:876–9.

Rao, M. A. and H. J. Cooley. 1994. Influence of glucose and fructose on high-methoxyl pectin gel strength and structure development. *J. Food Qual.* 17:21–31.

Redgwell, R. J. and R. R. Selvendran. 1986. Structural features of cell wall polysaccharides of onion. *Carbohydr Res.* 157:183–99.

Rees, D. A. 1982. Polysaccharide conformation in solutions and gels-recent results on pectins. *Carbohydr. Polym.* 2:254–63.

Rinaudo, M. 1974. Comparison between results obtained with hydroxylated polyacids and some theoretical models. In *Polyelectrolytes*, ed. E. Selegny, 157. Dordrecht: Reidel Publishing Company.

Rolin, C. and J. D. De Vries. 1990 Pectin. In *Food gels*, ed. P. Harris, 401–34. London and NY: Elsevier Applied Science.

Sakai, T., Sakamoto, T., Hallaert, J., and E. J. Vandamme. 1993. Pectin, pectinase and protopectinase-production, properties and applications. *Adv. Appl. Microbiol.* 39:213–94.

Vauquelin, M. 1790. Analyse du tamarin. *Ann. Chim.* (Paris) 5:92.

Yamaguchi, F., Shimizu, N., and C. Hatanaka. 1994. Preparation and physiological effect of low-molecular-weight pectin. *Biosci. Biotechnol. Biochem.* 58:679–82.

14

Starch

Introduction

Starch is suited to a vast array of food and industrial applications. Combined with its low cost, this has led to a remarkable increase in its utilization. In 1962, more than 150 million bushels (1 bushel = 35.24 L) of grain, generally corn, were processed for starch. Over half of the 5 billion pounds of cornstarch was converted to dextrose and corn syrup while the rest was marketed in raw form or modified to manufacture a variety of specialty starches and derivatives. Since then, there has been an average annual increase of ~5% in the quantity of processed grain. There are a vast number of applications for starch in the food industry, but, as stated by the French lawyer, politician, and author of a celebrated work on gastronomy, *Physiologie du Goût* (The Physiology of Taste) Anthelme Brillat-Savarin: "Starch is the basis of bread, of pastry, and of purées of all kinds, and thus to a great degree enters into the nourishment of nearly every nation."

Varieties of Starch

Starches are very important hydrocolloids, based on both their wide use and cheap price. In the United States (US), the most important starch crop is corn, accounting for more than 90% of the total manufacture of starch. In Europe, the predominant starch crop is potato, followed by wheat and then corn. Cornstarch is the most inexpensive of the common starches. It has a high gelatinization temperature and the ability to gel, and it also shows the phenomenon of retrogradation. Cornstarch, which includes inseparable amounts of oil, turns rancid during storage; as a consequence, the more expensive oil-free cornstarch was developed. When cooking cornstarch in water, a cloudy noncohesive solution is formed. Once it cools, this solution gels or sets back to produce viscous, fairly short and opaque pastes with a characteristic taste. Cornstarch is widely employed in sauces, gravies, salad dressings, puddings, pie fillings, canned foods, confections, and numerous additional formulations. It is not appropriate for frozen products whose pastes have a pronounced inclination to retrograde under freeze-thaw conditions. Sorghum starch is derived from sorghum grain. Its properties are fairly similar to those of cornstarch. Sorghum starch is sometimes converted into syrup and minute quantities are utilized as a replacement for cornstarch.

Wheat starch is manufactured as a byproduct of gluten production. Wheat includes a large quantity of protein and it is therefore difficult to extract the starch from the wheat flour. Several methods have been invented for this purpose. For instance, fermenting the gluten toward its conversion to allow washing out the starch with water is an option. This method is disadvantageous in the sense that the destroyed gluten has significant market value. Additional procedures deal with kneading wheat flour dough with water prior to extraction of the starch. This method permits separation of the wheat starch and gluten; nevertheless the starch contains leftover protein and consequently has a mealy taste and odor as well as foaming ability. The size of wheat starch granules is not consistent. Some are fairly large and similar to potato granules and others are quite small and round. At a given concentration, wheat starch pastes yield a softer gel than corn or sorghum. Some recipes for pastries, cakes, and cookies favor replacing part of the wheat flour with wheat starch for improved texture and sensory attributes as well as better volume.

Potato starch has the largest granules of any commercial starch. These granules have an oval shape and as a result of their size, they present fairly large interfaces for light reflection, giving dry potato starch its characteristic sheen. In many aspects, potato starch is similar to tapioca starch. Moreover, when it is modified by pregelatinization, it competes with tapioca starch for use in numerous food applications. Upon gelatinization, potato starch yields a paste that has long body, good clarity, and bland flavor. These attributes make it practical for many food applications such as baked goods, desserts, gravies, puddings, and soups. Both amylose and amylopectin from potato starch are commercially available. Tapioca starch has the lowest gelatinization temperature. In fact, tapioca starch has some unique properties; for instance, its cooked pastes have cohesive textures, with a clear, fluidic appearance, but it is important to note that in some food applications, the long cohesive texture is not desirable. Moreover, the noncereal flavor of tapioca starch is in high demand for various food products, especially when off-flavors are a problem. Rice starch is produced from damaged and broken rice grains. To achieve this, the gluten that holds the rice granules is treated with caustic soda that dissolves the protein and frees the starch. The starch is further purified by washing, sieving, setting, and drying. Rice starch granules are the smallest of the commercial starches. In addition to its small granule size, rice starch has soft textural properties, which are desirable for many nonfood uses as well as its use as a brewing adjunct. Arrowroot starch is derived from a plant that grows to a height of 0.6–1.5 m. The plant leaves are arrow shaped, giving it its name. It is similar to tapioca starch but less cohesive, typically cloudier and usually softer in texture. Arrowroot starch is manufactured in small quantities and is utilized mainly in baking for the manufacture of specialty biscuits and cookies. Lintnerized starch is obtained by mild acid hydrolysis of both α-1,4 and α-1,6 glycosidic linkages of amylose and amylopectin, preferentially in the amorphous regions of the granules (amylose and branched regions of amylopectin). Such a treatment leaves starch preparations with increased crystalline contents, which results in increased resistance to enzymatic hydrolysis. A bakery product showing moderate available starch and slow-release carbohydrate features was prepared using a resistant starch-rich powder (RSRP) from banana starch. The use of nutraceutical ingredients, such as RSRPs, may serve in the development of new products for population sectors with glycemia or a requirement for reduced calories. Sago starch is derived from the sago palm tree. Its inner turf consists of a soft, white starch material. The turf is processed by grating to achieve a powder. The powder is wetted over a strainer. The collected mixture of starch and water that passes through the strainer is kept in a trough, and the starch is precipitated out, washed, and then dried to get sago flour. The light brown flour can be processed later, by the importer. Raw sago starch yields a semiclear, cohesive fluid that maintains the viscosity level and consistency of cooked pastes stable over time. On the other hand, high-fluidity sago starches produce hard gels suitable for confectionary products. Waxy cornstarch is produced from the waxy maize plant. The name is derived from the waxy appearance of the cross section of the kernel. The waxy corn grains look microscopically similar to regular corn grains. Although the waxy cornstarch is extracted in a procedure similar to cornstarch, its properties differ due to differences in molecular structure. Waxy cornstarch is similar to tapioca starch in terms of their derived pastes, which are fluid, cohesive, and nongelling. In addition, waxy cornstarch stabilizes other starches by decreasing their tendency to gel. The clarity and nongelling properties of waxy cornstarch give its products uniqueness and encourage its utilization as a stabilizer and thickener in canned products, pie fillings, and salad dressings.

Structure and Composition

Starch is a reserve food supply for most plants. It is a carbohydrate polymer that is synthesized inside the plant by chemical interlinking of hundreds and thousands of individual glucose units to form long-chain molecules. These can be broken down by acidic or enzymatic treatments. When the plant draws on its reserve starch supply, dextrose is regenerated and utilized, as it is used as a source of energy by humans and animals. In the plant, two different mechanisms are used to synthesize starch. Amylose is produced by successive attachment of several hundred glucose units to form a long linear chain. A second mechanism involves attachment of a glucose unit to a linear chain in a branching position (amylopectin). Glucose groups are added to each of these branches until new branch points are started and amylopectin, that is, a large tree-like polymer, is synthesized.

Irrespective of their source, starches are small granules with intrinsic size and shape. Starch is a white granular material with different origins. The differences among the granules can be easily detected under a microscope. Sago and potato starches have fairly large oval granules. Cornstarch granules are smaller and have a polygon shape. Rice starch granules are polygonal but smaller than cornstarch granules. These granules have a complex internal structure, composed of successive, connected layers of starch placed via onion-type construction. The starch molecules are positioned in a systematic crystalline pattern. The granules are regarded as round crystals and the spherocrystalline character can be viewed through the microscope using polarized light as a Maltese cross pattern. Tapioca starch contains 17% amylose, 22% potato starch, and ~27% cornstarch. Waxy cereal grains are composed more or less wholly of amylopectin and the basic unit of both is D-glucose. Further information on the chemical structure and spatial arrangement can be located elsewhere.

Functional Properties of Starch Suspensions

Dry starch and processed or cooked starch have different functional properties. Raw starch has innate properties related to the variety and type of granule. Granules differ in their size, color, flavor, odor, moisture content, flowability, and dispersibility. Properties can be modified by physical and chemical treatments. During processing and cooking, properties such as gelatinization, rate of thickening, temperature, and time to attain maximal viscosity and control flow are most important. Following processing, properties such as clarity, paste and gel classification, flavor and texture, water retention, and storage stability become significant. When an aqueous suspension of starch is heated at a particular temperature, the hydrogen bonding that holds together the micellar network of molecules is weakened such that water can be absorbed by the granules. This point is the initial gelatinization temperature at which the granules swell and in parallel, lose their polarization crosses (birefringence). This process starts at the botanical center of the granule and spreads rapidly to its periphery. Gelatinization begins anywhere hydrogen bonding is weakest, that is, in the amorphous intermicellar areas of the granule, and hydrogen-bonding strength differs in the individual granules of each starch species. Therefore, granules gelatinize in a range of 60 to 82.2°C rather than at one specific temperature. When the heating temperature of the aqueous starch suspension is increased above the gelatinization range, more hydrogen bonds are disrupted, more water molecules attach to the liberated hydroxyl groups and the granules swell further.

As swelling progresses, the starch molecules become completely hydrated, disconnect from the complicated micellar network, and diffuse into the surrounding aqueous medium. Equilibrium is achieved when the concentration of soluble starch in the water inside the swollen granule is identical to that in the adjacent aqueous phase. Upon heating and cooling of a starch aqueous suspension, six stages occur: granules hydrate and swell to several times their size, they lose their Maltese crosses (birefringence), the mixture becomes clearer, viscosity increases rapidly, linear molecules dissolve and diffuse from the ruptured granules, and the mixture regresses to a paste-like mass or gel. The above-described changes are indicative of the functional properties of starch in food processing.

Starches gelatinize when their aqueous dispersions are heated over a range of temperatures. Gelatinization is characterized by tangential swelling of the granules and loss of their polarization crosses, as well as an increase in optical transmittance and a rise in viscosity. Cornstarch gelatinizes at a higher temperature than potato starch or waxy cornstarch. The potato and waxy corn starches reach a higher peak viscosity followed by considerable thinning out compared to cornstarch upon continued heating and stirring. Waxy maize and potato granules are in fact more fragile and more readily broken than cornstarch. This provides an explanation for the quick breakdown from initial peak viscosity.

Potato and waxy cornstarch thicken less than cornstarch paste; this is due to the lack or only trace amounts of amylose in waxy cornstarch, while the increased stability of potato starch relates to branching in the amylose and larger molecule size. When the temperature of an aqueous starch suspension is elevated to its pasting temperature, that is, it is higher than the gelatinization range, the solubility of the starch, as well as the paste's viscosity and clarity, increase. Potato starch swells quickly and markedly whereas cornstarch demonstrates comparatively slow and limited swelling. Waxy corn and sorghum starches show intermediate swelling capacity. Sorghum starch and cornstarch reveal two-stage swelling indicating two types of bonding forces within the granule. Waxy sorghum starch swells more freely than sorghum since it has no amylose to strengthen the molecular network in the granule. Potato and tapioca starches cook rapidly due to the breaking of their fragile, rapidly swelling granules. Corn and sorghum starches swell to a lesser degree and cook very slowly; nevertheless, they provide the user with more stable viscosity of the soluble portion. Starch derivatives create products that go through limited swelling and solubilization. Such products are suited for thickening in canned foods since their chemical cross-linking retains the formed viscosity during the heat treatment. The clarity of the created starch paste is directly related to the state of dispersion, the starch's tendency toward retrogradation, and the presence and type of other included materials. For example, sugar greatly increases the clarity of cornstarch; on the other hand, glycerol monostearate (emulsifier) tends to produce more opaque pastes.

Corn, sorghum, and wheat starches have a pronounced cereal flavor but it decreases during cooking and is masked or fades away at temperatures higher than the gelatinization temperature range of the starch. Potato, tapioca, and waxy corn starches have a less inherently cardboard-like flavor, perhaps due to their lesser lipid content. A few processes to eliminate off-flavors from treated starches have been recommended. The principal textural property of starch is the viscosity it creates upon heating and further cooling. The rheological parameters of starch can be estimated by the Brabender viscoamylograph. This instrument can measure changes in viscosity of stirred, heated, and then cooled (at a uniform rate) starch–water dispersions for specific times. Moreover, it is possible to hold any desired temperature for a specific period of time and then cool it at a uniform rate. The produced curves are characteristic of the functional properties of the starch under investigation. The procedure for Brabender measurements and their interpretation can be found elsewhere.

Starch Pastes and Gels

Starch behavior is dependent upon the affinity of the hydroxyl groups in one molecule for those in another. When the starch paste cools, the molecules become less soluble and aggregate, then crystallize out of solution. In amylose (linear polymer), straight chains orient themselves in a parallel alignment and promote an association via hydroxyl groups and bonds to form water-insoluble moieties. There is a difference between very dilute and concentrated solutions. In the former, aggregated chains of amylose precipitate, whereas in the latter they form a gel. This process, which is termed *retrogradation*, is essentially a crystallization phenomenon. It serves as a key factor in the amendment and use of starch. Since amylopectin has a highly branched structure, its alignment and crystallization are inhibited. Therefore, it tends to be soluble and form solutions that do not gel under normal conditions. It should be noted that under particular conditions (freezing, prolonged aging), some retrogradation as a result of water-molecule holding and attraction can be observed in amylopectin dispersions.

Amylose is regarded as the key component in gel formation. It shapes firm gels at as low as 1.5% concentration. It functions as a network that entraps and binds water and as a material that links intact starch granules or fragments. Starch gels are characterized by the size and type of their granules and their age and prior treatment, as well as time–temperature relationships of cooking. Starches that contain both amylose and amylopectin form gels readily at comparatively low concentrations. The exception to this is potato starch, which demonstrates fewer of the behaviors toward gel formation than would be expected from its amylose content. This is due to the atypically vast length of the linear molecules in potato starch or the minor degree of branching, which interferes with molecule alignment to shape micellar structures. A change in gel structure caused by amplified crystallization or retrogradation of the starch molecules is believed to be simply a continuation of those alterations that changed the thickened paste into a gel. Unwelcome changes as a result of retrogradation are represented by staling of bread and other baked goods, curdling of frozen sauces subsequent to thawing, and "skin" formation on cooked starch desserts and puddings. Upon standing, fluid tends to exude from starch gels. This is termed *syneresis* or *weeping*, which is an unwelcome occurrence in food items and results from slow shrinkage of the starch gel that decreases the network's ability to retain all of the entrapped fluid phase. This is followed by separation of the fluid from the gel. This effect is even larger when the gel is frozen and then thawed.

Effect of Food Ingredients on Starch Functionality

Starch gelatinization can be retarded by sugar, which inhibits its swelling in water. The sucrose preferentially bonds to water molecules, withholding them from the starch. The higher the sucrose concentration, the more inhibited the swelling becomes. Addition of sucrose also decreases the strength of the formed gel, as evidenced in confections such as gumdrops. Such confections are composed of 70% sugar, 20% water, and the rest starch. As a result of this composition, starch is not able to gelatinize fully due to lack of water. To bypass the problem, cooking is performed with more water for a longer time until the excess water evaporates. Pressure cooking is another option since starch can swell in the presence of high amounts of sugar at elevated temperature and pressure. Additional soluble constituents, for instance dry milk solids, egg, or fat, tend to retard gelatinization. A mixture of components can have intermediate or additive effects. For instance, a combination of vinegar and sugar is of interest given that decreasing the pH (increasing the acidity) speeds up gelatinization and viscosity breakdown, while sugar retards these processes. If fats and salt are added, both retard the swelling of starch granules. Within a pH range of 3 to 7, the gelatinization temperature of raw starches is not affected. Outside this range, the gelatinization temperature is a function of pH. At a pH level near 3, starch is subjected to hydrolytic degradation during manufacture and storage. In foods such as salad dressings that contain acetic acid as part of their formulation, the starches are cooked with the vinegar during their preparation and consequently are partially hydrolyzed, leading to thinning out. Thus the manufacturers of such commodities must control the processing parameters to achieve a desired texture. As in many other cases, addition of acid at the end of processing might contribute to a better texture.

In addition, modified starches should be used to reach the desired body and texture. Modified food starches have been developed as stabilizers, providing desirable consistency, texture, and storage ability. They are used primarily in strained and baby foods and, to a minor extent, in infant formulas. However, despite the increasing trend to introduce infants to solid foods at a very young age, there are a few long-term studies delineating the effect of starch feeding on the growth of young infants. Modified food starches used by the food industry for infants and young children are of concern: there is an urgent need for additional data regarding their bioavailability, effect on nutrient absorption, intestinal changes, and toxic, mutagenic, and carcinogenic effects. Therefore, the inclusion of modified food starches should be used prudently and sparingly. Various starches behave differently upon inclusion of electrolytes (salts) in the medium. Potato starch that contains ester-bonded phosphate groups and differs from cereal starches behaves as a polyelectrolyte and is sensitive to the addition of ions. In addition, the salt's concentration, type, and processing temperature are important. For example, sodium sulfate will repress granule gelatinization and raise the gelatinization temperature, whereas sodium nitrate increases the swelling and lowers the gelatinization temperature. Nonionic surfactants prevent gel formation in cooked farina and diminish tackiness in pasta-based manufactured goods. This performance is characteristic when amylose-surfactant complexes form, as these are able to inhibit the swelling and solubilization of the starch granules. In some cases, at higher temperatures, these complexes break down. Fats or fatty acids have been reported to repress the swelling of starch and its gelatinization. Other reports point to a contradictory effect of some fats and fatty acids and it is likely that these consequences rely on the nature of the systems; consequently, cases should be considered on an individual basis for particular products that contain different starches and fats.

Properties of Available Starches

Properties of Dry Starch

Native starch is made up of small granules. Their size and shape depend upon the particular botanical source. Rice starch granules have sizes of 3 to 8 μm, whereas potato averages 15 to 100 μm in diameter. Cornstarch ranges from 5 to 25 μm. Since the granules are small, their surface area and digestibility are major factors in their use in the food industry. The raw starch granules have a strong affinity for moisture due to numerous hydroxyl groups in the starch molecule. Absorbed moisture content depends on both relative humidity and temperature and is dissimilar for a variety of starch-producing species. Upon equilibration and under regular atmospheric conditions, most starches enclose 10 to 17% moisture. Cornstarch powder contains ~11% moisture and is able to adsorb moisture at high humidity. Cornstarch can potentially be dried to ~1.5% moisture for use in applications requiring the removal of moisture from other constituents, or the protection of active powders from the harmful effects of moisture. One such application is the use of dried starch in the confectionery industry to serve as a molding agent for jelly slices and gumdrops. In this application, the starch absorbs moisture and the confection sets under the proper conditions. Another use of dry starch is to separate active components of a powder formulation such as baking powder, enzyme preparation, and flavoring (see also discussion in **"Commercial Applications of Starches"** in **Chapter 14**).

Pregelatinized Starches

Pregelatinized starches are prepared by cooking and drying starch slurries. Via a precooking stage, starch granules swell and are preconditioned to go through thickening and gelatinization in cold liquid, with no need for subsequent heating. The manufacturer performs this starch cooking in the factory, shortening preparation times in the domestic kitchen. Pregelatinization of starches is designed to gain control of the requirements for a specific food product and to produce an item for consumption uniformity. Many convenience food items, such as cake and beverage mixes, dried soups, sauce and salad dressing mixes, desserts and puddings are in need of ingredients specifically suited to their preparation. The main requirement is ease of preparation and the option to achieve a uniform texture, especially for instant foods. The increased marketing of instant puddings has paved the way to broader applications for pregelatinized, precooked cold-swelling starches. These starches have replaced carrageenans (**Chapter 4**) and other water-soluble gums as stabilizers and thickeners since they are cheaper and have better hydration characteristics. Pudding desserts were prepared with κ-carrageenan, native cornstarch, skim milk powder (**Supplemental G**), sucrose, and water. The rheological properties of the pudding-like desserts seemed to be mainly governed by the exclusion effect of the swollen starch granules, thus concentrating κ-carrageenan in the continuous water phase. It is important to note that although instant starch puddings do not have the mouthfeel of cooked starch pudding, the convenience, ease of preparation, and time-saving attributes seem to be of top importance for today's consumer. Pregelatinized starches have found an important role in the production of fruit puddings, as well as meringues and icing mixes.

Modified Starches

Raw starches can be modified to obtain a combination of properties suitable for specific food products. Furthermore, starches can be treated to convey new properties, not just adjust and improve innate ones. These modifications involve the use of heat, acid, alkali, oxidizing chemicals, and other chemical agents to produce changes in the starch molecule. Such changes could be the introduction of a new chemical group, and changes in size, shape and/or structure. Several methods have been adopted for modification purposes. They include modification through plant breeding, modification through fractionation, modification by cross-linking, conversion by hydrolysis, and chemical derivatization. The genetic approach, that is, modification through plant breeding, is designed to develop hybrids that yield, for instance, starch in only one form, such as all-linear amylose or all-branched amylopectin. Another approach is to separate starch into amylose and amylopectin fractions. Potato starch fractionation was based on autoclaving a dispersion of potato starch in an aqueous solution of magnesium sulfate followed by cooling to 70°C, where amylose precipitated out and could be dried. Additional cooling to room temperature led to amylopectin precipitation. The colloidal properties and dispersibility of starches can be modified by mild chemical treatments. Changes in the swelling of the starch granule can be attained by reinforcing the bonds holding the granule together. Cross-linking reactions form reinforcing links between the molecules on or near the surface of the granules. The cross-linking leads to decreased swelling power of the granular starch and increased paste viscosity. The increased viscosity results from the strengthening of the granule structure. With an increase in cross-linking, granules become more resistant to swelling and the paste viscosity decreases. At even higher cross-linking levels, starch granules become totally unaffected by a regular heating cycle. In addition, cross-linking prevents the rapid drop in viscosity of the treated starch during continuous agitation, heating, or treatments with alkalis or acids. A small degree of cross-linking also helps provide a smooth creamy texture to starch pastes based upon potato, tapioca, or waxy corn starch instead of the rubbery cohesive texture of untreated starches.

Dextrins and thin-boiling starches are termed *converted starches*. They are manufactured by the degradation of ordinary starch molecules to yield manufactured goods of much lesser viscosity in comparison to raw untreated starches. These converted starches present a broad collection of practical properties, for instance, changes in solubility, gelling capacity and gel strength, and viscosity. As a result of such conversion, and at relatively high solids concentration, the viscosity of the solution is within workable limits and makes processing easier. Conversion means breaking, rearranging and/or recombining starch molecule chains. Transformation of functional groups into other kinds of functional groups is also an option. Conversion can be carried out by agents such as heat, oxidizing agents, enzymes, alkalis, and acids. Products can be prepared by treating aqueous slurries or dry starch granules. Through the conversion, the granule's structure grows weaker and disintegrates and this results in lower viscosity in the relevant products. Managed acid hydrolysis of raw starch could be valuable toward getting thin boiling or fluid starches in an extensive variety of viscosities. Thin-boiling starches have high gel strengths and good film-forming properties that are valuable in the production of gum candies.

Dextrins can be manufactured with different properties. In 1821, it was observed that starch turns brown upon heating and that this product dissolves easily in water to manufacture a viscous adhesive paste. Variations in roasting time and temperature and the use of minute amounts of acid in the roasting process explain the differences among different types. Dextrins with low to intermediate viscosity, good stability against gelling, and incomplete or complete cold-water solubility can be derived through different treatments. Three categories of chemical transformations prevail in the degradation–recombination reaction, specifically: hydrolysis of linkages between glucose units, rearrangement of the molecules to obtain increased branching, and repolymerization of minute fragments into larger molecules by catalytic action. The obtained products are determined by the extent of these three reactions. Common types are white, yellow, or other. White dextrins are manufactured at temperatures below ~150°C in the presence of moisture and high acidity. Their molecular weights range from comparatively high to a degree of polymerization of 20 glucose units or less per chain. White dextrins are soluble in cold water but require cooking to disperse. They set to a soft paste upon cooling and can be utilized in binders or adhesives. Yellow dextrins are produced by heating raw starches at temperatures of ~150–190°C with low moisture content and moderate acidity. The reaction results in highly branched molecules that do not retrograde effortlessly and thus have exceptional stability. Yellow dextrins have relatively low molecular weight with a degree of polymerization of ~20–50 glucose units. They are the most highly converted starches and are very soluble in cold water, producing solutions with low viscosities. As a result, they can be concentrated to 50–60%, are relatively fluid and may be used for adhesive applications. Other dextrins can be produced by bacterial and enzymatic action on starch. For example, with the help of amylase action on starch, various types of dextrin can be produced.

A starch derivative is formed when the chemical structure of some of the glucose units of the starch is chemically modified. The definition includes oxidized starches, not counting the previously produced and known hypochlorite-oxidized starch or the acid-modified starches. By derivatization, chemical or physical properties are modified to attain improved functionality under specific conditions. The derivatives of starch are characterized by the nature of their chemical modifications, their degree of substitution, degree of polymerization, and chemical form. Frequent forms of derivatization are substitution of hydrogen atoms (of hydroxyl groups) into starch esters and starch ethers; oxidation (at C-2, C-3, or C-6) to yield oxidized starches; attachment of polymer chains to starch molecules to get graft polymers, and cross-linking of starch by molecules with more than one reactive group to obtain intermolecular derivatives. Esterifying or etherifying some hydroxyl groups interferes with the parallel alignment of the linear amylose chains and thus the tendency to associate will be decreased or eliminated. Starch acetates are more stable than untreated starch and have fine hydrating capability and consistent viscosity in sol. These starches convey lower gelatinization temperatures than untreated starch. Special grades of such starches are used in processed foods for thickening purposes. Hydroxyethyl starch which has a degree of substitution ranging from 0.05 to 0.1 hydroxyethyl groups per anhydroglucose unit can be used in specific food applications. Carboxymethyl starch is dispersible in cold water, is an effective viscosity former, and can be used as an ice-cream stabilizer. Other ethers and esters are also available and further information about their virtues can be found elsewhere.

Commercial Applications of Starches

Starches have a remarkable variety of applications in foods. More than one type and variety of starch can serve a particular function and first choice depends on convenience, cost, and practical know-how. Thickening ability is starch's most significant property. Wheat starch and cornstarch give short-bodied and heavy pastes, whereas waxy corn and tapioca starches form long-bodied, cohesive, and stringy dispersions.

A common factor in thickening applications is that the utilized starch should provide a creamy and smooth texture and that retrogradation or gel creation is disadvantageous to the quality of the food. Amylose's tendency to gel can be decreased by combining tapioca or waxy corn starches or by plasticizing the starch with corn syrup where this is allowable. In general, tapioca and waxy corn starch do not need as much cooking and give notably clearer pastes than cornstarch due to their lower amylose content and the character of their linear fractions. Disadvantages of such applications are their cohesive stringiness, high peak viscosities, and tendency to thin out with continued cooking. In such cases, corn or wheat starches are added since their granules are more resistant to physical disintegration.

Starches are traditionally used in puddings. Such products can be manufactured from cornstarch that needs to be cooked over the inclusive gelatinization temperature range. After pouring into suitable containers, the gelatinized product cools and gels. Important factors are operating conditions and the type and concentration of the employed starch. To achieve a predetermined property, the use of starch blends or modified starches is required. For instant puddings, instructions for the consumer include mixing with water or cold milk to give a quick-setting, soft-gelled pudding. Pregelatinized and/or oxidized potato or tapioca starches are regularly used in this application. The main aim of the producers is to bring the instant-starch pudding texture as close as possible to that of the cooked-cornstarch product. Other puddings are Danish desserts (based on native starches or cross-linked noncereal starches), which resemble a gelled fruit juice or gelatin dessert, and aerated pudding which is composed of pregelatinized starch, often in combination with gelatin (**Chapter 9**).

The characteristic of pie filling is determined by the type of starch used. Modified cornstarches are used for fillings in Boston cream, coconut cream, and lemon cream pies. If acid is supposed to degrade the starch, as is the case in lemon pie fillings, encapsulation can be used to separate the acid from the other ingredients. Viscosity needs to be increased to decrease or eliminate fluid soaking into the crust or running out of the pie when it is cut. Starch used as a thickener in pie fillings must have sufficient body through the stages of processing, must not be degradable under acidic conditions, and must not set to a gel after cooking; the fluid should be clear, shiny, and flavorsome, the filling must not display syneresis upon standing and must be resistant to freeze thawing, in addition to being stable under storage at slightly above –17.8°C. For this application, cross-linked root surfaces, cross-linked waxy corn starch, cross-linked starch phosphates, and pregelatinized starches perform well.

Starch is used in the manufacture of confections. Jelly candies are manufactured by forming stiff starch gels enclosing high concentrations of sugar. A usual range of starch concentrations might be 7 to 12% and the gel's virtues are altered by sugar inclusion.

A preferred starch for such applications is a thin-boiling starch at the concentration necessary to produce the proper gel strength. The quality of the confection depends upon the viscosity of the gelatinized starch dispersion, the strength of the starch gel formed after cooking, the pH and the moisture content in the final product. Turkish delight is a gum confection that requires soft elastic sweetened and flavored starch gels. Raw thick-boiling cornstarch conveys suitable qualities to the product, as well as a high degree of water retention. Gumdrops are related to confections. They are produced from acid-modified cornstarch and contain a high dry-solids content, good water retention and clarity, as well as the desired firm and tender gel texture. In some cases, for the production of gumdrops, the traditionally used gum arabic (**Chapter 11**) is replaced with various starches to make the product cheaper. Gumdrops, gum slices, and jelly beans are produced with thin-boiling starches. The texture of the final product is influenced by the water content. Different textures, such as tough and chewy or soft, can be achieved. Qualities are influenced by the quality of the starch, cooking time and temperature, and starch-to-sugar ratio, among many other parameters. Starch can be used as the molding medium for numerous confections produced by casting in starch. Various candy confections are shaped by being deposited in starch. To cast starch jellies, the molding starch is dried and starch that has been reused a number of times acquires advanced molding properties owing to polishing, which makes the starch granules more spherical.

Starches are used as binders in various meat products. The aim of the effectual binder is to hold moisture during processing, which typically includes chopping, curing, smoking, cooking, and chilling, and eventually, storage. Another possible feature of the binder is its emulsification ability and the ability to hold well-mixed protein, fat, and moisture. The binder should also not be a source of off-flavors. Potato starch is regarded as a good binder for frankfurters and bologna. This is attributed to its low gelatinization temperature, which allows it to absorb water in the early stages of cooking and smoking. Full-fat and low-fat bolognas were compared to low-fat formulations supplemented with wheat flour (a standard binder) or pea starch. Fat reduction resulted in poor texture and binding, but this was overcome by binder addition. Use of pea starch and fiber restored texture-profile values of low-fat samples to full-fat levels. All pea ingredients lowered cooking and purge losses, and increased water-holding capacity compared to the low-fat product without binder. In general, pea starch and fiber performed as well as wheat flour in low-fat bologna, with little change in functionality and without compromising consumer acceptability. Turkey frankfurters with low levels of starch were more yellow, firmer, and not as cohesive as those with high starch levels. Frankfurters with high levels of water were softer, lighter in color, and juicier than those with less water. Increased levels of starch and water maximized organ meat/metallic flavor. For optimal sensory and physical attributes, the levels should be 2.3% starch and 33.6% added water. Partial replacement of fat with mixed konjac mannan–potato starch gels resulted in textural characteristics similar to those of reduced-fat frankfurters and an overall sensory quality comparable to that of reduced-fat or high-fat controls. Addition of mixed konjac mannan–potato starch gels to reduced-fat frankfurters not only provided the perception of a "healthy food," it also produced a product comparable to regular high-fat frankfurters in terms of textural and sensory traits. Therefore, it seems to be feasible and effective to incorporate mixed konjac mannan–potato starch gels at the tested levels to reduced-fat frankfurters for acceptable sensory merits with a reasonable shelf life. A similar effect can be obtained with tapioca flour or pregelatinized starches. Other starches (i.e., corn, wheat, or rice) are also utilized but do not perform as well due to poor absorption properties and high gelatinization temperatures.

Large quantities of starch are used in a variety of canned foods. Their virtues as stabilizers and thickening agents are much needed in baked beans, cream soups, pie fillings, and baby foods. When 5% (by weight) cornstarch in cream soups was replaced with varying levels of fenugreek gum (**Chapter 8**), 0.5% fenugreek gum gave the best taste and overall acceptability. Since canned foods are passed through heat processing for sterilization, the starches used in their formulations have to be previously heated or chemically modified and thus sterile. Special thermophile-free starches were invented for canned foods. Pregelatinized starches are used in bakery mixes. They increase water absorption of the batter and aid in the entrapment of air for production of moist baked goods using only moderate mixing. In the case of whole rye meal with low enzymatic activity, high bread volume is obtained when the pentosan fraction contains many water-extractable polymers, and there is an optimum ratio of water-extractable pentosans to starch, which should be in the range of 1:16 to 1:20. This is also accompanied by starch properties such as high amylose content, good swelling capacity (big swelling factor), and consequently high solubility and pasting temperature.

In cake frosting and fudge mixes, inclusion of pregelatinized starch offers tolerance to changeable additions of water, and simultaneously successfully restrains the growth of sugar crystals. Starches are also used as effective stabilizers in salad dressings and mayonnaise. To gain stability under low pH conditions, starches must be modified. Modified starches or mixtures of such (i.e., tapioca, waxy corn, and amioca) have effectively been used for such purposes. In frozen foods, such as the aforementioned pie fillings, stable thickening and reduced syneresis are achievable by using waxy-type starches. In addition, starch phosphates with a very low degree of substitution show resistance to freeze-thaw breakdown and are very effective as thickeners in frozen pies, freeze-thawed salad dressings, and gravies and sauces for frozen foods. In white sauce, the effect of the freeze/thaw cycle was lower in the native waxy cornstarch due to its lower amylose content, which reduced retrogradation. The texture of caramel sauces improved with an increase in the concentration of xanthan gum (**Chapter 15**), and such improvement was also observed in the sauces after storage. In a consumer survey of sauces prepared under laboratory conditions, those thickened with 0.3% potato starch and 0.02% xanthan gum received the highest score. Jackfruit seed starch is suitable as a thickener and stabilizer in chili sauce; this sauce showed the lowest serum separation and highest viscosity during storage compared with control chili sauce and sauce containing cornstarch. Dry starch is used in many applications. For example, in baking powder, starches are used as diluents for the acid and carbonate ingredients that must be kept dry and as separate as possible, even though they are mixed before their inclusion in a medium that includes water and that is going to be heated. The stability of baking powder depends largely on particle size. Moreover, cornstarch was observed to be more useful and less expensive than other starches for more efficient separation of the constituents in baking powder. Starch is also used as a carrier for organic peroxides, and as a dusting agent for baked goods and chewing gum. It can be used as a moisture-absorbing agent in condiments and powdered sugar. Starch flakes are used to entrap fat and protect it from oxidation. Starch is also utilized in the stabilization of dry milk products and pregelatinized starch has been found helpful in suspending cocoa solids in chocolate milk beverages. The viscosity of cocoa syrups thickened by a combination of starch and xanthan gum decreased with increasing xanthan gum concentration. The decrease in pseudoplasticity and hardness with increasing xanthan gum supplementation was a result of high sugar level in the syrup. This phenomenon was a result of competition between sugars (mono-, di-, and oligosaccharides) and hydrocolloids for water molecules in the studied systems. Other applications include starch incorporation in glaze formulations, as part of a batter mix for deep-fried foods, to achieve aerated textures, and to improve egg products, as well as many other foods.

Recipes with Starch

Blanc Manger (Figure 14.1)

Ingredients

(Serves 5)
30 g (5 Tbsp) cornstarch
50 g (a little under ⅓ cup) white sugar
300 mL (1⅕ cup) milk

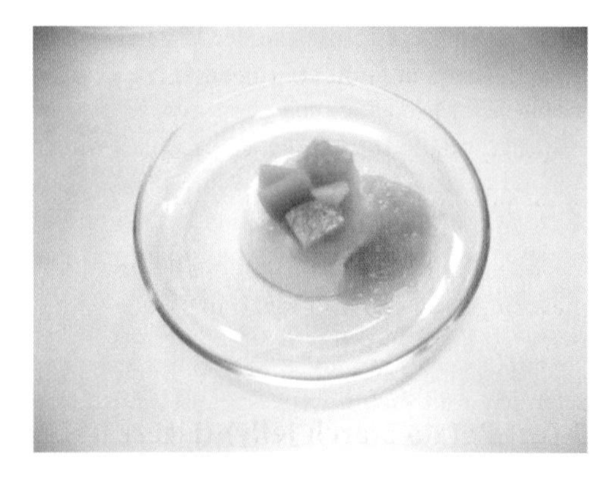

FIGURE 14.1 Blanc manger.

Vanilla flavoring

A {
50 g canned peaches
25 g canned syrup
2 g (½ tsp) rum
}

Preparation (Figure 14.2)

1. Mix cornstarch, white sugar, and milk in a pan, and heat with constant stirring with a spatula (**Figure 14.2A**) until it thickens (70–80°C) (**Figure 14.2B**) (**Hint 1**).
2. Reduce the heat, simmer for 2 minutes with stirring, remove from heat, and add vanilla flavoring.
3. Pour into molds (**Hint 2**), and cool to less than 20°C in a refrigerator or in an ice-water bath (**Figure 14.2C**) (**Hint 3**).
4. Make peach sauce using **Ingredients A**. Mix canned peaches, canned syrup, and rum in a blender for a few minutes.
5. Remove *blanc manger* (**3**) from the molds and garnish with peach sauce (**4**) (**Figure 14.1**).

pH = 6.46

FIGURE 14.2 Preparation of *blanc manger*. (**A**) Heat cornstarch, white sugar, and milk while stirring. (**B**) Continue until mixture thickens. (**C**) Pour into molds and cool.

Hints

1. The mixture can thicken suddenly, so it needs to be stirred constantly; otherwise it might burn.
2. Pour quickly because gelation occurs fairly rapidly.
3. The surface of the poured mixture should be flat.

Blanc manger is a sweet French dessert traditionally flavored with almonds. This recipe uses vanilla flavoring. You can also make the sauce with other fruits, such as kiwi, pineapple, strawberry, and so forth.

Warabi-Mochi (Sweet Potato Starch Jelly) (Figure 14.3)

Ingredients

(Serves 5)

75 g (a little over ½ cup) *warabi-mochi* powder (sweet potato starch)

300 mL (1⅕ cups) water

10 g (1 Tbsp) white sugar

10 g (1½ Tbsp) soybean flour

Preparation (Figure 14.4)

1. Mix *Warabi-mochi* powder and water in a pan, and heat with constant stirring with a spatula. When it starts boiling, turn the heat to low, and stir until it becomes transparent (at 80°C) (**Figure 14.4A**).

FIGURE 14.3 *Warabi-mochi* (sweet potato starch jelly).

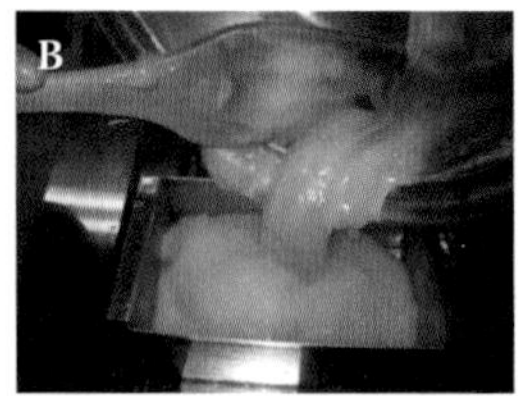

FIGURE 14.4 Preparation of *Warabi-mochi*. (**A**) Stir *Warabi-mochi* powder and water on low heat until mixture becomes transparent. (**B**) Pour into a moistened mold. (**C**) Cool.

2. Pour into a moistened rectangular mold (**Figure 14.4B**) (**Hint 1**), and cool it down to less than 20°C in the refrigerator or an ice-water bath (**Figure 14.4C**).
3. Mix white sugar and soybean flour.
4. Cut the *Warabi-mochi* (**2**) into bite-size pieces, put them on a dish or in a bowl, and sprinkle with the mixed white sugar and soybean flour (**3**) (**Figure 14.3**).

$$pH = 6.64$$

Hint

1. Any shape of mold will do.

Warabi-mochi is a traditional sweet Japanese dessert that is usually eaten in summer. Originally, bracken starch was used, but today, this starch is rare and expensive. Sweet potato starch is therefore used as a substitute. Sweet potato starch is also used as a substitute for *kudze* starch (**Supplemental L**).

Creamy Corn Soup (Figure 14.5)

Ingredients

(Serves 5)

150 g creamed sweet corn (canned)

30 g (a little over 3 Tbsp) all-purpose flour

FIGURE 14.5 Creamy corn soup. (**A**) Melt unsalted butter in a pot and stir in flour. (**B**) Pour soup stock, add sweet corn and milk, and season with salt and pepper. (**C**) Pour the soup into bowls, garnish, and serve.

20 g unsalted butter or margarine
500 mL (2 cups) soup stock
300 mL (1⅕ cups) milk
Salt, pepper, parsley, croutons

Preparation

1. Melt unsalted butter in a pan. Add all-purpose flour, and stir with a spatula (**Hint 1**) until it thins (at 70–80°C) (**Figure 14.5A**) (**Hint 2**).
2. Pour soup stock gradually into the pan (**Hint 3**).
3. Add sweet corn and milk, and season with salt and pepper (**Figure 14.5B**).
4. Serve the soup in a bowl, and garnish with finely chopped parsley and croutons (**Figure 14.5C**).

pH = 6.44

Hints

1. Be careful not to get burned.
2. At first the roux will thicken at around 60°C because the wheat starch gelatinizes and the starch granules swell; then the roux becomes thinner because the granules begin to fracture at over 70°C.
3. If you pour the soup stock in all at once, lumps will form.

This recipe is popular the world over. It is good for breakfast, lunch and dinner, and any time in between.

Sardine Meatball Soup (Figure 14.6)

Ingredients

(Serves 5)
150 g (5) sardines

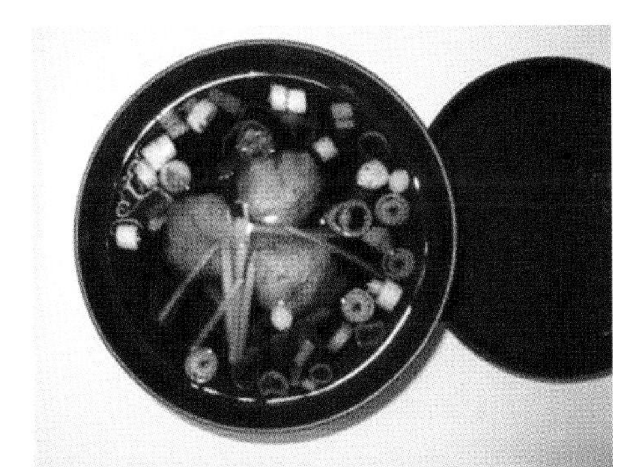

FIGURE 14.6 Sardine meatball soup.

FIGURE 14.7 Preparation of sardine meatball soup. (**A**) Remove sardine heads and debone. (**B**) Roughly chop sardines with a knife. (**C**) Grind. (**D**) Boil sardine meatballs.

15 g (a little under 1 Tbsp) *miso* paste
7.5 g (a little under 1 Tbsp) potato starch
5 g (a little over ⅔ tsp) grated ginger
750 mL (3 cups) water
25 g spring onion
25 g fresh ginger
Salt, soy sauce

Preparation (Figure 14.7)

1. Remove sardine heads and bones (**Figure 14.7A**) (**Hint 1**). Chop sardines roughly with a knife (**Figure 14.7B**), and grind in a blender or an earthenware mortar and pestle to make *surimi* (**Chapter 6**) (**Figure 14.7C**).
2. Add *miso* paste, potato starch, and grated ginger to the sardines and form meatballs.
3. Slice spring onion.
4. Cut fresh ginger into needle-like shapes, and soak in water (**Hint 2**).
5. Boil water in a pan, and put in sardine meatballs (**2**) (**Figure 14.7D**). When the meatballs float, season with salt and soy sauce, and add the sliced spring onion.
6. Pour the meatball soup into serving bowls, and garnish with the cut ginger (**Figure 14.6**).

pH of sardine meatball = 5.93

Hints

1. This is easily done by hand.
2. This removes the spiciness of the ginger.

This recipe can be made from *tofu* or potatoes instead of seafood. Sardines, prawns, or shrimp and squid (see below) are often used. Potato starch is used to make the *surimi* stick together.

Squid Meatball Soup (Figure 14.8)

Ingredients

(Serves 5)
100 g squid
1 g (1/6 tsp) salt
5 g (½ Tbsp) potato starch
15 g field peas
750 mL (3 cups) Japanese broth
Fresh ginger juice, salt, soy sauce, rind of *yuzu* (Japanese citrus fruit)†

†YOU CAN SUBSTITUTE ORANGE OR LIME RIND.

Preparation

1. Boil field peas, and cut thinly.
2. Slice the *yuzu* rind thinly.

FIGURE 14.8 Squid meatball soup. (**A**) Remove the internal organs and skin from the squid. (**B**) Chop finely with a knife. (**C**) Grind the chopped squid and season with salt. (**D**) Boil squid meatballs. (**E**) Put squid meatballs in serving bowls, add soup stock, field peas, rind of *yuzu*, and serve.

3. Remove the internal organs and skin from the squid (**Figure 14.8A**), and chop finely with a knife (**Figure 14.8B**).
4. Add 1 g salt to the chopped squid (**3**), and grind in a blender or food processor to make *surimi* (**Figure 14.8C**) (**Hint 1**), add fresh ginger juice and potato starch, and mix well. Then make meatballs (**Hint 2**).
5. Boil the squid meatballs (**Figure 14.8D**). When they float, take them out, and put them into serving bowls.
6. Boil Japanese broth, and season with salt and soy sauce.
7. Put the cut field peas in the serving bowls, pour in the broth, and garnish with sliced *yuzu* rind (**Figure 14.8E**).

pH of squid meatball = 6.23

Hints

1. It is harder to make squid *surimi* than sardine *surimi* because the stickiness of the protein in squid is weaker. If it is hard to make meatballs, add more potato starch.
2. The surface of the meatballs will not necessarily be smooth.

The crunchy texture of the squid and the smell of the *yuzu* rind make this recipe particularly appealing. Potato starch is also used to make the ground meat stick together in this recipe.

Goo Lou Yok (Sweet and Sour Pork) (Figure 14.9)

Ingredients

(Serves 5)

250 g pork thigh

A { 12 mL (⅔ Tbsp) soy sauce
12 mL (⅘ Tbsp) *sake*
5 g (1 tsp) fresh ginger juice }

23 g (2½ Tbsp) potato starch (or wheat flour)

100 g (1) carrot

100 g boiled bamboo shoots

200 g (1) onion

30 g *shiitake* mushroom

50 g bell pepper

10 g fresh ginger

10 g garlic

B
- 23 g (2½ Tbsp) white sugar
- 140 mL (2⅔ Tbsp) vinegar
- 113 mL (a little less than 1 Tbsp) sake
- 23 mL (1⅓ Tbsp) soy sauce
- 4 g (⅔ tsp) salt
- 150 mL (⅗ cup) Chinese soup stock

35 g (3 Tbsp) vegetable oil

C
- 12 g (1⅓ Tbsp) potato starch
- 30 mL (2 Tbsp) water

5 g (a little over 1 tsp) sesame oil

Oil for deep-frying

Preparation

1. Cut pork thigh into 2-cm cubes.
2. Mix **Ingredients A** in a bowl, and marinate the cut pork in it for 20 minutes.
3. Heat oil for deep-frying to 160°C; dust the marinated pork with 23 g potato starch, pat dry, and deep-fry for 5 minutes with turning.
4. Remove the deep-fried pork from the oil once, and deep-fry again at 180°C for 2 minutes (**Figure 14.9A**) (**Hint 1**).
5. Cut carrot into random shapes (**Figure 14.9B**), and boil until soft.
6. Cut boiled bamboo shoots into random shapes, cut onion into wedges, and thinly slice *shiitake* mushrooms (**Figure 14.9B**).
7. Slice bell pepper into quarters, and deep-fry at 140°C for 1 minute (**Figure 14.9C**).

FIGURE 14.9 *Goo lou yok* (sweet and sour pork). (**A**) Deep-fry the dusted pork twice. (**B**) Cut carrot, boiled bamboo shoots, and onion, and thinly slice mushroom. (**C**) Slice bell pepper and deep-fry. (**D**) Heat oil and stir-fry the chopped ginger and garlic. (**E**) Add boiled carrot, cut bamboo shoots, onion and mushroom, and deep-fried pork. (**F**) Add ingredients mix B to the stir-fried vegetables; when boiling, add fried bell pepper, ingredients mix C and sesame oil.

8. Chop fresh ginger and garlic finely.
9. Mix **Ingredients B** and **C**, respectively.
10. Heat vegetable oil in a frying pan, and stir-fry the chopped ginger and garlic (**8**) until they become odorous (**Figure 14.9D**); add the boiled carrot (**5**), the cut bamboo shoots, onion, and *shiitake* mushrooms (**6**), and the deep-fried pork (**4**) with stir-frying (**Figure 14.9E**) (**Hint 2**).
11. Add mixed **Ingredients B** to the stir-fry. Bring to boil and add the deep-fried bell pepper (**7**), pour mixed **Ingredients C** into the frying pan, and add sesame oil (**Figure 14.9F**).

pH = 3.85

Hints

1. Deep-frying twice produces a crispy texture.
2. Stir-frying leaves the vegetables with their crunchy texture.

This recipe is from China's Cantonese cuisine, and it is popular all over the world. Potato starch is used as a thickener for the sauce.

Mapo Tofu (Spicy *Tofu*) (Figure 14.10)

Ingredients

(Serves 5)
700 g (2 blocks) *tofu*

FIGURE 14.10 *Mapo tofu* (spicy *tofu*). (**A**) Heat oil, then stir-fry chopped garlic, ginger, ground pork, and *Toban jan*. (**B**) Season with white sugar, soy sauce and *sake*, add Chinese soup stock and cut *tofu*. (**C**) Thicken with potato starch and water and continue to stir with the addition of chopped spring onion and sesame oil. (**D**) Garnish with dried Szechuan pepper powder and the remaining spring onion.

30 g (2) spring onions
½ dried red pepper
15 g fresh ginger
10 g garlic
25 g (2 Tbsp) vegetable oil
200 g ground pork
10 g (1⅔ Tbsp) *Toban jan* or chili paste
3 g (1 tsp) white sugar
22 mL (1½ Tbsp) soy sauce
15 mL (1 Tbsp) *sake*
200 mL (⅘ cup) Chinese soup stock
A { 5 g (a little under 2 tsp) potato starch
15 mL (1 Tbsp) water }
2 g (½ tsp) sesame oil
Dried Szechuan pepper powder

Preparation

1. Wrap *tofu* in paper towels, and put it in a microwave oven (700 W) for 1–2 minutes to remove excess water (**Hint 1**), and cut it into 1.5-cm cubes.
2. Chop spring onion and red pepper roughly (**Hint 2**), and chop ginger and garlic finely.
3. Heat vegetable oil in a frying pan; stir-fry the chopped garlic, and add chopped ginger, ground pork and *Toban jan* in that order, and continue to stir-fry (**Figure 14.10A**).
4. Season with white sugar, soy sauce, and *sake*, add Chinese soup stock and the cut *tofu* (**1**) (**Figure 14.10B**), and simmer until *tofu* gets hot.
5. Add **Ingredients A** gradually to the frying pan, and heat until it thickens. Then, add half of the chopped spring onion (**2**) and sesame oil, and continue to stir-fry (**Figure 14.10C**), then remove from heat.
6. Garnish with dried Szechuan pepper powder and the remaining spring onion (**Figure 14.10D**).

pH = 5.75

Hints

1. Another way of removing water from *tofu* is to weigh down the *tofu*, boil it and squeeze dry with paper towels.
2. Kitchen scissors can be also used.

FIGURE 14.11 Deep-fried mackerel with vegetables.

This recipe is from China's Sichuan cuisine, and it is one of the best known Chinese dishes worldwide. Potato starch is also used as a thickener for the sauce.

Deep-Fried Mackerel with Vegetables (Figure 14.11)

Ingredients

(Serves 5)

600 g mackerel

A { 50 mL (1/5 cup) soy sauce; 30 mL (2 Tbsp) *sake* }

25 g (a little less than 3 Tbsp) potato starch

80 g carrot

30 g *shiitake* mushroom

60 g long onion

40 g boiled bamboo shoots

15 g green peas

50 g (1) egg

10 g (a little less than 1 Tbsp) lard

B { 200 mL (4/5 cup) Chinese soup stock; 24 mL (1 3/5 Tbsp) soy sauce; 20 mL (1 1/3 Tbsp) vinegar; 15 g (1 2/3 Tbsp) white sugar }

C { 10 g (a little over 1 Tbsp) potato starch; 10 mL (2/3 Tbsp) water }

Vegetable oil for deep-frying, vegetable oil, salt

Preparation (Figure 14.12)

1. Scrape off mackerel scales and remove the gills (**Figure 14.12A**); make a slit along the abdomen, and scoop out the innards with a knife (**Figure 14.12B**). Wash away any remaining innards, and dry with paper towels (**Figure 14.12C**). Make slits on both sides of the fish.
2. Mix **Ingredients A**, sprinkle over mackerel (**Figure 14.12D**), and let stand for 20–30 minutes.
3. Dust the seasoned mackerel with 25 g potato starch (**Figure 14.12E**); heat vegetable oil for deep-frying to 160°C, and deep-fry both sides of the mackerel by pouring oil onto the top side with a ladle until the fish becomes brown (**Figure 14.12F**); remove from oil. Raise the temperature to 180–190°C, and put the deep-fried mackerel in oil again for 2 minutes (**Figure 14.12G**).
4. Slice carrot, *shiitake* mushrooms, bamboo shoots, and long onion thinly (**Figure 14.12H**).
5. Boil green peas, and cool them in water (**Hint 1**).
6. Beat egg and a pinch of salt, heat vegetable oil in a rectangular frying pan, pour in the beaten egg, and make a paper-thin omelette (**Figure 14.12I**). Cool and cut into julienne strips (**Figure 14.12J**).

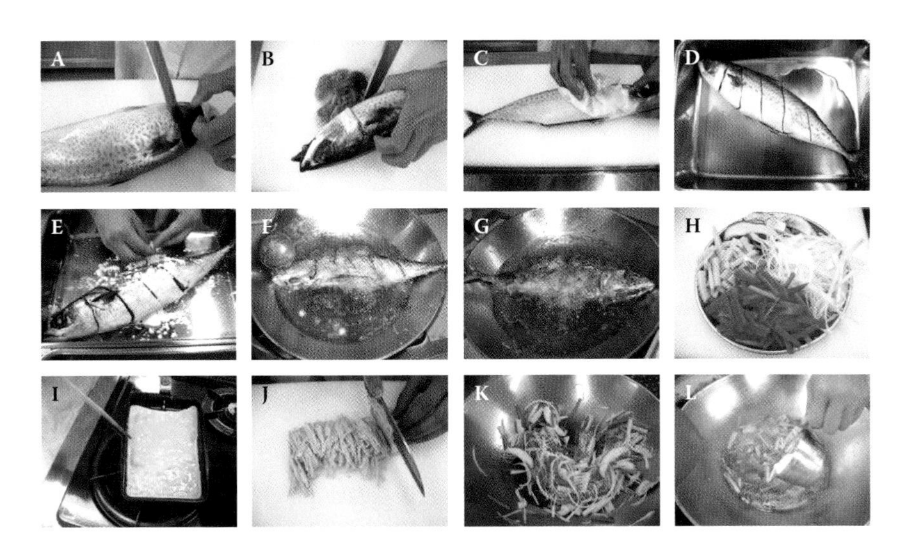

FIGURE 14.12 Preparation of deep-fried mackerel with vegetables. (**A**) Remove the gills. (**B**) Make a slit along the abdomen and scoop out the innards with a knife. (**C**) Dry with paper towels. (**D**) Make slits on both sides of the fish and sprinkle with spices. (**E**) Dust the seasoned mackerel with potato starch. (**F**) Deep-fry both sides of mackerel. (**G**) Refry the mackerel. (**H**) Slice carrot, *shiitake* mushrooms, boiled bamboo shoots, and onion. (**I**) Make a paper-thin omelette. (**J**) Cut it into julienne strips. (**K**) Stir-fry carrot, mushrooms, boiled bamboo shoots, and onion, and add seasoning. (**L**) Add potato starch and water to the vegetable sauce. (***Continued***)

FIGURE 14.12 (***Continued***) Preparation of deep-fried mackerel with vegetables. (**M**) Add the boiled green peas. (**N**) Pour the vegetable sauce over the deep-fried mackerel and garnish.

7. Mix **Ingredients B** and **Ingredients C**, respectively.
8. Heat lard in a frying pan, and stir-fry the sliced carrot, *shiitake* mushrooms, bamboo shoots, and long onion (**4**) (**Figure 14.12K**), and add mixed **Ingredients B**. When it boils, pour mixed **Ingredients C** into the vegetable sauce (**Figure 14.12L**). Then add the boiled green peas (**5**) (**Figure 14.12M**).
9. Pour the vegetable sauce over the deep-fried mackerel (**3**) (**Figure 14.12N**) (**Hint 2**), and garnish with julienned egg (**6**).

pH of vegetable sauce = 4.37

Hints

1. Blanching green vegetables gives them a bright green color.
2. Pour vegetable sauce quickly, because it is easier to pour when hot.

This recipe is also from Chinese cuisine; it was originally made with carp, but other fish, such as mackerel, horse mackerel, or sea bream are also good.

Tips for the Amateur Cook and Professional Chef

- Wheat starch only thickens during heat treatment; consequently, pregelatinized starches can be used to bind the limited available water early on.
- High-amylose starches require higher temperatures.
- Extremely high levels of dextrins or partially gelatinized, high-amylose starches may cause stickiness.
- Achieving the needed texture for desserts relies upon optimal cooking of the starch.
- Acidic products (i.e., $pH < 4.5$) will necessitate a higher degree of cross-linking than neutral products under the same heat treatment.
- Potato starch has the highest swelling capacity combined with a low gelatinization temperature.
- The heated dispersions of potato, tapioca, sweet potato, and sago starches are more transparent than those of other starches.
- Low-calorie or low-fat dressings benefit from hydroxypropylated starches.
- Adding salt (sodium) decreases the viscosity of the potato starch dispersion.

REFERENCES AND FURTHER READING

Aparicio-Saguilan, A., Sayago-Ayerdi, S. G., Vargas-Torres, A., Tovar, J., Ascencio-Otero, T. E., and L. A. Bello-Perez. 2007. Slowly digestible cookies prepared from resistant starch-rich lintnerized banana starch. *J. Food Comp. Anal.* 20:175–81.

Arocas, A., Sanz, T., and S. M. Fiszman. 2009. Influence of corn starch type in the rheological properties of a white sauce after heating and freezing. *Food Hydrocolloids* 23:901–7.

Baldwin, A. B. and W. Hach. 1949. Caramel-flavored starch powder. U.S. Patent 2,484,543.

Beggs, K. L. H., Bowers, J. A., and D. Brown. 1997. Sensory and physical characteristics of reduced-fat turkey frankfurters with modified corn starch and water. *J. Food Sci.* 62:1240–44.

Brautlechat, C. A. 1953. *Starch—Its sources, production and uses*. New York: Reinhold Publishing Corp.

Buksa, K., Nowotna, A., Praznik, W., Gambuś, H., Ziobro, R., and J. Krawontka. 2010. The role of pentosans and starch in baking of wholemeal rye bread. *Food Res. Int.* 43:2045–51.

Buleon, A., Colonna, P., Planchot, V., and B. Ball. 1998. Starch granules: Structure and biosynthesis. *Int. J. Biol. Macromol.* 23:85–112.

Bus, W. C., Muetgeert, J., and P. Hiemstra. 1956. Process for fractionating starch into components with branched and linear chains. U.S. Patent 2,914,410.

Debet, M. R. and M. J. Gidley. 2006. Three classes of starch granule swelling: Influence of surface proteins and lipids. *Carbohydr. Polym.* 64:452–65.

Hirashima, M., Takahashi, R., and K. Nishinari. 2012. The gelatinization and retrogradation of cornstarch gels in the presence of citric acid. *Food Hydrocolloids* 27:390–3.

Jobling, S. 2004. Improving starch for food and industrial applications. *Curr. Opin. Plant Biol.* 7:210–8.

Kao W. T. and K.W. Lin. 2006. Quality of reduced-fat frankfurter modified by konjac–starch mixed gels. *J. Food Sci.* 71:S326–32.

Kerr, R. W. 1950. *Chemistry and industry of starch*. 2nd ed. New York: Academic Press.

Kraybill, H. R. 1955. Sugars and other carbohydrates in meat processing. *Adv. Chem. Ser.* 12:83–8.

Krystyjan, M., Sikora, M., Adamczyk, G., and P. Tomasik. 2012. Caramel sauces thickened with combinations of potato starch and xanthan gum. *J. Food Eng.* 112:22–8

Lanciers, S., Mehta, D. I., Blecker, U., and E. Lebenthal. 1998. Modified food starches in baby foods. *Indian J. Pediatr.* 65:541–6.

Larch, H. W., McCowen, L. D., and T. J. Schoch. 1959. Structure of the starch granule. I. Swelling and solubility patterns of various starches. *Cereal Chem.* 36:534–44.

Minkema, W. H. 1964. Upgrades product and process. *Food Eng.* 36:94.

Moe, O. A. 1953. The physical and chemical characteristics of starch and its derivatives. In *Starch—Its sources, production and uses*, ed. C. A. Brautlecht, 354–89. New York: Reinhold.

Morris, E. 1994. Rheological and organoleptic properties of food hydrocolloids. In *Hydrocolloids, structures, properties and functions*, ed. K. Nishinari and F. Doi, 201–10. New York: Plenum Press.

National Starch Co. 1953. *The story of starches*. Plainfield, NJ: National Starch Co.

Pietrasik, Z. and J. A. M. Janz. 2010. Utilization of pea flour, starch-rich and fiber-rich fractions in low fat bologna. *Food Res. Int.* 43:602–8.

Radley, J. A. 1963. Technical properties of starch as a function of its structural chemistry. *Recent Adv. Food Sci.* 3:310–8.

Ravindran, G. and L. Matia-Merino. 2009. Starch–fenugreek (*Trigonella foenum-graecum* L.) polysaccharide interactions in pure and soup systems. *Food Hydrocolloids* 23:1047–53.

Rengsutthi, K. and S. Charoenrein. 2011. Physico-chemical properties of jackfruit seed starch (*Artocarpus heterophyllus*) and its application as a thickener and stabilizer in chilli sauce. *LWT-Food Sci. Technol.* 44:1309–13.

Rutgers, R. 1958. The chemistry of starch-milk puddings. II. Influence of cooking and cooling conditions and type of starch. *J. Sci. Food Agric.* 9:69–77.

Sandstedt, R. M. and R. C. Abbot. 1964. A comparison of methods for studying the course of starch gelatinization. *Cereal Sci. Today* 9:13–26.

Schoch, T. J. and A. L. Elder.1955. Starches in the food industry. *Adv. Chem. Ser.* 12:21–34.

Schopmeyer, H. H. 1947. Waxy starch granules. U.S. Patent 2,431,512.

Schopmeyer, H. H. and G. E. Felton. 1940. Thermophilic starch. U.S. Patent 2,218,221.

Sikora, M., Juszczak, L., Sady, M., and J. Krawontka. 2003. Use of starch/xanthan gum combinations as thickeners of cocoa syrups. *Nahrung/Food* 47:106–13.

Srichuwong, S. and J. L. Lane. 2007. Physiochemical properties of starch affected by molecular composition and structures, a review. *Food Sci. Biotechnol.* 16:663–74.

Stoloff, L. 1958. Polysaccharide hydrocolloids of commerce. *Adv. Carbohydr. Chem.* 13:265–87.

Stumbo, C. R. and R. Heinemann. 1966. Sour cream dressing. U.S. Patent 3,235,387.

Vandeputte, G. E. and J. A. Delcour. 2004. From sucrose to starch granule to starch physical behaviour: A focus on rice starch. *Carbohydr. Polym.* 58:245–66.

Verbeken, D., Thas, O., and K. Dewettinck. 2004. Textural properties of gelled dairy desserts containing κ-carrageenan and starch. *Food Hydrocolloids* 18:817–23.

Waldt, L. 1960. Pregelatinized starches for the food processor. *Food Technol.* 14:50–3.

Whistler, R. E. and E. F. Paschall. 1965. *Starch: Chemistry and technology, vol. I. Fundamental aspects*. New York: Academic Press.

Whistler, R. E. and E. F. Paschall. 1967. *Starch: Chemistry and technology, vol. II. Industrial aspects*. New York: Academic Press.

Wollermann, L. A. and E. W. Makstell. 1958. Properties of pregelatinized starches. *Cereal Sci. Today* 3:244–6.

15

Xanthan Gum

Introduction

Xanthan gum is manufactured by biotechnological processes. The hydrocolloid is produced by the bacteria *Xanthomonas campestris*, classified under the name B-1459, and can successfully replace other natural gums. Numerous species of *Xanthomonas* produce extracellular polysaccharides and ordinarily, extracellular polysaccharides are produced by various species of microorganisms. Following their manufacture, they do not form covalent bonds with the microorganism's cell walls, and are exuded into the culture media instead. Xanthan gum is manufactured in the United States (US), Europe, China, and Japan. The preferred production route is fermentation since it does not depend on variable factors, for instance weather conditions, and consequently a more consistent product is obtained, at a cost which is less subject to political or economic whim. The gum is regarded as a harmless food additive for, among other aims, thickening, when its use follows practical manufacturing practices. In the early 1960s, the Kelco Company in San Diego, California, started producing xanthan gum under the registered trademark Kelzan, and its use was permitted by the Food and Drug Administration (FDA) in 1969.

Processing

The manufacture of xanthan gum requires growing pure cultures of *X. campestris* by means of submerged aerobic fermentation within a sterilized medium containing carbohydrates, a nitrogen source, phosphates, and trace minerals. The process continues with incubation at 30°C for 3 days in a manufacturing-size fermenter, followed by a thermal treatment to eradicate viable microorganisms. This culture is then precipitated in isopropyl alcohol and the fibers are separated out by centrifugation, then dried, milled, sieved, and packaged. A technologically practical fermentation medium was formulated for *X. campestris* composed of glucose, sucrose, or sirodex A serving as the only carbon source, corn-steep liqueur as a combined nitrogen-phosphate foundation, and supplementary phosphate and citrate, depending on the application to which the gum is destined. Xanthan gum yields of 16.2, 15.1, and 16.5 g/kg were obtained after 96 hours fermentation on glucose (2%), sucrose (2%), or sirodex A (2.8%), respectively. Addition of citrate enhanced the pyruvate content of the xanthan gum, with a concomitant reduction in viscosity. Additional descriptions of medium formulations for xanthan gum production can be found elsewhere. The resultant products are regular grades of xanthan gum with a particle size of 80 mesh, a fine-mesh grade (not easy to disperse although it is quickly hydrated) with a particle size of 200 mesh, a granulated powder with advanced dispersibility, and a transparent xanthan gum powder for clear solutions at low concentrations. The gum powder contains a very small total bacterial count due to sterilization.

Chemical Structure

Xanthan gum is a microbial polysaccharide composed of a 1,4-linked β-D-glucose backbone, with side chains including two mannoses and one glucuronic acid. The pyruvic acid residues carried on half of the terminal mannose units correspond to ~60% of the molecule and provide the gum with many of its unique qualities, for example, its unusual opposition to hydrolysis and its consistent chemical and physical properties. Marketable xanthan gums have a substitution degree of ~30 to 40% for pyruvate and 60 to 70% for acetate, even though a considerable difference in these proportions occurs. The covalent structure of xanthan gum is associated with its purpose as a thickening, suspending, or gelling agent. The chief amino acids in its proteinaceous constituent are alanine, glutamic acid, aspartic acid, and glycine. Xanthan gum has a molecular weight of $\sim 2.5 \times 10^6$. Hydration in water is complete owing to its side chains. Xanthan gum's molecular conformation includes a helix having a pitch of 4.7 nm and stabilization by means of hydrogen bonds. It has been suggested that the xanthan gum macromolecule be regarded as a rigid helix in solution, while not discarding the possible presence of double or triple helices. Xanthan gum passes from a rigid ordered state to a more flexible disordered state under the influence of temperature. Xanthan gum can be in one of two ordered conformations: native and renatured. The latter is more highly viscous at an identical concentration, although its molecular weight is identical to that of the former. The transition from native to denatured state is irreversible, whereas the transition from denatured to renatured state is reversible. The transition temperature depends on gum concentration, ionic strength, and pyruvic and acetic acid contents of the xanthan gum. The transition can be checked by optical rotation, calorimetry, circular dichroism, or viscometry. The thermal transition at low concentrations, up to 0.3% in distilled water, usually takes place at near 40°C; however, the transition temperature can increase to >90°C when minute quantities of salt are incorporated.

Xanthan Gum Solutions

Xanthan gum is sold as a yellowish powder. The gum is soluble in cold or hot water. It dissolves readily in 8% solutions of acetic, nitric, and sulfuric acids, 10% hydrochloric acid and 25% phosphoric acid. Such solutions are stable at ambient temperature for more than a few months. Up to 50% solvent, such as ethanol and propylene glycol, can be included in aqueous xanthan gum solutions. Glycerol at 65°C can also be used to solubilize the gum. Xanthan gum's dissolution in water produces a very viscous and turbid solution. This solution shows pseudoplastic characteristics, stemming from both its conformation (rod-like) in solution and its high molecular weight. The viscosity of xanthan gum is dependent upon its concentration in the dispersion.

The relationship between xanthan gum concentration and viscosity is similar to that of other natural gums, for example, gum tragacanth (**Supplemental H**), guar gum (**Chapter 8**) and alginate (**Chapter 3**). The pseudoplasticity of xanthan gum solutions offers an advantage over other gums in that amplification in shear rate causes a decrease in viscosity values, making it easier to treat the food during processing. The pseudoplasticity is relevant for some foods' sensory qualities. A number of equations connecting apparent viscosity, concentration, shear rate, and temperature have been described. Upon degradation of the molecular structure of xanthan gum by high shear rates, a decrease in functional properties is expected.

Xanthan gum solutions show higher viscosities than other gums at similar low concentrations. Shear thinning of xanthan gum solutions is more pronounced than with other gums as a result of its semirigid conformation (as compared to the random coil conformation observed in other gums). The configuration of a weak network in solution results in high yield values. The interactions are shear reversible and as a result, temporary. At a 1% concentration in a 1% KCl solution, the yield values for xanthan gum, guar gum, hydroxymethycellulose (HMC) (**Chapter 5**), locust bean gum (LBG) (**Chapter 8**), carboxymethylcellulose (CMC) (**Chapter 5**), and sodium alginate (**Chapter 3**) are 11300, 4000, 830, 360, 410, and 210 mPa, respectively. Xanthan gum displays a yield value of 500 mPa at a concentration of 0.3%, while the other mentioned gums have no noteworthy yield values at this concentration. This feature accounts for xanthan gum's ability to stabilize dispersions and emulsions.

Stability in Different Media, under Different Technological Treatments

Due to its side chains, xanthan gum is fairly stable against freeze-thaw cycles, enzyme activity, extensive mixing, and degradation by acids, bases, and heat treatments. Xanthan gum solutions are stable over a broad range of pHs, being affected only at pHs >11 and <2.5. At ca. pH 9 or higher, xanthan gum progressively deacetylates. Its stability depends on hydrocolloid concentration: the higher the concentration the more stable the solution. High levels of strong oxidizing agents such as hydrogen peroxide, persulfates, and hypochlorite degrade gum solutions. Xanthan gums are well matched with elevated concentrations of a variety of salts. In the range of 10 to 90°C, xanthan gum viscosity is more or less unchanged in the presence of salts. Following sterilization (30 minutes at 120°C) of food items for consumption, including different gums, only 10% of the viscosity is lost in the case of xanthan gum as compared with other hydrocolloids, for example, guar gum (**Chapter 8**), alginate (**Chapter 3**), and CMC (**Chapter 5**). Enzymes such as amylase, pectinase, and cellulase, found intrinsically in the food or put in afterward, do not degrade xanthan gum. In microwaveable foods, the addition of xanthan gum eliminates moisture separation. Furthermore, at a concentration of 0.4%, the viscosity of xanthan gum is not influenced by electrolytes, while at 1% there is a major increase in viscosity. With sugar concentrations as high as 60%, xanthan gum hydration is unmodified. Interactions and precipitation can occur in acidic or heat-processed dairy-protein systems.

Solution Preparation

The manufacture of xanthan gum solutions relies on their dispersibility and solubility. Dispersion occurs when the gum particles are suspended in the liquid and detached from one another. Solubility is the process whereby particles swell and the fluid becomes viscous. Hydration relies on successful dispersion, and the size of the gum particles relative to other ingredients in the mix. Lumping can be prevented by high-shear mixing and bringing the gum powder (previously mixed with supplementary dry constituents of the mix if feasible) through the top of the vortex. In commercial preparations, dispersion funnels or cyclone chambers are employed to achieve quick dispersion.

Xanthan Gum Interactions

Xanthan gum can interact with other polysaccharides with synergistic consequences, for instance increased viscosity or gelation with galactomannans (**Chapter 8**) and glucomannans (**Chapter 12**). Such synergistic interactions can occur with guar gum (**Chapter 8**), LBG (**Chapter 8**), and tara gum (**Chapter 8**). Most significant are the guar– and LBG–xanthan gum interactions. LBG reacts more strongly because of its 4:1 proportion of mannose to galactose, as compared to the 2:1 mannose-to-galactose ratio in guar gum. The interaction occurs between xanthan gum molecules and the "smooth" region of galactomannans, explaining the strong interaction with LBG, which is not as branched as guar gum and has a more favorable galactose distribution. Xanthan gum also displays specific reactivity with tara gum. The scientific literature offers a number of thoughts on the function of pyruvate and acetate within the xanthan gum in the synergy.

Xanthan–guar gum interactions can amplify solution viscosity and elastic modulus. Salad dressings need a high elastic modulus for stabilization, whereas in sauces and soups, a high apparent viscosity is needed while a high degree of elasticity is undesirable. Maximal viscosity can be achieved for mixtures with a total gum content of 1.0%, composed of no more than 20% xanthan gum. The synergy decreases in the presence of salts, and is also influenced by gum concentration: the higher the latter, the stronger the synergy. Xanthan–guar gum blends improve guar gum's heat stability, yet xanthan gum on its own remains more thermally stable. Xanthan–guar gum mixtures have an extensive variety of applications owing to their fine sensory qualities and lower price compared to xanthan gum alone. Only a minute proportion of xanthan gum is needed to change guar gum's rheological properties, and this is why the mixture is extensively utilized in foods even though xanthan gum alone provides high-quality thermal stability and stabilizing properties.

Through xanthan gum–LBG interactions, at gum concentrations of >0.2%, strong and cohesive thermoreversible gels displaying very modest syneresis can be manufactured. The most favorable gum ratio is between 40:60 and 60:40. In view of the fact that the gel is elastic, cohesive, and nonbrittle, in terms of its sensory evaluation, the product is unattractive. On the other hand, these properties can be improved by a suitable addition of starch (**Chapter 14**) or protein. Such gels are especially utilized in pet foods. The effect of pH, addition of xanthan gum and presence of NaCl on the formation of whey protein isolate (**Supplemental G**)–xanthan gum gels was studied in the hope of obtaining a synergistic effect. A synergistic effect on gel strength was indeed observed at pH 6.5 and 6.0 for all concentrations of xanthan gum added. At these pHs, phase separation was also observed between the denatured whey proteins and the xanthan gum when heat was applied. A better knowledge of the microstructure of these systems and of the type of mixed gels obtained under different conditions (pH, ionic strength, etc.,) will allow for a more optimal use of mixed protein–polysaccharide systems in food formulations.

Interactions between xanthan gum–konjac mannan (glucomannan) (**Chapter 12**) are strong at concentrations higher than 0.2%. The most favorable ratio is between 40:60 and 30:70 (xanthan gum to konjac mannan). The presence of salt decreases the interaction. The highest gel strength is achieved at ~1.0% total gum concentration, with no additional increase in this property with increasing gum concentration. The mechanical properties of xanthan gum–LBG are also comparable to those of xanthan gum–konjac mannan. Minute quantities of xanthan gum are adequate for stabilizing starch solutions during storage. Xanthan gum at 0.1–0.2% prevents retrogradation of the starch dispersion and increases its stability, particularly at low pH (around 3). These small quantities of xanthan gum also improve starch stability during freeze-thaw cycles. Reactions of xanthan gum with specific dextrins (**Chapter 14**) and incompatibility between xanthan gum and gum arabic (**Chapter 11**) at pH below 5 have been reported. Reactivity is also noticeable with casein (**Supplemental G**) at lower than its isoelectric point. Interactions with proteins and other hydrocolloids are described elsewhere.

Information about the performance of starch pastes during frozen storage is required to understand more complex systems such as sauces, dressings, and desserts. Consequences of storage at subzero temperatures on the properties of 10% cornstarch pastes, manufactured in the presence or absence of 0.3% xanthan gum, were reported. Pastes were frozen at different rates (0.3–270 cm/h) and stored at –5°, –10°, and –20°C. Ice recrystallization, amylopectin retrogradation, syneresis, and rheological properties were examined. Samples stored at –5°C or at elevated temperature deteriorated as a result of amylose and amylopectin retrogradation and ice recrystallization. Starch retrogradation was not noticed in samples stored at –10 or –20°C, even though ice recrystallization was observed. The addition of xanthan gum improved the value of frozen pastes by preventing the formation of a spongy structure connected with amylose retrogradation and by preserving the smooth textural features of unfrozen samples. Nevertheless, the presence of xanthan gum did not prevent ice recrystallization or amylopectin retrogradation.

Food Applications

Once it was authorized as a food additive, xanthan gum found many applications in food processing, given that at small concentrations it offers storage stability, water-binding ability, and esthetic appeal. Xanthan gum is employed as a stabilizer for salad dressings because it imparts high yield values and strong pseudoplasticity. These, in turn, allow the suspension of spices, herbs, and vegetables in a dressing that adheres to the salad in addition to having the desired texture. Due to its stability, xanthan gum can be applied in products with low pH (~3.5 in some dressings) and high-salt content (>15%), or those that undergo thermal treatment. The consistent viscosity of xanthan gum from 5 to 70°C also assists in yielding a consistent texture and superior stability. The required quantity of xanthan gum depends on oil content, with lower oil inclusion requiring a higher amount of xanthan gum in the product. In fact, very similar flow properties can be attained with dissimilar oil levels by adjusting the gum level. Addition of xanthan gum at concentrations of up to 0.5% to citrus and other fruit-flavored beverages helps improve stability and mouthfeel. The heat tolerance and outstanding stabilization and suspension properties of xanthan gum are significant in canned foods. In foamed creams and mousses, the high yield values of xanthan assist in stabilizing air bubbles and whipping is performed more easily due to the gum's pseudoplasticity. In instant mixes such as breakfast drinks, soups, and low-calorie and aerated desserts, instant milkshakes and sauces, xanthan gum is employed as a thickener, suspension and/or body agent, at levels of ~0.1–0.2% by weight; in such beverages, xanthan gum blends with CMC (**Chapter 5**) or guar gum (**Chapter 8**) can also be used. Xanthan gum and LBG (**Chapter 8**) were added to a white sauce, and viscosity was decreased compared to the control sauce. All samples performed as weak gels, but the control samples were more structured. In these samples, an aqueous upper phase separated out during storage, while in samples containing hydrocolloids, oil, granules, and gum were also observed. Gum aggregates were concentrated locally in a continuous starch polymer network. The pseudoplastic performance of xanthan gum is vital in baked goods throughout the preparation stages of dough, that is, pumping, kneading and molding.

Xanthan gum prevents lump formation throughout kneading and improves dough consistency. It also enhances product volume and standardizes the distribution and size of the air cells in baked goods. Xanthan gum is employed in syrups as a particle suspender, and in toppings and fillings it improves texture, controls syneresis, and improves freeze-thaw stability. In frozen desserts and confections such as ice cream and sherbets, the inclusion of ~0.2% xanthan gum results in products with improved mouthfeel and heat-shock resistance. Ice cream is commonly stabilized by a mixture of xanthan gum, LBG, and guar gum at a final concentration of ~0.1%. Xanthan gum is also used to coat frozen confections. Ice-cream viscosity can be controlled with xanthan gum as well as with marketable mixes together with stabilizers based on mixtures of guar gum, LBG (**Chapter 8**) and CMC (**Chapter 5**). Xanthan gum has been used in frozen desserts to protect against heat shock and to control the development of a coarse, icy texture. Nevertheless, its functionality is limited in lower-pH manufactured milk goods since long, narrow fibers appear, resulting from a complex formed between xanthan gum and whey proteins. This reaction has been utilized to yield fat replacers for use in a number of foods. Stabilization of chocolate milk, yoghurt-based beverages, and low-fat spreads using galactomannans (**Chapter 8**), pectins (**Chapter 13**), alginates (**Chapter 3**), carrageenans (**Chapter 4**), and xanthan gum has been reported. The effect of xanthan gum and mixtures of it with other gums at different concentrations on the properties of yoghurt and soy yoghurt were also studied. The addition of xanthan gum or its mixtures had no noticeable effect on pH values, total solids content, or lactic acid bacterial counts in yoghurt and soy yoghurt samples when fresh or during storage. The viscosity values of cows' milk yoghurt or soy milk yoghurt throughout the fermentation period were noticeably increased when xanthan gum was added either by itself or in combination with other gums. Soy yoghurt treated with 0.005% xanthan gum exhibited high resistance to syneresis during storage. The microstructures of yoghurt with 0.01% and soy yoghurt with 0.005% xanthan gum showed the smallest decreases in the dimensions of the pores, as well as the dimensions of the casein (**Supplemental G**) particles. Yoghurt and soy yoghurt treated with 0.01 and 0.005% xanthan gum, respectively, gained the highest sensory scores of all treatments. Additional uses consist of stabilizing creamy cottage-cheese emulsions with a combination of xanthan gum–LBG at levels of

0.10–0.12%. A similar blend at ~0.5% has been described in the production of starch-jelly candies, to minimize their setting time. The inclusion of xanthan gum in meat batters, at levels of up to 1.0%, augments their consistency properties. Substitution of 50% of the normal fat content with gums allowed the production of cakes that were light with respect to their fat content. Those made with xanthan gum were considered better than controls with respect to overall appreciation and texture in the acceptability test. A comparison of the characteristics of gluten-free cakes prepared with rice and corn flours and with different concentrations of xanthan gum was performed. Cakes formulated with xanthan gum displayed improved quality characteristics, such as increased specific volume, enhanced texture in terms of decreased firmness, and delayed staling. Xanthan added to 0.3% and 0.4% of the formulated cakes presented desirable quality characteristics that resembled the physical, chemical, and sensory attributes of traditional cakes formulated with wheat flour. Xanthan gum also produces flexible tortillas because it retards retrogradation due to decreased water loss and increased starch gelatinization. Further utilizations of xanthan gum consist of the manufacture of pourable aerated dairy desserts and ready-to-serve frozen desserts.

Toxicity

The toxicity of xanthan gum was examined in experiments in which it was fed to rats and dogs. At 5%, little effect was noted in rats. When levels of 7.5 to 15% were included in the diet, diarrhea was noted but there was no verification of a pathological effect. Dogs were more sensitive to the dietary inclusion of xanthan gum. At 1%, no effect was noted but at 2%, persistent diarrhea occurred, but no unusual pathological findings were reported. For food-thickening purposes, less than 1% is used and therefore no toxic effects are expected. Clearance for the use of this gum in foods was obtained in 1969. Xanthan gum was approved for use in accordance with food manufacturing practice as a stabilizer, emulsifier, thickener, suspending agent, bodying agent, and foam stabilizer.

Recipes with Xanthan Gum

Italian Dressing (Figure 15.1)

Ingredients

(Makes 150 g)

72 g (⅓ cup) olive oil (**Hint 1**)

30 mL (2 Tbsp) vinegar†

15 mL (1 Tbsp) lemon juice

4.5 g (1½ tsp) white sugar

6 g (1 tsp) salt

0.5 g (¼ tsp) xanthan gum powder

30 mL (2 Tbsp) water
0.5 g (1 tsp) dried parsley or basil
0.5 g (⅕ tsp) celery seed
0.5 g (¼ tsp) garlic powder
Dried oregano, red pepper, black pepper

†YOU CAN USE ANY TYPE OF VINEGAR YOU LIKE.

Preparation

1. Mix xanthan gum, white sugar, and salt in a bowl, and stir in water, vinegar and lemon juice with a hand mixer (**Figure 15.1A**); mix in dried parsley, celery seed and garlic powder (**Figure 15.1B**).
2. Add olive oil to the mixture gradually with stirring (**Figure 15.1C**).
3. Add dried oregano, red pepper, and black pepper (optional) (**Figure 15.1D**) (**Hint 2**).

$$pH = 2.63$$

Hints

1. When making dressing with a high concentration of oil, add 0.2–0.3% xanthan gum, and when medium or low concentrations of oil are used, add 0.3–0.4% and 0.4–0.6% xanthan gum, respectively.
2. When the dressing contains extra virgin olive oil and is refrigerated, the oil phase may separate from the water phase. Before use, let the dressing stand at room temperature, or warm up the dressing a little, and shake it to emulsify.

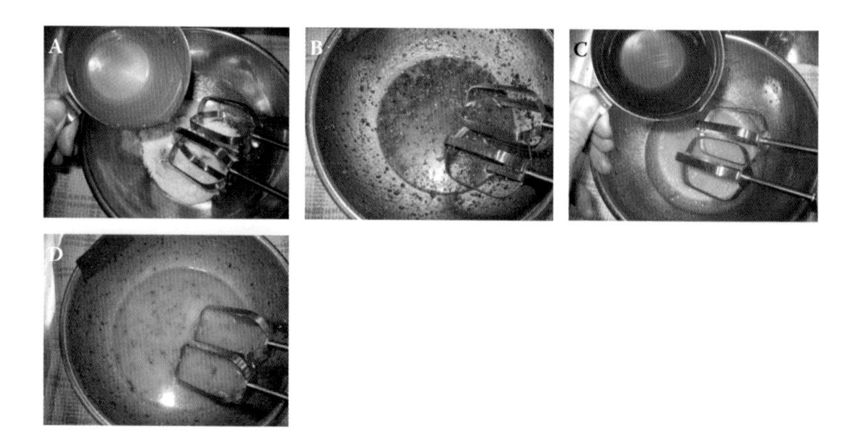

FIGURE 15.1 Italian dressing. (**A**) Mix xanthan gum, white sugar, and salt; stir in water, vinegar, and lemon juice. (**B**) Mix in dried parsley, celery seed, and garlic powder. (**C**) Add olive oil gradually to the mixture while stirring. (**D**) Add spices to the formulation (optional).

This recipe is good for any type of salad, and is as easy to make as French dressing (**Chapter 3**).

Creamy Italian Dressing (Figure 15.2)

Ingredients

(Makes 300 g)
190 g (a little under ¾ cup) mayonnaise
45 mL (3 Tbsp) red wine vinegar
9 g (1 Tbsp) white sugar
2 g (⅓ tsp) salt
0.7 g (⅓ tsp) xanthan gum powder
60 mL (4 Tbsp) water
1 g (½ tsp) garlic powder
0.5 g (⅕ tsp) black pepper
0.5 g (⅙ tsp) onion powder

Preparation

1. Mix xanthan gum, white sugar, and salt in a bowl, and stir in water and vinegar with a hand mixer (**Figure 15.2A**); stir in mayonnaise (**Figure 15.2B**).
2. Add garlic powder, black pepper, and onion powder to the dressing and mix (**Figure 15.2C**).

pH = 3.50

This recipe is good for any salad and also goes well with pasta.

FIGURE 15.2 Creamy Italian dressing. (**A**) Mix xanthan gum, white sugar, and salt, and stir in water and vinegar. (**B**) Stir in mayonnaise. (**C**) Add garlic powder, black pepper, and onion powder to the dressing and mix.

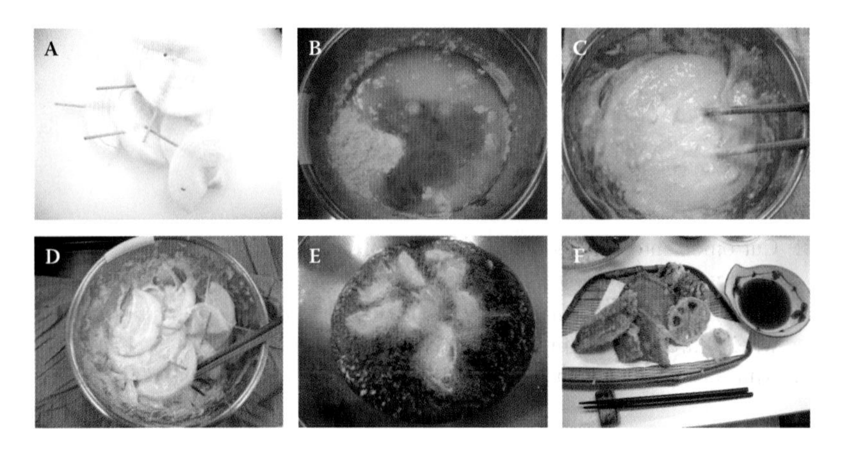

FIGURE 15.3 *Tempura*. (**A**) Cut onion into half rounds. (**B**) Add ice water and beaten egg to powder mixture of xanthan gum, wheat flour, wheat starch, and baking soda. (**C**) Stir briefly. (**D**) Coat each piece with the batter. (**E**) Deep-fry vegetable and fish pieces. (**F**) Garnish and serve with dipping sauce.

Tempura (Figure 15.3)

Ingredients

(Serves 4)

A:
- 1 egg
- 50 g (a little over ⅓ cup) weak flour
- 50 g (⅖ cup) wheat starch†
- 120 mL (a little under ½ cup) ice water (0–5°C)
- 0.5 g baking soda
- 3 g (1½ tsp) xanthan gum powder

4 fillets of white fish‡
200 g (1) onion‡
200 g (2) eggplants
80 g (¼) lotus root
120 g (¼) pumpkin

B:
- 160 mL (⅔ cup) Japanese broth
- 40 mL (2 Tbsp and ⅔ tsp) soy sauce
- 40 mL (2 Tbsp and ⅔ tsp) *mirin*

100 g Japanese radish
10 g ginger
Vegetable oil for deep-frying, weak flour

†**YOU CAN REPLACE WHEAT STARCH WITH OTHER STARCH OR WHEAT FLOUR.**

‡YOU CAN USE OTHER SEAFOOD, SUCH AS PRAWNS OR SQUID, AND OTHER VEGETABLES, SUCH AS BELL PEPPERS, MUSHROOMS, ASPARAGUS, SWEET POTATO, AND SO FORTH.

Preparation

1. Cut onion into rounds or half rounds (**Figure 15.3A**), cut eggplants into quarters, slice lotus root and pumpkin.
2. Make *tempura* batter using **Ingredients A**. Mix xanthan gum, wheat flour, wheat starch and baking soda in a bowl, and beat egg lightly.
3. Add ice water and the beaten egg to the powder mixture (**Figure 15.3B**), stir briefly with chopsticks or fork (**Figure 15.3C**) (**Hint 1**).
4. Heat vegetable oil for deep-frying to 160–180°C, and coat each vegetable and fish piece with the batter (**3**) (**Figure 15.3D**) (**Hint 2**).
5. Deep-fry the vegetable and fish pieces until the color changes at 160–180°C (**Figure 15.3E**) (**Hint 3**).
6. Make a dipping sauce using **Ingredients B**. Mix Japanese broth, soy sauce, and *mirin* in a pan, heat to boiling. Grate Japanese radish and ginger.
7. Garnish the *tempura* (**5**) with the grated Japanese radish and ginger, and serve alongside the dipping sauce (**6**) (**Figure 15.3F**).

pH of batter = 6.97

Hints

1. Do not stir vigorously and use cold water so that gluten does not form.
2. If the batter does not stick well to the vegetables or seafood, sprinkle light wheat flour over the pieces before you dip them in the batter.
3. Deep-fry vegetables and then fish to keep the smell and condition of the oil fresher for longer.

This is a famous Japanese recipe, similar to fritters. This is made throughout the year.

Chocolate Soufflé (Figure 15.4)

Ingredients

(Serves 5)
120 g (a little over ½ cup) granulated sugar
28 g (¼ cup) cocoa
23 g (a little over 2½ Tbsp) tapioca starch
23 g (a little under 4 Tbsp) cornstarch

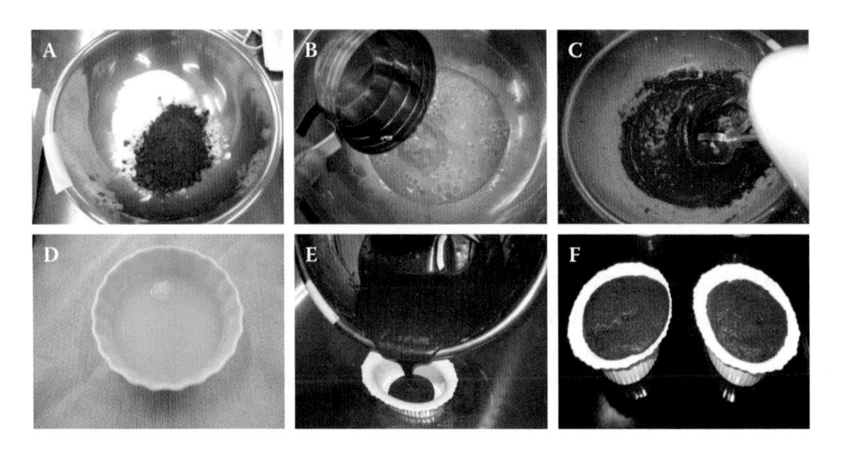

FIGURE 15.4 Chocolate soufflé. (**A**) Mix xanthan gum, granulated sugar, cocoa, tapioca starch, cornstarch, salt, and baking soda. (**B**) Beat eggs and mix in water. (**C**) Stir the powder mixture into the eggs and water. (**D**) Melt butter. (**E**) Pour the batter into cups. (**F**) Bake the soufflé.

1 g (½ tsp) xanthan gum powder
0.8 g salt (a pinch)
0.5 g baking soda
2 eggs
60 mL (4 Tbsp) water
40 g butter

Preparation

1. Mix xanthan gum, granulated sugar, cocoa, tapioca starch, cornstarch, salt, and baking soda in a bowl (**Figure 15.4A**) (**Hint 1**).
2. Beat eggs in a bowl, and mix in water (**Figure 15.4B**).
3. Stir the powder mixture (**1**) into the eggs and water (**2**) with a hand mixer for 3 minutes (**Figure 15.4C**).
4. Melt butter in a microwave oven (**Figure 15.4D**) and add it to the batter (**3**) (**Hint 2**) with stirring for 30 seconds.
5. Pour the batter into heat-resistant cups (**Figure 15.4E**).
6. Preheat the oven to 180°C, and bake soufflé for 25 minutes (**Figure 15.4F**).

pH of batter = 6.88

Hints

1. Commercial soufflé mix is made from wheat flour, baking soda, and xanthan gum.
2. Add the butter last so that the soufflé batter will be smooth.

This recipe can be made in a variety of flavors. You can make it with instant coffee, green tea powder, tea leaves, and so forth. It is a gluten-free food.

Tips for the Amateur Cook and Professional Chef

- Xanthan gum dissolves in hot or cold water to form viscous nonthixotropic solutions.
- Xanthan gum concentrations of 0.1–1.0% give a pH in the range of 6.0–7.0.
- Temperature has a very slight effect on xanthan gum solution viscosity.
- Xanthan gum solution viscosity is in essence independent of pH and the dried polymer powder can be dissolved directly in acid solutions.
- Aqueous solutions of 1% xanthan gum can produce unsupported films of good quality; inclusion of glycerol reduces their brittleness.
- Stable edible oil-in-water emulsions can be prepared by using as little as 0.05% xanthan gum.
- Stable cheese dressings and flavored oil emulsions that are stabilized by the inclusion of xanthan gum have improved freeze-thaw resistance.
- Xanthan gum is an effective foam stabilizer for beer and other malt beverages.
- Xanthan gum is useful as a thickening agent in pickled sauces and relishes.

REFERENCES AND FURTHER READING

Anderson, D. M. W. 1986. The amino acid components of some commercial gums. In *Gums and stabilisers for the food industry*, vol. 3, ed. G. O. Phillips, D. J. Wedlock, and P. A. Williams, 79–86. London: Elsevier Applied Science Publishers.

Anon. 1969. Xanthan gum. *Federal Register* 34:5376–7.

Bertrand, M.-E. and S. L. Turgeon. 2007. Improved gelling properties of whey protein isolate by addition of xanthan gum. *Food Hydrocolloids* 21:159–66.

Carnie, G. A. R. 1964. Evaluation of a new polysaccharide gum. *Australian J. Pharm.* 45:580–3.

Colarow, L., Dalan, E., and A. Kusy. 1994. Composition enabling proteins to remain heat-stable and product thus obtained. European Patent Application, EP 0,628,256 A1.

Cornan, J., Tsuchiya, H. M., and C. S. Stringer. 1956. Enzymatic preparation of high levulose food additives. U.S. Patent 2,742,365.

da Matta Jr., M. D., Sarmento, S. B. S., de Oliveira, L. M., and S. S. Zocchi. 2011. Mechanical properties of pea starch films associated with xanthan gum and glycerol. *Starch* 63:274–82.

de Vuyst, L. and A. Vermeire. 1994. Use of industrial medium components for xanthan production by *Xanthomonas campestris*. *Appl. Microbiol. Biotechnol.* 42:187–91.

Edlin, R. L. 1972. Method of producing a dehydrated food product. U.S. Patent 3,694,236.

El-Sayed, E. M., Abd El-Gawad, I. A., Murad, A. H., and S. H. Salah. 2002. Utilization of laboratory-produced xanthan gum in the manufacture of yogurt and soy yogurt. *Eur. Food Res. Technol.* 215:298–304.

Farkas, E. 1974. Method for preparing a low-calorie aerated structure. U.S. Patent 3,821,428.

Ferrero, C., Martino, M. N., and N. E. Zaritzky. 1994. Corn starch-xanthan gum interaction and its effect on the stability during storage of frozen gelatinized suspensions. *Starch* 46:300–8.

Foegeding, E. A. and S. R. Ramsey. 1987. Rheological and water-holding properties of gelled meat batters containing iota-carrageenan, kappa-carrageenan, or xanthan gum. *J. Food Sci.* 52:549–53.

Glicksman, M. and H. E. Farks. 1973. Pudding compositions. U.S. Patent 3,721,571.

Glicksman, M. and H. E. Farks. 1974. Gum gelling system: Xanthan-tara dessert gel. U.S. Patent 3,784,712.

Goff, H. D. and V. J. Davidson. 1994. Controlling the viscosity of ice cream mixes at pasteurization temperatures. *Modern Dairy* 7:12, 14.

Hart, B. 1988. Fixing formulations. *Food Proc.* (UK) 57:15–6, 21.

Jansson, P. E., Kenne, L., and E. B. Lindberg. 1977. Protein-lipid and protein-carbohydrate interactions in flour-water mixtures. *Carbohydr. Res.* 45:275–82.

Jeanes, A. R., Pittsley, J. E., and F. R. Samtis. 1961. Polysaccharide B-1459: A new hydrocolloid polyelectrolyte produced from glucose by bacterial fermentation. *J. Appl. Polymer Sci.* 5:519–26.

Jenkenson, T. J. and T. P. Williams. 1973. Gel-coated frozen confection. U.S. Patent 3,752,678.

Jordan, W. A. and W. H. Lester. 1973. Blends of xanthomonas and guar gum. U.S. Patent 3,765,950.

Katz, M. H. 1971. Edible mix composition for producing an aerated product. U.S. Patent 3,582,357.

Kelco Company. 1975. Xanthan gum/Keltral/Kelsan. In *Natural biopolysaccharides for scientific water control*, 2nd ed. San Diego: Kelco Co.

Kelco Company. 1992. Gum synergy K.O. competition. *Dairy Foods* 93:8, 78.

Kovacs, P. and K. S. Kang. 1977. In *Food colloids*, ed. H. P. Graham, 500–21. Westport, CT: AVI Publishing Co. Inc.

Kovacs, P. and B. D. Titlow. 1976. Stabilizing cottage cheese emulsions with a xanthan gum blend. *Am. Dairy Rev.* 38:34J–M.

Lilly, V. G., Wilson, H. A., and J. G. Leach. 1958. Bacterial polysaccharides. II. Laboratory-scale production of polysaccharides by species of *Xanthomonas*. *Appl. Microbiol.* 6:105–8.

Malone, M. J. and J. G. Sage. 1994. Ready to serve frozen dessert for soft-serve dispensing. U.S. Patent 5,328,710.

Mandala, I. G., Savvas, T. P., and A. E. Kostaropoulos. 2004. Xanthan and locust bean gum influence on the rheology and structure of a white model-sauce. *J. Food Eng.* 64:335–42.

McNeely, W. H. and K. S. Kang. 1973. Xanthan and some other biosynthetic gums. In *Industrial gums*, ed. R. L. Whistler and N. J. BeMiller, 473–521. New York: Academic Press Inc.

Melton, L. D., Mindt, L., Rees, D. A., and G. R. Sanderson. 1976. Covalent structure of the extracellular polysaccharide from *Xanthomonas campestris*: Evidence from partial hydrolysis studies. *Carbohydr. Res.* 46:245.

Milani, F. X. and R. L. Bradley Jr. 1992. Modification of induced complex formation between xanthan gum and whey proteins at reduced pH. *J. Dairy Sci.* 75:121.

Moorhouse, R. 1992. In *Thickening and gelling agents for food*, ed. A. Imeson, 202–26. Bishopbriggs, Glasgow: Blackie Academic & Professional, an imprint of Chapman and Hall.

Moorthy, P. R. S. and R. Balachandran. 1992. Certain non-conventional stabilizers—Their effect on whipping ability of ice cream mix. *Cheiron* 21:180–2.

Nussinovitch, A. 1997. *Hydrocolloid applications: Gum technology in the food and other industries*. London: Blackie Academic & Professional.

Nussinovitch, A. 2003. *Water soluble polymer applications in foods*. Oxford, UK: Blackwell Publishing.

Nussinovitch, A. 2010. *Polymer macro- and micro-gel beads: Fundamentals and applications*. New York: Springer.

Pettitt, D. J. 1982. Xanthan gum. In *Food Hydrocolloids*, vol. 1, ed. M. Glicksman, 127–49. Boca Raton: CRC Press Inc.

Preichardt, L. D., Vendruscolo, C. T., Gularte, M. A., and A. da S. Moreira. 2011. The role of xanthan gum in the quality of gluten free cakes: Improved bakery products for coeliac patients. *Int. J. Food Sci. Technol.* 46:2591–7.

Román-Brito, J. A., Agama-Acevedo, E., Méndez-Montealvo, G., and L. A. Bello- Perez. 2007. Textural studies of stored corn tortillas with added xanthan gum. *Cereal Chem.* 84:502–5.

Rossi, E. A., Faria, J. B., Borsato, D., and F. L. Baldochi. 1990. Optimization of a stabilizer system for a soya-whey yoghurt. *Alimentos e Nutricao* 2:83–92.

Sanderson, G. R. 1982. The interaction of xanthan gum in food systems. *Prog. Food Nutr. Sci.* 6:77–87.

Sandford, P. A. and J. Baired. 1983. Industrial utilization of polysaccharides. In *The Polysaccharides*, vol. 2, ed. G. O. Aspinalli, 411–90. New York: Academic Press.

Schmidt, K. A. and D. E. Smith. 1992. Milk reactivity of gum and milk protein solutions. *J. Dairy Sci.* 75:3290–5.

Schuppner, H. R. 1968. Non-alcoholic beverage. U.S. Patent 3,413,125.

Schuppner, H. R. 1971. Heat-reversible gel. U.S. Patent 3,507,664.

Shatwell, K. P., Sutherland, I. W., Ross-Murphy, S. B., and I. C. M. Dea. 1991. Influence of the acetyl substituent on the interaction of xanthan with plant polysaccharides I. Xanthan-locust bean gum systems. *Carbohydr. Polym.* 14:29–51.

Smith, I. H., Symes, K. C., Lawson, C. J., and E. R. Morris. 1981. Influence of the pyruvate content of xanthan on macromolecular association in solution. *Int. J. Biol. Macromol.* 3:129–34.

Symes, K. C. 1980. The relationship between the covalent structure of the *Xanthomonas* polysaccharide (xanthan) and its function as a thickening, suspending and gelling agent. *Food Chem.* 6:63–76.

Tako, M. 1991. Synergistic interaction between xanthan and tara-bean gum. *Carbohydr. Polym.* 16:239–52.

Thomas, A. W. and H. A. Murray. 1928. Gum arabic. *J. Phys. Chem.* 32:676–97.

Tilly, G. 1991. Stabilization of dairy products by hydrocolloids. In *Food Ingredients Europe: Conference proceedings*, 105–21.

Tilly, G. 1995. Stabilizer composition enabling the production of a pourable aerated dairy dessert. European Patent Application EP 0,649,599 A1.

Urlacher, B. and B. Dalbe. 1992. Xanthan gum. In *Thickening and gelling agents for food*, ed. A. Imeson, 206–226. Bishopbriggs, Glasgow: Blackie Academic & Professional, an imprint of Chapman and Hall.

Whiting, R. C. 1984. Addition of phosphates, proteins and gums to reduced-salt frankfurter batters. *J. Food Sci.* 49:1355–7.

Wilkenson, J. F. 1958. The extracellular polysaccharides of bacteria. *Bacteriol. Rev.* 22:46–113.

Yaki, K. and T. Okamura. 1969. Manufacturing method for ice cream and sherbet. Japanese Patent 44–10149.

Zambrano, F., Despinoy, P., Ormenese, R. C. S. C., and E. V. Faria. 2004. The use of guar and xanthan gums in the production of 'light' low fat cakes. *Int. J. Food Sci. Technol.* 39:959–66.

16

The Use of Multiple Hydrocolloids in Recipes

Synergistic Combinations

Mixtures of hydrocolloids can be employed to achieve novel and improved characteristics in a variety of food products. Sometimes, such combinations might lead to a reduction in cost. The synergy might be caused by the association of different hydrocolloid molecules or phase separation. In solutions that contain mixtures of hydrocolloids, precipitation or gelation might occur. Hydrocolloids with opposite charges associate and produce either soluble or insoluble complexes, depending upon the ionic strength of the solution, pH, and mixing ratio. If there is no association, at "low" concentrations hydrocolloids form a single homogeneous phase. At higher concentrations, the hydrocolloids separate into two liquid phases, with each phase being enriched in one of the hydrocolloids.

The phase separation creates a situation in which droplets enriched in one hydrocolloid are dispersed in a continuous phase that is enriched in the other. The relative concentration is determined by whether the hydrocolloid is present in the continuous or dispersed phase. When either one or both hydrocolloids are able to produce gels, gelation, and phase separation occur simultaneously and the resultant gel properties depend upon the relative rates of each process. Since such combinations lead to different gel textures, this is an important experimental and practical field.

Protein–Polysaccharide Interactions: Conjugates and Complexes

The effect of added and interacting polysaccharides on the stability of an emulsion can be described simply. A small amount of polymer is introduced to an initially stable emulsion. The addition leads to bridging flocculation. When the droplet interface is saturated by sufficient polymer, the emulsion is stabilized electrostatically and steric stabilization of droplets with saturated coverage of polysaccharide occurs. When the hydrocolloid concentration in the aqueous phase is high, the emulsion droplets are entrapped within a polymer gel network. Experimental evidence of protein–polysaccharide interactions at the oil–water interface can be provided by rheological determinations. The structure of the composite interfacial protein–polysaccharide layer depends to a certain extent on the preparation method. There are two approaches. In the first, after preparation of the biopolymer solutions, the protein–polysaccharide complex is added by homogenization. The second method involves the preparation of an emulsion with the protein as an emulsifying agent followed by the addition of the interactive polysaccharide to an aqueous phase to obtain an emulsion containing droplets coated by a protein–polysaccharide bilayer.

Applications

Locust Bean Gum (LBG) Interactions with Carrageenan and Xanthan

The galactomannan LBG (**Chapter 8**) has a level of substitution of one part mannose with four parts galactose. Carob gum has a "block" structure, that is, branching units clustered mainly in blocks of ~25 residues ("hairy" region), followed by even longer blocks of unsubstituted β(1, 6)-D-mannopyranosyl units ("smooth" region). The latter are important in forming interchain associations. Interactions of galactomannans with additional hydrocolloids can be utilized to construct rigid, thermally reversible gels, as is the case when hot 1:1 solutions of LBG and xanthan gum (**Chapter 15**) are mixed together and then cooled. The outcome, which cannot be attained using either hydrocolloid by itself, is most probably the consequence of junction-zone formation between the galactose-deficient segments of LBG and xanthan gum. In support of this presumed mechanism is the fact that an increase in galactose content results in a reduced interaction. The supermolecular features of xanthan gum–LBG gels were studied by electron microscopy and rheology. Results from the latter demonstrated that the xanthan gum–LBG network is formed from xanthan gum supermolecular strains, and that the addition of LBG does not influence xanthan gum structure. Observed structural features of the gels were independent of heat treatment and LBG fraction. Structural similarities and rheological differences observed between xanthan gum and LBG fractions were compared to existing interaction models at the molecular level.

Evidence for heterotypic binding in the synergistic gelation of LBG or konjac mannan (KM) (**Chapter 12**) with xanthan gum was reported. Gels incorporating KM showed evidence of structural rearrangement after their initial formation. No such effects were seen for LBG. In addition, the effects of LBG and KM on the conformation and rheology of agarose (**Chapter 2**) and κ-carrageenan (**Chapter 4**) were studied. It was concluded that LBG and KM chains promote conformational ordering of carrageenan and agarose by binding to the double helix as it forms. Comparable gel textures were obtained when LBG was mixed with κ-carrageenan to produce a strong elastic gel, in contrast to the common brittle texture. Carob gum is also capable of producing cohesive gels with carboxymethylcellulose (CMC) (**Chapter 5**) at a total gum concentration of 0.5%. Equal amounts of both gums led to maximal synergy. Thermodynamic incompatibility, when mixing proteins and polysaccharides above a critical concentration, could result in separation into two phases. This occurs at concentrations at which molecular overlap starts to occur and the two molecular species are no longer totally miscible. The two phases include one rich in protein and the other containing mostly polysaccharides. Phase separation is favored by reducing the solubility of the protein and increasing the intrinsic viscosity of the galactomannan.

Carrageenan and Protein Synergy

One of the features of carrageenans' reactivity with milk is gelation. This takes place when κ-casein (**Supplemental G**) and carrageenan interact to form a complex which aggregates into a 3D gel network. Specific interactions with κ-casein occur at pH values above the isoelectric point of the protein. Nonspecific interactions can take place below this point. Carrageenans may interact with αs1- and β-caseins (**Supplemental G**), but binding is much weaker than with κ-casein and does not result in gel formation. The electrostatic attraction flanked by negatively charged sulfate groups and the positively charged region in the peptide chain of κ-casein has been described, demonstrated, and discussed in many studies. κ-casein isolated from a milk-protein system occurs independently of the spherical submicelles which are ~20 nm in diameter. When carrageenan is present, the submicelles become aligned into thread-like structures ~20 nm in width. In comparison, the carrageenan chains are ~2 nm wide. The κ-casein particles are connected to the hydrocolloid chain by electrostatic attraction, first and foremost with κ- and ι-carrageenans. Those carrageenans (and not λ-carrageenan) are efficient for suspending cocoa particles in chocolate milk, possibly because formation of a gel network is essential, in parallel to the fact that suspension ability only holds for high-molecular-weight carrageenans. κ-casein composes 8–15% of the casein micelle. A fraction of the external surface of the micelle is exposed and can interact with carrageenan. The ultrastructure of carrageenan–milk sols and gels has been described. In the micelle, κ-casein protects the calcium-sensitive caseins from contact with Ca^{2+} ions, almost certainly protecting the αs1- and β-caseins. This defensive task is shared with the carrageenans. Electron-microscopy studies revealed a κ-carrageenan/αs-casein system in which αs-casein in aggregate sizes of 100 to 500 nm was entrapped by interconnecting strands of κ-carrageenan. Additional specific information on the interaction of sulfated polysaccharides with proteins can be located elsewhere. The option of no complex formation between αs-casein and carrageenan in the absence of Ca^{2+} was suggested following electrophoresis and sedimentation studies, and verified by physicochemical techniques. In fact, results suggested that Ca^{2+} ions serve as mediators in ionic interactions. Nevertheless, it was proposed that because the stability of the complex is independent of ionic strength, the cations neutralize charged ester sulfates of carrageenan and promote aggregation of the gum chains into a double-helix formation.

λ-carrageenan is not effective in the stabilization of αs-casein systems, possibly owing to its structure, although, alkaline modification of nongelling θ-carrageenan alters these circumstances, albeit not to the extent attained by κ-carrageenan. Additional proteins such as β-casein, para-casein and those from soy, peanut, cottonseed, and coconut can be stabilized against precipitation with Ca^{2+} by carrageenans. Studies of additional sulfated polysaccharides as stabilizers of αs-casein can be found elsewhere. Nonspecific interactions, such as the precipitation reaction of carrageenan with gelatin (**Chapter 9**) and the interaction between carrageenan and β-lactoglobulin (**Supplemental G**) have also been reported.

Recipes with Multiple Hydrocolloids

Coffee Jelly (Figure 16.1)

Ingredients

(Serves 5)
500 mL (2 cups) brewed coffee[†]
3 g (1½ tsp) xanthan gum powder
2 g (2 tsp) locust bean gum (LBG) powder
60 g (a little over ⅖ cup) granulated sugar
Light cream

[†]PREPARED INSTANT COFFEE CAN ALSO BE USED.

Preparation

1. Mix xanthan gum, LBG and granulated sugar in a pan (**Figure 16.1A**), stir in coffee with a whisk and heat until the powders dissolve (**Figure 16.1B**).
2. Pour the mixture into moistened molds (**Hint 1**), and cool them in a refrigerator or ice-water bath (**Figure 16.1C).**
3. Cut jelly into pieces and serve with light cream (**Figure 16.1D**).

pH = 4.76

Hint

1. You can use any mold shape.

This recipe is refreshing on hot summer days. You can serve it with condensed milk, whipped heavy cream, or ice cream instead of light cream.

FIGURE 16.1 Coffee jelly. (**A**) Mix xanthan gum, LBG, and granulated sugar. (**B**) Stir in coffee and heat the mixture until the powders dissolve. (**C**) Pour the mixture into molds and cool. (**D**) Serve with light cream.

Hot Savory Jelly (Figure 16.2)

Ingredients

(Serves 5)
3.5 g (1¾ tsp) κ-carrageenan powder
0.8 g (⅙ tsp) konjac flour (*seiko*)
0.2 g (1/10 tsp) xanthan gum powder
500 mL (2 cups) soup stock
Salt, black pepper

Preparation

1. Mix carrageenan and konjac in a bowl, and stir in soup stock with a hand mixer in several steps (**Figure 16.2A**) (**Hint 1**).
2. Heat the mixture while stirring with a hand mixer until it boils, season with salt and black pepper, and continue to boil for a few more minutes (**Figure 16.2B**).
3. Cool the dispersion down to 50°C, stir in xanthan gum with a hand mixer (**Figure 16.2C**) (**Hint 2**), and cool to 35–40°C (**Figure 16.2D,E**).

$$pH = 6.13$$

Hints

1. Stir vigorously and add soup stock little by little to prevent the formation of lumps.
2. Changes in the amount of xanthan gum can alter the viscosity or texture of the jelly.

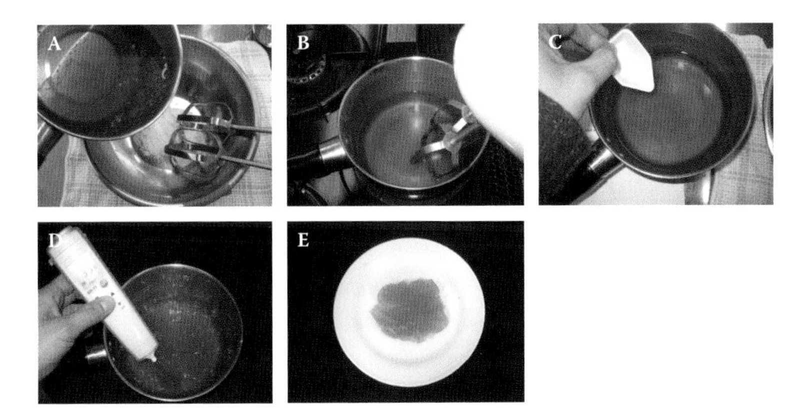

FIGURE 16.2 Hot savory jelly. (**A**) Gradually add soup stock to a mixture of carrageenan and konjac while stirring. (**B**) Heat the mixture while stirring until it boils. (**C**) Cool the dispersion and stir in xanthan gum. (**D**) and (**E**) Cool to 35–40°C.

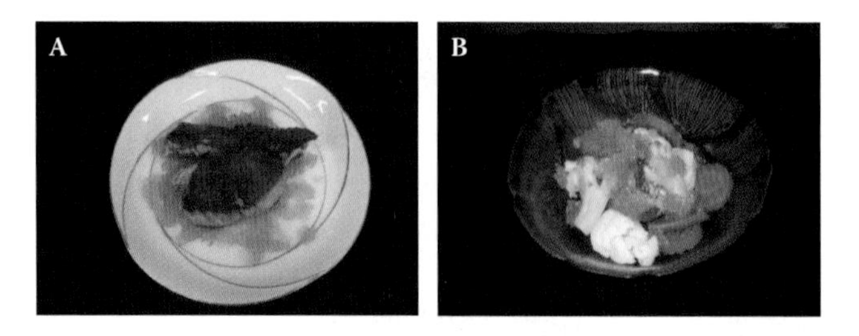

FIGURE 16.3 Hot savory jelly. (**A**) Served with hot seafood. (**B**) Served on hot vegetables.

This jelly is heat-resistant. It is suitable on hot seafood (**Figure 16.3A**), hot meats and hot vegetables (**Figure 16.3B**).

Dulce de Batata (Sweet Potato Jam) (Figure 16.4)

Ingredients

(Serves 5)
400 g sweet potato
180 g (1⅓ cups) white sugar
100 g (a little over ½ cup) corn syrup
1.5 g (1½ tsp) locust bean gum (LBG) powder
200 mL (⅘ cup) water
0.9 g (a little over ⅓ tsp) agar–agar powder
200 mL (⅘ cup) water
Vanilla flavoring

Preparation

1. Peel sweet potato and cut it into rounds; boil in enough water to cover until the sweet potato can be easily pierced with a fork (**Figure 16.4A**).
2. Strain the boiled sweet potato (**Figure 16.4B**) (**Hint 1**) and purée it (**Figure 16.4C**).
3. Mix white sugar and LBG in a bowl; add to 200 mL water with stirring with a whisk (**Figure 16.4D**) until sugar and LBG dissolve, then add corn syrup (**Figure 16.4E**).
4. Mix agar–agar and 200 mL water in a pan, and heat until agar–agar dissolves (**Figure 16.4F**).
5. Place the sweet potato purée (**2**) in another pan, add the sugar and LBG mixture (**3**) to it in a few steps while stirring with a whisk (**Figure 16.4G**), add the agar–agar solution (**4**) to it, and simmer with occasional stirring until it the mixture has thickened, about 1 hr. (**Figure 16.4H**).

FIGURE 16.4 *Dulce de batata* (sweet potato jam). (**A**) Boil cut sweet potato until soft. (**B**) Strain the boiled sweet potato. (**C**) Mix boiled sweet potato to a purée. (**D**) Add white sugar and LBG to water while stirring. (**E**) Add corn syrup to the sugar and LBG solution. (**F**) Mix agar–agar and water and heat until agar–agar dissolves. (**G**) Add the sugar and LBG to the sweet potato purée while stirring. (**H**) Add the agar–agar solution, stir, and simmer until the mixture thickens. (**I**) Pour the paste into molds and cool.

6. Remove the sweet potato paste from the heat and mix in vanilla flavoring.
7. Pour the sweet potato paste into molds, and cool in a refrigerator or ice-water bath (**Figure 16.4I**).

$$pH = 5.80$$

Hint

1. Strain the boiled sweet potato quickly before it becomes cold, because the cooled pectin in the potatoes will become sticky, and working with the potatoes will become difficult.

This recipe makes a traditional Argentinian, Paraguayan, and Uruguayan dessert, originally made using agar–agar. It is also good for spreading on bread.

Low-Fat Ice Cream (Figure 16.5)

Ingredients

(Serves 5)

0.2 g ($^{1}/_{10}$ tsp) κ-carrageenan powder

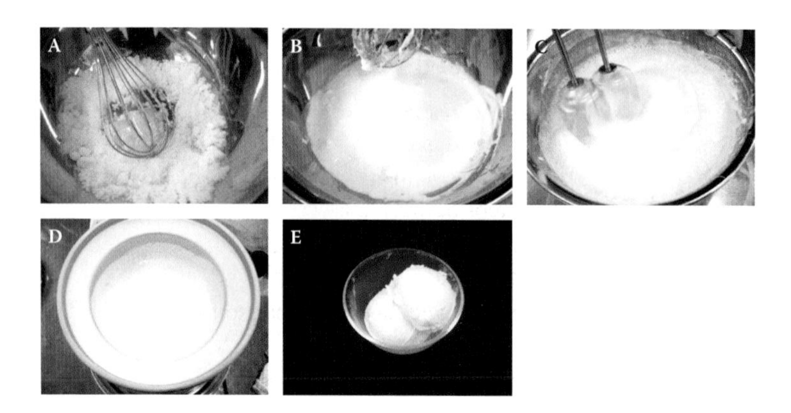

FIGURE 16.5 Low-fat ice cream. (**A**) Stir corn syrup into a mixture of carrageenan, xanthan gum, LBG, guar gum, and white sugar. (**B**) Stir in heavy cream and milk. (**C**) Gradually add lecithin and water while stirring. (**D**) Put the ice-cream mix in an ice-cream maker. (**E**) Serve.

0.5 g (¼ tsp) xanthan gum powder
0.3 g (⅓ tsp) locust bean gum (LBG) powder
0.2 g (1/15 tsp) guar gum powder
100 g (a little under ¾ cup) white sugar
35 g (a little over 1½ Tbsp) corn syrup
120 g (a little under ½ cup) heavy cream
200 mL (⅘ cup) milk
3 g (⅗ tsp) lecithin
450 mL (1⅘ cups) water
Vanilla flavoring

Preparation

1. Mix carrageenan, xanthan gum, LBG, guar gum, and white sugar in a bowl, and stir in corn syrup with a whisk (**Figure 16.5A**).
2. Add heavy cream and milk to the mixture, stirring with a hand mixer (**Figure 16.5B**); add lecithin and continue to stir for 5 minutes (**Hint 1**); gradually stir in water in several steps (**Figure 16.5C**), and add vanilla flavoring.
3. Put the ice-cream mix in an ice-cream maker (**Figure 16.5D**), and follow the manufacturer's instructions (**Figure 16.5E**) (see **Chapter 5**).

$$pH = 6.78$$

Hint

1. Stirring vigorously for a longer time will introduce more air bubbles and produce a softer texture.

This recipe contains more water than traditional ice cream, and it does not contain eggs, making it suitable for an egg-free diet.

Tips for the Amateur Cook and Professional Chef

- Hot solutions of κ-carrageenan and LBG form strong, elastic gels with low syneresis when they are cooled to below 50–60°C.
- Maximum interaction and maximal peak rupture strength occur at κ-carrageenan-to-LBG ratios between 60:40 and 40:60.
- Maximum interaction and maximal peak rupture strength occur at a xanthan gum-to-LBG ratio of 60:40.
- Konjac flour interacts strongly with κ-carrageenan to form cohesive elastic gels that are stronger than κ-carrageenan gels.
- Very low levels of 150–250 ppm carrageenan might be sufficient to prevent whey separation in various dairy products.
- Xanthan gum–LBG gels set and melt thermoreversibly at ~55–60°C.

REFERENCES AND FURTHER READING

Cairns, P., Miles, M. J., Morris, V. J., and G. J. Brownsey. 1987. X-ray fibre-diffraction studies of synergistic binary polysaccharide gels. *Carbohydrate Res.* 160:411–23.

Dea, I. C. M. 1979. Interactions of ordered polysaccharide structures synergism, and freeze-thaw phenomena. In *Polysaccharides in food*, ed. J. M. V. Blanshard and J. R. Mitchell, 229–40. London: Butterworths.

Dickinson, E. and K. Pawlowsky. 1997. Effect of i-carrageenan on flocculation, creaming and rheology of a protein-stabilized emulsion. *J. Agric. Food Chem.* 45:3799–806.

Fox, J. E. 1992. Seed gums. In *Thickening and gelling agents for food*, ed. A. Imeson, 153–70. Glasgow: Blackie Academic & Professional, an imprint of Chapman & Hall.

Glicksman, M. 1969. *Gum technology in the food industry*. New York and London: Academic Press.

Glicksman, M. 1983. *Food hydrocolloids*, vol. 1, 2 & 3. Boca Raton, FL: CRC Press Inc.

Goycoolea, F. M., Morris, E. R., and M. J. Gidley. 1995. Viscosity of galactomannans at alkaline and neutral pH: Evidence of 'hyperentanglement' in solution. *Carbohydr. Polym.* 27:69–71.

Goycoolea, F. M., Richardson, R. K., Morris, E. R., and M. J. Gidley. 1995. Stoichiometry and conformation of xanthan in synergistic gelation with locust bean gum or konjac glucomannan. *Macromolecules* 28:8308–20.

Goycoolea, F. M., Richardson, R. K., Morris, E. R., and M. J. Gidley. 1995. Effect of locust bean gum and konjac glucomannan on the conformation and rheology of agarose and κ-carrageenan. *Biopolymers* 36:643–58.

Guzey, D. and D. J. McClements. 2006. Formation, stability and properties of multilayer emulsions for application in the food industry. *Adv. Colloid Interface Sci.* 128–130:227–48.

Hansen, P. M. T. 1982. Hydrocolloid-protein interactions: Relationship to stabilization of fluid milk products. A review. In *Gums and stabilizers for the food industry*, ed. G. O. Phillips, D. J. Wedlock, and P. A. Williams, 127–38. Oxford: Pergamon Press.

Hidalgo, J. and P. M. T. Hansen. 1969. Interactions between food stabilizers and β-lactoglobulin. *Agric. Food Chem.* 17:1089–92.

Hollingworth, C. S. 2010. *Food hydrocolloids: Characteristics, properties and structures*. New York: Nova Science Publishers, Inc.

Hood, L. F. and J. E. Allen. 1977. Ultrastructure of carrageenan-milk sols and gels. *J. Food Sci.* 42:1062–5.

Kaletunc, G. and M. Peleg. 1986. Rheological characteristics of selected food gum mixtures in solution. *J. Texture Studies* 17:61–70.

Laaman, T. R. 2010. *Hydrocolloids in food processing*. Oxford: Wiley-Blackwell.

Lin, C. F. 1977. Interaction of sulfated polysaccharides with proteins. In *Food colloids*, ed. H. D. Graham, 320–46. Westport, CT: Avi Publishing Co.

Lundin, L. and A. M. Hermannson. 1995. Supermolecular aspects of xanthan-locust bean gum gels based on rheology and electron microscopy. *Carbohydr. Polym.* 26:129–40.

McCandless, E. L., and M. R. Gretz. 1984. Biochemical and immunochemical analysis of carrageenans of the Gigartinaceae and Phyllophoraceae. *Hydrobiologia* 116/117:175–8.

Nussinovitch, A. 1997. *Hydrocolloid applications, gum technology in the food and other industries*. London: Blackie Academic and Professional.

Nussinovitch, A. 2003. *Water soluble polymer applications in foods*. Oxford, UK: Blackwell Publishing.

Nussinovitch, A. 2010. *Plant gum exudates of the world sources, distribution, properties, and applications*. Boca Raton: CRC Press, Taylor and Francis Group.

Payens, T. A. J. 1972. Light scattering of protein reactivity of polysaccharides especially of carrageenans. *J. Dairy Sci.* 55:141–50.

Snoeren, T. H. M. 1976. κ-Carrageenan. A study on its physiochemical properties, sol-gel transition and interaction with milk proteins, Ph.D. Thesis, Nederlands Instituut voor Zuivelonderzoek, Ede, The Netherlands.

Snoeren, T. H. M., Payenes, T. A. J., Jeunink, J., and P. Both. 1975. Electrostatic interaction between κ-carrageenan and κ-casein. *Milchwissenschaft* 30:393–6.

Stainsby, G. 1980. Proteinaceous gelling systems and their complexes with polysaccharides. *Food Chem.* 6:3–14.

Tolstoguzov, V. B. 1991. Functional properties of food proteins and role of protein-polysaccharide interaction. *Food Hydrocolloids* 4:429–68.

Urlacher, B. and B. Dalbe. 1992. Xanthan gum. In *Thickening and gelling agents for food*, ed. A. Imeson, 164. London: Blackie Academic and Professional.

Glossary

Bamboo shoots: Bamboo sprouts or young shoots. They are eaten as vegetables in the spring in East Asia, Southeast Asia, and India.

Bazzar: Traditional North African spice blend. The main ingredients are 2 sticks cinnamon, 4 g dried chilies, 3 g cloves, 2 g turmeric, 2 g ginger powder, 4 g black pepper and 8 g cumin seeds. All spices are mixed, then ground in a coffee grinder. The powder keeps in a jar for several months.

Bittern: A coagulant for *tofu*. It is a ca. 0.4% magnesium chloride aqueous solution.

Blender: Electrically powered standing mixer. It is used to mix, purée, emulsify, and so on.

Brine: Solution of salt, sugar, and spices in which meat is soaked to produce ham or sausage.

Calcium hydroxide: $Ca(OH)_2$. This is widely used in the food industry, serving as, for example, a *tofu* coagulant, sugar-extraction agent, and so on.

Cassonade: Nonrefined sugar made from 100% sugarcane.

Celery seeds: Used as a spice or herb. It is also used as medicine for pain relief.

Chinese soup stock: There are many kinds of Chinese soup stock. The following is a recipe for simple clear soup stock.

Ingredients: 2400 mL water, 200 g lean pork, 200 g chicken with bone, 1 leek, 20 g ginger.

Preparation:

1. Cut leek into 10-cm long pieces; crush ginger with a knife.
2. Put water, pork, chicken, the cut leek, and crushed ginger in a pan, and simmer for 1–2 hours.
3. Strain the soup stock.

Cinnamon: A spice obtained from the dried inner bark of cinnamon trees, and used as a flavoring in foods.

Citric acid: An organic acid found mainly in citrus fruit, and various other kinds of fruits and vegetables.

Cup: In this book, 1 cup = 250 mL. The volume of a United States (US) cup is 240 mL, so if you are using a US cup you need to adjust the volume a little bit. Similarly, a Japanese cup = 200 mL, so 1 cup in this book is equivalent to 5/4 Japanese cups.

Dried bonito: A fermented fish product called *katsuobushi* in Japan. It is an extremely hard food, which is shaved into flakes for use. It has the taste of *umami* and is used in Japanese broth.

Ginkgo nuts: Fruit of the maidenhair tree. These nuts are surrounded by shells that must be cracked open. They are eaten in China and Japan in the autumn.

Glucose: A type of sugar and one of the monosaccharides, called *blood sugar*. It has 60–80% the sweetness of sucrose. It is the only energy source used by our brains.

Glycerol monostearate: A kind of emulsifier known as GMS.

Hand mixer: Electrically powered hand-held mixer. It usually has one or two beaters and is used to mix, stir, beat, whip, and so on.

Ice-cream maker: Machine to make small quantities of ice cream at home. It is also called an *ice-cream machine* or an *ice-cream freezer.*

Japanese broth: Kelp (*Laminaria* spp.) and dried bonito are usually used for making Japanese broth called *dashi*. The recipe follows.

Ingredients: 10 g kelp, 10 g dried bonito flakes, 1000 mL water.

Preparation:

1. Soak kelp in water in a pan for 30 minutes, and then heat to boiling.
2. Remove kelp from the pan, and add dried bonito; boil for 1–2 minutes and remove from heat.
3. Strain the broth.

Japanese pepper: Seedpod of the Japanese prickly ash. It is ground and used as pepper.

Judas ear: A kind of mushroom (Figure 7.4). It is sold as a dried product, and used in Chinese and Japanese cuisines. It has a gelatinous texture.

Lecithin: A phospholipid used as an emulsifier. Egg yolk and soybeans contain large amounts of lecithin.

Lily roots: Edible lily roots. These are crunchy, with a light and dry texture. They are used in Chinese and Japanese cuisines.

Mannitol: A sugar alcohol. This is reduced mannose; it is 60% as sweet as sucrose with 50% the calories.

Mirin: Sweet *sake* (alcohol) made from rice in Japan. It contains about 14% alcohol.

Nutmeg: A spice. It is a tree seed that is used as a flavoring agent, particularly with minced meats.

Okara: Soybean bran or pulp (also known as *tofu* lees). It is a byproduct of the *tofu*-making process. It is not tasty, but contains a great deal of fiber.

Piment d'Espelette: Sweet pepper cultivated in Espelette, France. It is used in traditional Basque dishes.

Potassium sorbate: Sorbic acid. It is used as a food preservative.

Red bean paste: A filling for Japanese sweets.

***Sake*:** Rice wine containing about 15% alcohol.

***Shiitake* mushrooms:** A representative of Japanese mushrooms. They have a distinct *umami* flavor.

***Shiratama* rice flour:** Waxy rice flour (**Figure 2.6**). It is used for Japanese sweets.

Shortening: A type of fat, which was developed as a substitute for lard.

Sodium monoglutamate: An *umami* seasoning. It is widely used in East Asian and Southeast Asian cuisines.

Sorbitol: A sugar alcohol. This is reduced glucose, and it is 60% as sweet as sucrose with 30% of the calories.

Sorghum: A kind of cereal crop, which originated in Africa. Today, there are many types of sorghum cultivated throughout the world. It is used in porridges and couscous.

Soup stock: Bouillon. There are many kinds of bouillon cubes or powders. A simple recipe for soup stock follows.

Ingredients: 2500 mL water, 200 g beef shin, 1 chicken bone, 100 g carrot, 100 g onion, 100 g celery, 1 bay leaf.

Preparation:

1. Cut beef shin, chicken bone, carrot, onion, and celery roughly.
2. Put the cut meat and vegetables, water, and bay leaf in a pan, and cook for 2 hours.
3. Strain the soup stock.

Soy sauce: A basic seasoning in Japanese dishes made from soybeans, wheat, and salt. Heavy soy sauce, which is termed *soy sauce* in this book, is the most widely used variant, although light soy sauce is also used in cooking.

Szechuan pepper: A kind of Chinese spice. The husk or hull around the seeds is usually consumed.

Tablespoon: 1 tablespoon = 15 mL. The abbreviation for tablespoon is "Tbsp."

Teaspoon: 1 teaspoon = 5 mL. The abbreviation for teaspoon is "tsp."

***Toban jan*:** A fermented seasoning in China. It is a spicy and salty paste.

Tomato purée: A type of processed tomato food. It usually consists of only tomato.

Trefoil: A Japanese herb used in many kinds of Japanese cuisines. This is a spring herb.

Trisodium citrate: Known as sour salt. Trisodium citrate is used as a flavoring agent and preservative.

Whisk: Hand-operated mixer. It is used to mix, stir, beat, whip, and so on.

***Yuzu*:** A Japanese citrus fruit. Its peel is used for zest. It has a distinct citrus flavor. It is a winter fruit.

Alphabetical List of Hydrocolloid Manufacturers and Suppliers

(Last checked on Jun 1, 2013)

ABHS Aboul-Hassanein
Products: gum arabic
E-mail: abhs.info@almahrigroup.com
http://www.almahrigroup.com/pages/abhs_us.htm

Ace Gum Industries Pvt Ltd.
Products: guar gum
http://www.acegumindustries.com/contactus.html
http://www.acegumindustries.com/

Agar del Pacifico S.A.
Products: agar
http://www.agarpac.com/

Agarmex S.A. DE C.V.
Products: agar and carrageenan
E-mail: info@agarmex.com
http://www.agarmex.com/

Algas Marinas S.A. (Algamar)
Products: agar
Phone: +56 (2) 205 5086
Fax: +56 (2) 205 5184

Algina Inversion SA
Products: carrageenan
http://www.gelymar.com/contacto_gerencia.htm
http://www.gelymar.com/

Altrafine Gums
Products: Altrafine gums
http://altrafine.com/contact_us.html
http://altrafine.com/

Andi Johnson

Products: konjac, agar, and carrageenan

http://andi-johnsongroup.com/

Arthur Branwell

Products: acacia gums, stabilizers for frozen foods, and ingredients for the meat industry

E-mail: gums@branwell.comhttp://www.branwell.com/

http://www.branwell.com/

Avebe

Products: solutions based on potato starch

E-mail: Info@AVEBE.COM

http://www.avebe.com/

B&V Agar Producers

Products: agar

http://www.agar.com/old/index.htm

BD Biosciences

Products: agar

http://www.bd.com/support/contact/international.asp

http://www.bd.com/

Beneo-Orafti

Products: inulin

http://www.beneo-orafti.com/

Bhansali International

Products: guar gum

http://www.bhansaligums.com/contact.aspx

http://www.bhansaligums.com/

Bunge Foods

Products: food and feed ingredients

E-mail: millingsales@bunge.com

E-mail: bna.hrbenefits@bunge.com

http://www.bungenorthamerica.com

Carbomer Inc.

Products: carbohydrate and polymer technology

http://www.carbomer.com/

Cargill Texturizing Solutions

Products: alginates, carrageenans, guar gum, locust bean gum, pectins, scleroglucan, xanthan gum, and starch and derivatives

https://www.contacts.cargilltexturizing.com/texturizing/emailtracker.nsf/ctsemailformExt?OpenForm

http://www.cargilltexturizing.com/

Carob, S.A. Balearic Islands (Spain)

Products: gums

E-mail: comercial@carob.es

http://www.carob.es/

Ceamsa

Products: carrageenan and pectin

E-mail: ceamsa@ceamsa.com

http://www.ceamsa.com/

Clariant

Products: food additives

http://www.clariant.com/corpnew/internet.nsF/vwWebPagesByID/27F67cF40ecc9de4c12579ecϕϕ2Fd929?OpenDocument&TableRow=3.2.1#3.2.

http://www.clariant.com

E-mail: info@clariant.com

Colloides Naturels International

Products: gum acacia

E-mail: information@cniworld.com

http://www.cniworld.com/

Continental Ingredients Canada Inc.

Products: design and application of functional food ingredient systems

http://www.cic-can.net/

E-mails: sales.support@cic-can.net
technical.support@cic-can.net
customer.support@cic-can.net

CP Kelco

Products: carrageenan, cellulose gum/CMC, diutan gum, gellan gum, microparticulated whey protein concentrate, pectin, refined locust bean gum, welan gum, and xanthan gum

http://www.cpkelco.com/

Daicel Co.

Products: cellulosics and CMC

http://www.daicel.com/en/contact/form.html

http://www.daicel.com/en

Daiwa Pharmaceutical Co., Ltd.

Products: arabinoxylan

http://www.daiwa-pharm.com/eng/index.html

Danisco

Products: alginate, carrageenan, cellulose gum, guar gum, locust bean gum, microcrystalline cellulose, pectin, and xanthan

http://www.danisco.com/contact

http://www.danisco.com

Dansa Food Processing Co., Ltd.

Products: gum arabic

Email: gme3209974@aol.com

http://www.dansagum.com/main.html

Dinesh Enterprises

Products: guar gum powder and guar gum splits

E-mail: info@dineshgum.com

http://www.dineshgum.com

Dow Wolff Cellulosics

Products: cellulosics

http://www.dow.com/dowwolff/en/contact/

http://www.dow.com/dowwolff/en/index.htm

DSP Gokyo Food & Chemical Co., Ltd.

Products: guar gum, xanthan gum, and xyloglucan

http://www.dsp-gokyo-fc.co.jp/ (in Japanese)

Eastman

Products: cellulosics

http://www.eastman.com

Est-Agar AS

Products: furcellaran

E-mail: info@estagar.ee, hugo@estagar.ee, tiit@estagar.ee

http://www.estagar.ee

Exandal

Products: tara gum
http://www.exandal.com/portal/en/contact-us
http://www.exandal.com

FMC

Products: hydrocolloid and cellulosics
http://www.fmcbiopolymer.com/Home.aspx

Fufeng Group

Products: xanthan gum and starch sweetener
http://www.fufeng-group.com/html/en/
http://www.fufeng-group.com/html/en/contact

Fuji Nihon Seito Co.

Products: inulin
http://www.fnsugar.co.jp/index.html (in Japanese)

Fuji Oil Co., Ltd.

Products: soluble soybean polysaccharide
http://www.fujioil.co.jp/fujioil_e/index.html

Gelymar SA

Products: carrageenan
http://www.gelymar.com/contacto_gerencia_ing.htm
http://www.gelymar.com/contacto_gerencia_ing.htm

Haji Dossa Nutralgum Ltd.

Products: guar
E-mail: faroukdossa@cyber.net.pk
http://www.guargum-hajidossa.com/index.html

Haresh Overseas Pvt. Ltd.

Products: guar gum
http://www.hareshoverseas.com/contactus.html
http://www.hareshoverseas.com

Hayashibara

Products: glucose obtained enzymatically from starch, and pullulan
http://www.intl.hayashibara.co.jp
Contact page: http://www.intl.hayashibara.co.jp/contact_en.php

HBGum Ind Pvt Ltd.

Products: guar

E-mail: deepakh@icenet.net, hbgum@icenet.net

http://www.hbgum.com/

Hebei Xinhe

Products: gums

E-mail: sales@xinhe-biogums.com

http://www.xinhe-biogums.com

Herbstreith & Fox

Products: pectin

E-mail: info@herbstreith-fox.de

http://www.herbstreith-fox.de/

Hindustan Gum & Chemicals Ltd.

Products: oxidized guar gum powder, carboxymethyl derivative of guar gum, carboxymethyl TKP, cold water-soluble TKP, and carboxymethyl starch.

E-mail: hichem@vsnl.com

http://www.hindustangum.com/

Hispanagar

Products: agar and agarose

http://www.hispanagar.com/contact.htm

http://www.hispanagar.com

Huey Shyang

Products: agar and carrageenan

E-mail: vin@hueyshyang.com

http://www.21food.com/showroom/3258/aboutus/puning-huey-shyang-sea-weed-industrial-co.ltd.html

Iberagar S.A.

Products: agar

http://www.iberagar.com/contact.html

http://www.iberagar.com

Ina Food Industry Co., Ltd.

Products: agar, carrageenan, guar gum, and xanthan gum

E-mail: privacy@kantenpp.co.jp

www.kantenpp.co.jp

Industrias Roko, S.A.

Products: agar

http://www.rokoagar.com/contacto.php

http://www.rokoagar.com

Japan Corn Starch Co., Ltd.

Products: cornstarch, modified starch, and rice starch

http://www.nihon-cornstarch.com/english/tabid/141/Default.aspx?language=en-US

Jianlong Biochemical Co., Ltd.

Products: xanthan gum and cornstarch

E-mail: liangdm@jianlonggroup.com/Liang171717@163.com; jin@jianlonggroup.com

http://www.jianlonggroup.com/English/profile.asp

JRS

Products: cellulosics

E-mail: info@jrs.de

http://www.jrs.de/wEnglisch/index.shtml

Jungbunzlauer International AG

Products: xanthan gum

http://www.jungbunzlauer.com/contact/contact-us.html

http://www.jungbunzlauer.com

Kewpie Egg Co.

Products: frozen egg, egg yolk and egg white, and dried egg, egg yolk and egg white

http://www.kewpie.co.jp/english/our_business/index04.html

Khartoum Gum Arabic Processing

Products: gum arabic

E-mail: info@baghlaf.com

http://www.baghlaf.com/en/Companies.aspx?ID=24

Kimica Corporation

Products: alginate, chitin, and chitosan

http://www.kimica-alginate.com/contacts/

http://www.kimica-alginate.com

Kirin Kyowa Foods Co., Ltd.

Products: curdlan and polysaccharide thickener/stabilizer
http://www.kirinkyowa-foods.co.jp/english/

Kitozyme

Products: chitosan and chitin-glucan, as well as all of their derivatives
http://www.kitozyme.com/index.php?page=contact&hl=en_US
http://www.kitozyme.com

Lamberti Spa

Products: cellulose, guar, and starch derivatives
http://www.lamberti.com/contacts/contacts.cfm
http://www.lamberti.com

Larex

Products: larch arabinogalactan
http://www.larex.com/larex/welcome/contact_us.html
http://www.larex.com

LBG Sicilia

Products: locust bean gum
http://www.lbg.it/contact-us.html
http://www.lbg.it

Lubrizol

Products: food hydrocolloids
http://www.lubrizol.com/ContactUs.aspx
http://www.lubrizol.com

Lucid Colloids, Ltd.

Products: hydrocolloids
E-mail: admin@lucidgroup.com
http://www.lucidgroup.com

Lunawat Industries

Products: guar gum
http://lunawatindustries43808_ww.indiabizclub.com/

Marcel Carrageenan

Products: carrageenan products
http://www.marcelcarrageenan.com/contactus.html
http://www.marcelcarrageenan.com

Marokagar

Products: agar

Phone: +212 2 623 611

Fax: +212 2 614 895

Maruyama Shouten Co., Ltd.

Products: konjac powder and agar–agar powder

http://www.maruyama-syouten.com/(in Japanese)

Matsuki Agar-Agar Industrial Co., Ltd.

Products: agar

Phone: +81 (266) 724 121

Matsutani Chemical Industry Co., Ltd.

Products: tapioca starch and dextrin

http://www.matsutani.com/

http://www.matsutani.co.jp/

Meron Biopolymers

Products: agar and chitosan

E-mail: info@meron.com

Contact page: http://meronbiopolymers.com/contacts.aspx

Mingtai Chemical Co., Ltd.

Products: cellulosics

E-mail: sales@mingtai.com, cindy.yuan@mingtai.com, alice.lu@mingtai.com william.tseng@mingtai.com

http://www.mingtai.com

Mitsubishi Shoji Food Tech

Products list: http://www.mcft.co.jp/mcft-e/product/01.html

http://www.mcft.co.jp/mcft-e/company/04.html

E-mail: customer@mcft.co.jp

http://www.mcft.co.jp

MRC Polysaccharide Co., Ltd.

Products: carrageenan, guar gum, tara gum, and xyloglucan

http://www.mrc.co.jp/mrcps/(in Japanese)

Myeong Shin Chemical Ind. Co., Ltd.

Products: agar and carrageenan

http://www.miryangagaragar.com

Natural Colloids Food Gums

Products: food gum and stabilizer systems
http://www.natural-colloids.com/contactus.html
http://www.natural-colloids.com/index.php

New Zealand Coast Biologicals, Ltd.

Products: agar
http://www.coastbio.co.nz/Contact_Us.htm
http://www.coastbio.co.nz

Nitta Gelatin, Inc.

Products: gelatin
http://www.nitta-gelatin.co.jp/english/

Norevo

Products: natural hydrocolloids, gum arabic, and agar–agar
E-mail: contact@norevo.de, a.benech@norevo.de
http://www.norevo.de

Norevo GmbH

Products: agar–agar
E-mail: contact@norevo.de
http://www.norevo.de/en/Home.html

NovaMatrix

Products: alginate, chitosan, and hyaluronan
http://www.novamatrix.biz/ContactUs.aspx
http://www.novamatrix.biz

Orafti

Products: inulin and oligofructose ingredients
http://www.orafti.com/contact
http://www.orafti.com

Pektowin Jaslo

Products: pectin
E-mail: marketing@pektowin.pl
http://www.pektowin.com.pl/

Penn Carbose

Products: cellulosics
http://www.penncarbose.com/message.asp
http://www.penncarbose.com/#

P L Thomas & Co.

Products list: http://www.plthomas.com/all-products
http://www.plthomas.com/combined.php
http://www.plthomas.com

Polygal

Products: carob, guar, tamarind, alginate, starch, and tara gum
http://www.polygal.ch/site/content.asp?typ=contact&lang=EN
http://www.polygal.ch/

Polypro International, Inc.

Products: guar
E-mail: polypro@polyprointl.com
http://www.polyprointl.com/index.html

Premcem Gums

Products: guar
Email: info@premcemgums.com, Chandresh.Joisher@btinternet.com
http://www.premcemgums.com/

Prodoctora de Agar S.A. (Proagar S.A.)

Products: agar
E-mail: gm@proagar.cl

Provisco AG

Products: xanthan, agar–agar, carrageenan, alginate, cellulose, pectin, and starch
http://www.provisco.ch/kontakt.php
http://www.provisco.ch
http://www.provisco.ch/english/produkte.html

P.T. Agarindo Bogatama

Products: agar–agar
http://www.swallow-globe.co.id/
http://www.swallow-globe.co.id/

P.T. Electrindodaya Agar

Products: agar
E-mail: elsa-agar@nusaweb.com

P.T. Surya Indoalgas

Products: agar
http://indoalgas.co.id/index.php/msg/cat/contact
http://indoalgas.co.id/
E-mail: sales@indoalgas.co.id

Qingdao Jiaonan Bright Moon Seaweed Industrial Co.

Products: alginate

http://yssi.en.ecplaza.net/

Quimica Amtex LTDA

Products: carboxymethylcellulose

E-mail: qac@amtex.com.co

http://www.amtex.com.co/

Rajasthan Gum Industries

Products: guar

http://www.tradeget.com/free_list/p37481/index.html

Rama Industries

Products: guar

http://ramagum.com/contact.aspx

http://ramagum.com

Rhodia

Products: xanthan

http://www.rhodia.com/en/info/contacts.tcm

http://www.rhodia.us/

Samsung

Products: cellulosics

http://anycoat.samsung-cellulose.com/ENG_DOCU/Contact/contact.asp

http://www.samsung-cellulose.com/

San-Ei Gen F.F.I., Inc.

Products: bacterial cellulose carrageenan, gellan gum, and xanthan gum

http://www.saneigenffi.co.jp/english/index.html

San-Ei Sucrochemical Co., Ltd.

Products: cornstarch and dextrin

http://www.sanei-toka.co.jp/(in Japanese)

Sanming Mingfu Agaragar Co., Ltd.

Products: agar

www.agar.cn

Sansho Co., Ltd.

Products: pectin, carrageenan, locust bean gum, and guar gum

http://www.sansho.co.jp/(in Japanese)

http://www.sansho.co.jp/profile-E.htm

Sanwa Cornstarch Co., Ltd.

Products: cornstarch and dextrin

http://www.sanwa-starch.com/english/index.html

Sensus

Products: inulin

E-mail: info@sensus.nl, contact@sensus.us,info.asia@sensus.nl

http://www.sensus.nl

Setexam SA

Products: agar

E-mail: setex@setexam.com

http://www.setexam.com/

Shandong Deosen Corporation

Products: xanthan

Phone: 86-533-7220838

Fax: 86-533-7216024

Shandong Gold Millet Biologicals

Products: xanthan

http://www.goldmillet.com/

Shanghai Beilian Foodstuff Co.

Products: konjac, locust bean gum, xanthan, and carrageenan

E-mail: info@sh-blg.com

Shemberg

Products: carrageenan

http://www.shembergcorp.com/#!contact

http://shembergcorp.com

Shimizu

Products: konjac glucomannan

http://www.shimizuchemical.co.jp/eng/contact/index.html

http://www.shimizuchemical.co.jp/

Shin Etsu

Products: cellulosics

http://www.shinetsu.co.jp/en/company/office.html

http://www.shinetsu.co.jp/en

Shiraimatsu Pharmaceutical Co., Ltd.

Products: sodium alginate

http://www.shiraimatsu.com/index.html (in Japanese)

Shree Nath Gum and Chemical

Products: guar gum split, guar gum powder, psyllium husk powder

http://www.shrinathgum.com/Contact-Us.aspx?SubjectID=Contact-Us

http://www.shrinathgum.com/

Shree Ram Gum and Chemicals

Products: organic guar gum

http://www.shreeramgum.com/contact.htm

http://www.shreeramgum.com/

Shubhlaxmi Industries

Products: guar

E-mail: info@shubhgum.net, info@shubhgum.com

http://www.shubhgum.com/

Sichem

Products: cellulosics

E-mail: info@sichemuae.ae

http://www.sichem.ae/

Simosis International

Products: gum karaya, gum arabic, tamarind kernel powder

E-mail: simosis@simosis.com, simosis9@gmail.com

http://www.simosis.com/

SNAP Natural & Alginate Prod., Ltd.

Products: alginic acid, carrageenan, and seaweed extract

http://www.snapalginate.com/contactus.html

http://www.snapalginate.com

Sobigel S.A.

Products: agar–agar

E-mail: info@sobigel.com

http://www.sobigel.com/

Soriano S.A.

Products: agar, carrageenan, and seaweed gums

http://www.soriano-sa.com.ar/

Sosa Ingredients s.l.

Products: agar and carrageenan

http://www.sosa.cat/?lang=en

Supreme Gums Pvt., Ltd.

Products: guar

http://supremegums.com/contact_us.html

http://supremegums.com

Tacara

Products: carrageenan

http://www.tacara.com.my/home.html

http://www.tacara.com.my/

Taiyo Kagaku Co., Ltd.

Products: egg (frozen, refrigerated, dried and concentrated)

http://www.taiyointernational.com/

Tate and Lyle

Products: locust bean gum

http://www.tateandlyle.com/contactus/Pages/Contactus.aspx

http://www.tateandlyle.com/Pages/default.aspx

TBK Manufacturing Corporation

Products: carrageenan

http://www.tbk.com.ph/ContactUs/tabid/59/Default.aspx

http://www.tbk.com.ph

The Nissin Oillio Group, Ltd.

Products: soybean protein

http://www.nisshin-oillio.com/english/company/index.shtml

Tic

Products: pectin: apple, citrus, grapefruit, and sugar beet, citrus fiber, specialty fiber for fat replacement, stabilizer systems, customized hydrocolloid blends, xanthan, guar, LBG, tragacanth, carrageenan, barley starch, specialty starch for mouthfeel in beverages

http://www.theingredients.co.uk/tic/Contact_Us.html

http://www.theingredients.co.uk

http://www.theingredients.co.uk/tic/product_alliances.html

TIC Gums

Products: gums: acacia, agar, carrageenan, guar, konjac, locust bean gum, pectin, tara gum

http://www.ticgums.com/contact.html

http://www.ticgums.com/

Tokai Denpun Co., Ltd.

Products: starch

http://www.tdc-net.co.jp/(in Japanese)

Toshniwal Associates PVT

Products: guar

Phone: 91-291-2577-144

Fax: 91-291-2577-144

Ugur Selluloz Kimya

Products: cellulosics

http://www.uskcmc.com/eng/index.htm

http://www.uskcmc.com/

Unipektin

Products: guar

http://www.unipektin.ch/kontakt/index.asp?pmid=76

http://www.unipektin.ch

Vikas Granaries Limited (Adarsh Derivatives Limited)

Products: guar gum

E-mail: investor.grevience@vikagranaries.in

http://vikasgranaries.in/

Vinyak Industries

Products: guar

http://vinayakgum.com/feedback.htm

http://vinayakgum.com

V V Eco Gum

Products: gums

Phone: +(91)-(22)-26360400/26363237

Fax: +(91)-(22)-26368419

Willpowder

Products: agar, carrageenan, gellan gum, methylcellulose, pectin, locust bean gum, tapioca starch, xanthan gum, guar gum

E-mail: spice@willpowder.com

http://www.willpowder.net/locustBeanGum.html

Wolff cellulosics GmbH & Co

Products: cellulosics

http://www.dow.com/about/contact/

http://www.wolff-cellulosics.de/

Yantai Andre Pectin Co., Ltd.

Products: pectin

http://www.andrepectin.com/contact.aspx

http://www.andrepectin.com

Yantai Xinwang Seaweed Industry Co., Ltd

Products: alginate

http://www.yantaimarket.com/en/companyhtml/109774.htm

Zibo Zhongxuan Biological Product Co, Ltd.

Products: xanthan

Phone: 0533-7220834 0533-7212171

Fax: 0533-7216024

Subject Index

A

Acacia senegal, 195–196
Acacia seyal, 196
Acid hydrolysis, 67, 244, 251
Acid processes
  agar extraction, 14
  gelatin manufacture, 161
  pectin extraction, 221, 222–224
  starch conversion, 224, 250–251
Acid stability
  carboxymethylcellulose, 86
  gel vs. sol state, 62
  gum arabic, 197
  gum tragacanth, 5
  pectins, 43, 221, 229
  xanthan gum, 273
Acyl groups, 179
Agar
  additives, 18
  as antistaling agent, 31
  applications, 13, 17, 18–21, 30, 31; *See also specific foods*
  bar form, 14–16, 21
  in carrageenan-based gels, 66
  cooking tips, 30–31
  as culture medium, 13
  defined, 13, 30
  extraction processes, 13–14
  flakes, 15–16
  gelation, 17, 18, 31
  and gellan gum, 182
  gel properties, 16, 17, 18, 30–31, 63
  and gum arabic, 7
  history of, 1, 13
  and konjac mannan, 209
  and locust bean gum, 7, 18–19, 31
  packaging, 14
  powdered, 14–16, 19, 26, 28–29
  products from, 14–15
  properties of, 16–18
  purity, 14
  recipes with, 21–30
  regulatory status, 16, 18
  seaweed sources, 3, 13, 18
  sols, 18
  solution properties, 16–17
  string form, 14–16
  structure, 16
  syneresis, 18
  terminology, 2
  toxicity studies, 16
Agaropectin, 14, 16
Agarose
  gelling temperature, 17
  interactions with other hydrocolloids, 288
  microparticles, 21
  physical gels, 17
  sols, 18
  stability, 19
  structure, 16
Alanine, 141, 161, 272
Albumin, 8, 9, 122–123, 162, 164, 165, 166
Alcaligenes faecalis var. *myxogenes* strain *10C3*, 101
Alcoholic beverages, *See also* Beer; Wine
  alginate beads, 43
  carboxymethylcellulose in, 82–83, 86
  gellan gum in, 183, 184
  gum arabic in, 199
Alcohol precipitation
  carrageenan extraction, 57, 59
  pectin manufacture, 224
Algae. *See* Seaweed and *individual algae*
Algin
  cross-linking reaction, 50
  in frozen foods, 40, 42–43, 51
  isolation of, 35
Alginate(s)
  applications, 5, 10, 19, 40–43, 85, 166, 183, 184, 199; *See also specific foods*
  calcium conversion and gel texture, 38
  cooking tips, 50–51
  degree of polymerization, 38
  enzyme stability, 50
  gelation, 38–40, 50
  gel properties, 38–40, 63
  G-type, 38
  history of, 35
  manufacture of, 36–37
  M-type, 38
  -pectin gels, 38–39, 229
  in pet foods, 183
  propylene glycol alginate production, 35–36

recipes with, 44–50
regulatory status, 36–37
solution properties, 37–38, 50
sources, 3, 35–36
structure, 36
thermostability, 43
world market, 3
Alginate beads, 39, 43
Alginate casings, 41–42
Alginate-high methoxy pectin gels, 38
Alginate-peach gels, 40
Alginate-phosphate, 43
Alginic acid, 35–37, 39
Alkaline stability, xanthan gum, 273
Alkali processes
agar extraction, 14, 15
alginate extraction, 35, 36
carrageenan, 63
gelatin manufacture, 162–163
konjac flour manufacture, 209–210
Alkaloids, fenugreek gum, 141
Alkoxylation reactions, 77–78, 79
Alkylation reactions, 77–78
All-purpose flour, 47, 202, 258, 259
Amidated pectin, 224–225, 227, 230
Amino acids
in collagen, 161
egg white, 120
in locust bean gum, 141
in xanthan gum, 272
Ammonium alginates, 37
Amorphophallus konjac, 207, 208
Amylase
and alginates, 50
dextrin production, 252
starch replacement, 145–146, 252
xanthan gum resistance, 273
Amylopectin
Dispersion, 247
from potato starch, 244, 250
retrogradation, 275
separation methods, 250
starch structure, 245, 247, 248
Amylose
Dispersion, 246, 247
in plants, 245
from potato starch, 244
retrogradation, 247, 252, 275
separation methods, 250
in starch gels, 248, 268
starch structure, 244–245, 248–249, 254–255
Angel food cakes, 121
Antimicrobial packaging, 42, 88
Antistaling agents, 31, 88, 146, 277
Anthocyanins, 228
Apple
cooked, 39
filling, 153
jam, 231
juice, 29, 87, 222
pectins from, 222–223
pomace, 221, 222–224
tannic acid in, 31
Arabinogalactan, 2–3, 196
Arabinose, 196, 222
Arako, 208
Arginine, 141
Arrowroot starch, 244
Arsenic, 16, 141, 164, 209
Ascophyllum nodosum, 35
Ascorbic acid, 41, 184
Ash content
agar, 14, 16
cellulose products, 78, 79
eggs, 119, 120
gelatin, 163, 164
gum arabic, 196
konjac flour, 208, 209
pectins, 225
tara gum, 141
Aspartic acid, 141, 272
Avidin, 120

B

Baby foods, 249, 254
Bacteria
and cellulose solutions, 84
and curdlan, 101
and dextrin, 252
and gellan gum, 179
and xanthan gum, 271–272
in gelatin gels, 164
in gum arabic, 196
in konjac flour, 209
in pectins, 225
as polysaccharide source, 179
poultry products, 42
Baked goods
agar in, 18–19, 31
alginates in, 40, 43
cellulose derivatives in, 86
furcellaran in, 64, 147
gelatin in, 164, 166–167
gellan gum in, 183
guar gum in, 146, 147
gum arabic in, 198
konjac flour in, 217–218
locust bean gum in, 146

pectins in, 229
xanthan gum in, 276–277
Bakery fillings
alginates in, 43
gelatin in, 164
gellan gum in, 185
pectins in, 229
Bakery toppings
agar in, 18–19
alginates in, 43
gelatin in, 164
xznthan in, 276
Baking powder, 47, 111,112, 250, 255
Bamboo shoots, 130, 262–264, 266–268, 297
Batters, 85, 146, 184, 277
Bazzar, 157, 297BBQ,
okara konjac, 215–216
sauce, 215
Beef products
agar in, 20
calcium-alginate films, 42
carrageenans in, 68
Beer
fining agents
gellan gum, 183
gum, 5
foam stabilizers
alginates, 43, 199
gum arabic, 199
xanthan gum, 283
Beverages
alginates in, 43
carboxymethylcellulose in, 85–86
gum arabic in, 197–198
gums in, 5
konjac mannan in, 209
pectin in, 229–230
xanthan gum in, 275–276
Binding agents
carrageenans, 64, 66, 68, 183
cellulose derivatives, 80, 85–87
curdlan, 103
dextrins, 251
galactomannans, 146, 158
gelatin hydrolysates, 175
gums, 5
konjac flour, 217
starches, 254
Biscuits, 86, 146, 147
Bittern, 112–113, 297
Blanc manger pudding, 66, 195, 255–257
Blanching, 224, 268
Bleaching agents, 14
Blender, 26, 44, 47, 49, 70–71, 135, 148, 150–151, 155, 256, 260, 262, 297
Bloom gelometer, 163, 164, 165, 166, 175
Bodying agents
alginates, 41
carboxymethylcellulose, 86
carrageenans, 65, 66
gums, 5, 8
starches, 249, 253
xanthan gum, 276, 277
Bologna, 68, 146, 254
Boucher Electronic Jelly Tester, 228
Bovine serum albumin, 122–123
Bovine spongiform encephalopathy (BSE), 4, 168
Brabender viscoamylograph, 247
Breads. *See* Baked goods
Breadings, 184
Brewing, 197, 244
Brillat-Savarin, Anthelme, 243
Brine, 68, 72, 184, 297
Brown seaweed, 2, 35
Bulking agents
agar, 20
cellulose, 86
gums, 5
konjac mannan, 209
Butter, with gelatin, 165–166

C

Caesalpinia spinosa, 140
Cakes
agar in, 31
cellulose derivatives in, 85, 86
coating for, 19
curdlan in, 103, 104
decoration, 64, 95, 115
eggs in, 121, 146
frosting, 254
gelatin in, 166, 167
gluten free, 277
guar gum in, 146, 147, 158
konjac flour, 217
low-calorie, 86
mixes, 86, 158, 166, 167, 250
starches in, 244, 250
traditional, 277
xanthan gum in, 277
Calcium
in alginate gels, 3, 35, 37, 38–40, 41, 50, 183
in gellan gum, 191
in low-methoxy pectin gels, 3, 224, 226–229, 238
Calcium alginates, 10, 37, 42
Calcium hydroxide, 162, 209, 211–212, 214–215, 218, 297
Calcium hypochlorite, 14

Candy
    bars, 19
    confections, 253
    fruit-gummy, 169–170
    sugar-free, 200–201
Canned foods, 20, 25, 40, 43, 146, 158, 166, 243, 145, 247, 254, 276
Caramel(s), 165
    custard, 70, 130–132, 134
    flavor, 158
    sauces, 255
Carbohydrates
    eggs, 119, 120
    konjac, 208–209, 217
    starches, 245
Carboxymethylcellulose (CMC)
    applications, 10, 65, 78, 885–86, 96; *See also specific foods*
    and alginate, 8
    and carob gum, 288
    and carrageenans, 8
    and hydroxypropylcellulose, 80
    clump-prevention techniques, 83
    food-grade, 82
    gelation, 84
    history of, 81–82
    manufacture of, 77–78, 82, 83
    in microcrystalline cellulose dispersions, 80–81
    mixed with other gums, 84
    and pectin gels, 7–8
    properties of, 78
    solution properties, 82–85
    technical-grade, 82
    world market, 3
Carboxymethyl starch, 252
Carob fruit, 139
Carob gels, 143
Carob gum, 139, 142, 146, 288
Carob tree, 139
Carrageenans
    applications, 8, 9, 13, 19, 57, 64–68, 73, 85, 145, 183, 213–214, 250, 276, 230, 295; *See also specific foods*
    cooking tips, 73
    extraction methods, 57, 58–59, 63
    food-grade, 60
    food safety, 60
    function in seaweed, 57
    and gelatin, 289
    gelation, domain model, 62
    gel properties, 7, 17, 62–63, 73, 295
    and guar gum, 67, 85, 145
    history of, 57
    and konjac mannan, 209, 295
    with β-lactoglobulin, 289
    locust bean gum interactions, 7, 73, 144, 288, 295
    marketable forms, 59–60
    molecular weight, 60, 61, 62
    packaging, 60
    in pet foods, 67, 183
    protein stabilization, 8, 63–64, 295
    protein synergy, 289
    recipes with, 68–73
    regulatory status, 60
    salts of, 59
    solution properties, 8, 19, 60–62
    sources, 3, 13, 35, 57, 58–59
    structure, 58, 59
    types of, 58
    water applications, 66–67
    world market, 3, 4
ι-Carrageenan
    gel formation, 62
    gel properties, 62–63
    in milk products, 64
    salts of, 60, 61
    solubility, 60–61, 67
    sources, 58, 59
    structure of, 58
    water gelation, 66
κ-Carrageenan
    applications, 19, 20, 45–46, 64, 65–68, 213, 250
    cryoprotective effects, 64
    gel formation, 62
    and gellan gum, 182–183
    gel properties, 62–63
    and guar gum, 67
    and konjac mannan, 209, 288, 295
    and locust bean gum, 288, 295
    in milk products, 64–66
    in pet foods, 60
    salts of, 60, 61
    solubility, 60–61, 67
    sources, 58–59
    structure, 58
    whey-off prevention, 65–66
λ-Carrageenan
    cryoprotection, 65
    in milk products, 65–66, 289
    solubility, 60–61
    sources, 58–59
    structure, 58
Carriers
    caseins, 167
    cellulose, 80
    gelatin, 167, 175
    starch, 255

Casein(s)
carboxymethylcellulose and, 86
carrageenan and, 63–66, 289
as carriers, 167
emulsifying properties, 122
foaming ability, 164
pectin and, 229
types of, 63, 289
-xanthan gum interactions, 275, 276
αs-Casein, 289
β-Casein, 63, 289
κ-Casein, 63–64, 289
Caseinate, and gellan gum, 184–185
Casings
alginate, 41–42
cellulose, 87
gelatin, 42
pectin, 42
Cassia gum, and gellan gum, 183
Cassonade, 131–132, 297
Celery seeds, 297
Cellulases, 84, 273
Cellulose/cellulose derivatives
applications, 78, 79, 80, 81, 82, 85–88; *See also specific foods*
cooking tips, 96
degree of polymerization, 77
degrees of substitution, 77
flocculation temperature, 79
manufacture of, 77–78, 82
in plants, 77
purity of, 78
recipes with, 88–96
solubility, 78
structure, 77, 80, 82
types of, 78
Cellulose gums, 64, 82, 83, 84–85, 146
Cereal,
breakfast, 20
flavor, 244, 247
grains, 1, 245
starches, 249, 253
Ceramium spp., 14
Ceratonia siliqua, 139
Cheese
agar in, 20
carrageenans in, 66, 73
cellulose in, 80, 88
curdlan in, 103
folic acid in, 229
galactomannans in, 145
gelatin in, 145
gellan gum in, 66, 184
gums in, 5, 10
locust bean gum in, 140, 158
sodium alginate in, 41
spread, 10, 41, 73
Cherries, artificial, 40
Cherry balls, 40
Chicle, 1, 3
Chiffons
agar in, 18
galactomannan blends in, 145
gelatin in, 145, 166
guar gum in, 147
Chinese soup stock, 127–128, 263–266, 297
Chocolate-milk drinks
alginate in, 43, 276
carboxymethylcellulose particles, 85
carrageenans in, 8, 64, 65, 276, 289
cellulose in, 87
furcellaran in, 65
galactomannans, 276
gums in, 5, 184
pectin, 276
starch in, 255
xanthan gum in, 255, 276
Cholesterol, egg yolk, 120
Chondrus chrispos, 57, 58
Cinnamon, 150, 297
Citric acid, 40, 65, 186–187, 189–190, 200–201, 224, 226, 237, 269, 297
Citrus juice products, 9, 158, 222, 275–276
Citrus peel, 222–223, 224, 229
Clarifying agents
agar, 20
alginates, 43
gellan gum, 183
gum arabic, 199
gums, 5
Clouding agents
glycerol, 198
gums, 5,
locust bean gum, 158
pectins, 229–230
vegetable oil, 198
Clump-prevention techniques
carboxymethylcellulose, 83
carrageenan solutions, 61
gelatin, 175
sodium alginate, 41
xanthan gum solutions, 274
CMC. *See* Carboxymethylcellulose
Coatings, *See also* Films
agar, 20
alginate, 10, 40, 41, 42, 50, 184
antimicrobial, 42, 88
ascorbic acid in, 41, 184
cellulose derivatives, 10, 82, 87, 88
curdlan, 103

gelatin, 10, 166–167
gellan gum, 41, 183, 184
guar gum, 145
gum, 4, 5, 8
gum arabic, 198
konjac mannan, 217
locust bean gum, 145
pectins, 10, 229, 230
sunflower oil in, 41, 184
xanthan gum, 283
Cocoa. *See* Chocolate-milk drinks
Cohesive texture, 244, 251
Collagen amino acids, 161
defined, 161
in sausage casings, 166
structure, 161
Color-stabilization, 42, 43, 228
Confections
agar in, 19
cellulose gum, 64
carrageenan, 164
furcellaran in, 57, 64
gelatin in, 164–165, 167
gellan gum in, 184
gum arabic in, 198, 198
gums in, 5, 8, 10
konjac mannan, 209
pectins in, 221, 228, 230, 238
starch in, 243, 245, 248, 249, 253
xanthan gum, 278
Converted starches, 251
Cooper, Peter, 164
Corned beef, 20
Cornstarch
and alginate, 42
and konjac, 217
and PGA, 43
and xanthan gum, 275
applications, 243, 250, 252–255
flavor of, 247
gelatinization, 246–247
products, 243
properties of, 243, 249
solutions, 243
structure, 245
recipes, 91–92, 93, 127, 173–174, 255–256, 281–282
Corn syrup, 19, 40, 85, 94–95, 151, 153, 164, 169–170, 173–174, 234, 235–236, 243, 252, 292–293, 294
Cottage cheese
carrageenan in, 66
locust bean gum in, 66
xanthan gum in, 276–277
Cotton, cellulose concentration, 77
Cream
and agar, 20
and alginate, 41
and λ carrageenan, 65
and cellulose derivatives, 79, 80, 86
and furcellaran, 66
and locust bean gum, 65
recipes, 93, 95–96, 131–132, 172–173, 290, 294
Cream,
Bavarian, 172
Boston, 253
boiled, 131
butter fat, 165
cheese, 20
coconut, 253
custard, 131, 133, 134
foamed, 276
fresh, 172, 173
frozen, 175
heavy, 93, 95, 96, 125, 131, 132, 290
lemon, 253
low-fat whipped, 95
puffs, 133, 134
soups, 254
synthetic, 41
thicken whipped, 41
whipped dairy, 65, 86, 173
Cream cheese, 20
Creaming stability
eggs, 122
gellan gum, 185
gum arabic, 198
Cross-linking agent, 6, 39, 84, 143
Croutons, 259
Cryoprotection, carrageenans, 64, 65
Crystalline domains, 77, 102
Culture media,
agar, 13
xanthan gum, 271
Curdlan
applications, 102, 103–104; *See also specific foods*
conformational changes, 102
cooking tips, 115
crystalline forms, 103
degree of polymerization, 101, 102
gelation, 102–103
gel properties, 103, 115
high-set gels, 103, 115
history of, 101
low-set gels, 102, 115
production of, 101
recipes, 105–115
regulatory status, 102

solution properties, 102
structure, 101–102
toxicity, 102
water suspensions, 102
Curdling, 248
Custard, 64, 65, 70, 119, 128–130
Cyamopsis psoraloides, 140
Cyamopsis tetragonolobus, 140

D

DA. *See* Degree of amidation
Dairy products, *See also* Milk products and *individual products*
agar, 20
alginates in, 41, 43
carboxymethylcellulose in, 65, 86, 96
carrageenans in, 63–64, 65–66, 289, 295
furcellaran in, 64, 65, 66, 73
galactomannans in 144, 145
gelatin in, 164, 165–166
gellan gum in, 184
guar gum in, 65,
locust bean gum in, 65
pectin, 221, 228, 229, 230
starch in, 250, 253, 255
xanthan gum in, 276
Danish agar, 2, 57
Danish desserts, 253
DE. *See* Degree of esterification
Deacylated gellan, 19
Deep-frying, 47, 104, 111–112, 215–216, 263–264, 266–267, 280–281
Deformability modulus, 63
Degree of acetylation, 223
Degree of amidation (DA), 225, 227
Degree of esterification (DE), 223, 225, 226, 227, 229
Degree of polymerization (DP)
alginates, 36, 38
cellulose, 77
curdlan, 101, 102
dextrin, 251
galactomannans, 141
starch derivatives, 251, 252
Degrees of substitution (DS)
carboxymethylcellulose, 82, 83, 84
cellulose, 77, 78, 85
Derivatization reactions,
cellulose manufacture, 7778
starch, 250, 252
Depolymerization, 59–60, 62, 144
Dessert gels
alginates in, 8, 41
cellulose in, 8
carrageenan, 145
furcellaran in, 64
galactomannans in, 145
gelatin in, 164
gellan gum in, 184
gums in, 5
Dextran, 179
Dextrins, 199, 251, 252, 268, 275
Dextrose, 164, 243, 245
Diabetes, and fenugreek, 141, 147
Dietary supplements, 21, 141, 167, 175, 197
Diffusion setting, 39
Domain model, 62
Double emulsions, 167
Double helices, 62–63, 181
Doughnuts, 19, 86, 104, 111–112
DP. *See* Degree of polymerization
Dried bonito, 89–91, 297–298
Drug delivery, pectin formulations, 225
Drum drying, 10, 59
Dry starch, 246, 249–250, 251, 255
DS. *See* Degrees of substitution
Dusting agents, 87, 255

E

Edible casings. *See* Casings
Edible films. *See* Films
Egg-box structure
alginate gels, 38
carrageenan gels, 62
pectin gels, 227
Eggs
composition of, 119–121
cooking tips, 135
gelling properties, 123–124
history of, 119
nutritional value, 121
recipes with, 124–135
separating, 135
structure of, 119
uses, 119, 121–122, 123
Egg white
composition of, 119, 120–121
foams, 122, 135
gels, 123–124
Egg yolk
composition of, 119, 120
emulsions, 121–122
gels, 123
Eklonia cava, 35
Eklonia maxima, 35
Emulsifying agents/stabilizers
alginates, 37, 43
carrageenan, 67

cellulose, 79, 80, 81, 82, 86, 87
curdlan, 102
eggs, 121–122
gelatin, 165, 167, 175
gellan gum, 184–185
guar gum, 158
gum arabic, 197–199, 238
gum karaya, 238
gums, 5, 8
gum tragacanth, 238
konjac mannan, 217
locust bean gum, 276
pectins, 230, 238
protein-polysaccharides, 287–288
starches, 254
xanthan gum, 273, 276, 277, 283
Emulsion, defined, 121
Encapsulating agents
agar, 21
alginate/pectin polymers, 229
calcium/alginate films, 10
fenugreek, 141
gelatin, 164, 167
gum arabic, 10, 197, 198–199
gums, 5, 9
starch, 253
with hydrogel beads, 21
Endosperm, 139, 140, 141, 142
Enzyme stability
alginates, 40, 50
galactomannans, 144
lintnerized starch, 244
xanthan gum, 273
Epigallocatechin gallate, 167
Escherichia coli, 164, 225
Est-Agar AS, 57, 59
Ethanol production, alginate beads, 43
Eucheuma spp., 58
Extruded food products, 42, 81, 87, 146, 147
Extrusion processes
carrageenan production, 59
cellulose derivatives, 87
pasta, 43
reformed fruit, 40
Exudate gums, 1, 3, 4, 195

F

Fabricated foods, 103, 184
Fats, and starch gelatinization, 248, 249
Fat stabilizers, 65, 66
Fat substitutes
alginates, 276
carrageenan, 67, 68, 230, 276, 294
cellulose, 95–96
curdlan, 103, 111
galactomannans, 276, 294
gellan gum, 183, 184, 209
guar gum, 294
konjac mannan, 209, 211
pectins, 230
starch, 67, 254, 268
xanthan gum, 276
FDA, 16, 18, 60, 82, 102, 140, 217, 225, 305
Fenugreek gum, 141, 146, 147, 157, 254
Fenugreek seeds, 141, 147, 156–157
Fermentation, gum production, 180, 271
Fiber sources
fenugreek gum, 141
guar gum, 147
gum arabic, 198
konjac flour, 209
microcrystalline cellulose, 80, 86, 87
pectin, 225
starch, 254
Films, *See also* Coatings
Agar, 20
alginate, 10, 40, 41, 42, 50, 184
antimicrobial, 42, 88
cellulose, 10, 80, 82, 87, 88
curdlan, 103
gelatin, 10, 166–167
gellan gum, 41, 183
gum, 4, 5, 8
gum arabic, 198
konjac mannan, 217
locust bean gum, 145
pectin, 10, 229
sunflower oil in, 41, 184
xanthan gum, 283
Fining agents
agar, 20
alginates, 43
gellan gum, 183
gum arabic, 199
hydrocolloids as, 5
Fish fillets, 106, 107, 146
Fish gelatin, 4, 67, 166, 167
Fish-oil emulsions, stabilizers, 197
Fish products
agar in, 20
in alginate jelly, 40, 41–42
carrageenans in, 67–68, 73
curdlan, 103, 106–110
gelatin in, 166
Flavonoids, 167
Flavor-fixation, 9–10, 167
Flavor oils
encapsulating agents, 10, 197, 198–199
and gelatin, 166–167

Flocculation, 5, 30, 79, 810, 145, 185, 197, 198, 229, 287
Fluid gel, 21, 189–190
Foaming agents/stabilizers
alginates, 43, 199
carrageenans, 8, 65
gelatins, 164–165, 175
guar gum, 147
gums, 5, 9
karaya gum, 199
locust bean gum, 8
microcrystalline cellulose gel, 80, 81, 87
proteins, 8
xanthan gum, 276, 277, 283
Foams
defined, 122
egg white, 121, 122, 135
gelatin, 166
-mat drying, 87
Folic acid, encapsulated, 229
Frankfurters
carrageenan in, 230
gellan gum in, 184
high-fat, 211
konjac mannan in, 209, 211
pectin gels in, 230
reduced-fat, 184, 209, 211, 230
starches in, 254
Freeze drying process
agar production, 14
flavor fixation, 10
Freeze thaw process
agar extraction, 13–14
carrageenan extraction, 59
Freeze-thaw stability
Carrageenan, 62, 65
Cellulose, 87
curdlan, 103
galactomannans, 146
hydrocolloid blends, 182
starches, 253, 255
xanthan gum, 273, 275, 276, 283
Freeze-thaw cycles, 87, 273
French
dessert, 257
dressing, 8, 49, 279
meringue, 135
vinaigrette, 50
Fried foods
alginates in, 43
cellulose in, 86, 87
curdlan in, 104
gellan gum in, 184
guar gum in, 147
starch in, 255
Frozen foods
alginate in, 40, 42, 43, 47–48, 51
carrageenan, 64, 65
cellulose derivatives, 78, 82, 87, 96
curdlan, 103, 104
galactomannans in, 146
gelatin, 165, 175
gum, 5
pectins in, 230
starches in, 255, 275
xanthan gum in, 275, 276, 277
Frozen yoghurt, 66
Fructose, 227
Fruit
fresh-cut, coatings for, 41, 73, 184
reformed, black currants, 40
Fruit products
agar in, 13, 19–20
alginates in, 39, 40–41, 43, 51
carboxymethylcellulose in, 85, 86, 96
carrageenan, 64, 67
and curdlan, 103, 115
dried texturized, 20
furcellaran in, 73
gelatin, 165, 166
gellan gum in, 185, 191
guar gum in, 158
gums in, 5
pectin in, 221, 228, 230
simulated, 19, 40
starch, 250
Fruit pulp, 19, 20, 21, 40, 43, 73, 85, 221, 227, 228
Fruit purées, 40, 96, 151
Furcellaran
applications, 64, 65, 66, 73; *See also specific foods*
cooking tips, 73
extraction methods, 59
gel properties, 63
history of, 57
recipes with, 68–73
regulatory status, 60
solubility, 61
sources, 59
structure, 58
terminology, 2
Furcellaria fastigiata, 2, 57
Furcellaria lumbricalis, 57

G

Galactomannans
alginate in, 276
applications, 144–147, 230; *See also specific foods*

carrageenans in, 276
cooking tips, 158
degree of polymerization, 141
interactions with other hydrocolloids, 288
pectin in, 230, 276
proteins in, 145
recipes with, 147–158
solution properties, 142–143
stability, 144, 146
structure, 141–142
types of, 139; *See also* Fenugreek; Guar gum; Tara gum and Locust bean gum
-xanthan gum interactions, 274, 276
D-Galactopyranosyl, 141, 143
Galactose, 58, 60, 140, 141, 143, 196, 222, 274, 288
Galacturonic acid, 221, 222, 223, 225
Gefilte fish, 67
Gel(s)
brittle, 7, 18, 38, 63, 64, 179, 180, 182, 191
chewy, 7
properties of, 7, 17, 38, 123, 181, 228, 287
strength, defined, 17, 39, 62, 63
textures, 7
Gelatin
and agar, 18
as a carrier, 167, 175
applications, 4, 5, 8, 10, 20, 145, 164–167; *See also specific foods*
Bloom strength, 163, 165
with carrageenan, 67, 289
and gellan gum, 183
cooking tips, 175
defined, 161
gelatinization, 163
gel properties, 7, 163, 175
and gum arabic, 198
history of, 161
manufacture of, 162–163
nonfood applications, 161, 164, 167
properties of, 163–164
recipes with, 168–175
regulatory status, 167–168
solubility, 163
sources, 4, 162
and starch, 253
type-A, 162, 163
type-B, 162, 163
world market, 3
Gelatin casings, 42
Gelatin desserts. *See* Dessert gels
Gelatin hydrolysates, 165, 167, 175
Gelatin sponges, 167
Gelation, cross-linking mechanism, 6–7
Gelidium spp., 13, 14, 15
Gelidium amansii, 14
Gellan gum
applications, 41, 66, 183–185; *See also specific foods*
benefits of, 184
cooking tips, 191
gelling abilities, 179, 181, 184
gel properties, 181–182
high-acyl (HA), 180, 181, 182, 183, 189, 190, 191
hydration of, 180–181, 191
low-acyl (LA), 180–184, 186, 187, 188, 191
manufacture of, 179–180
and other hydrocolloids, 182–185
purity, 180
recipes with, 186–191
regulatory status, 180
solution viscosity, 182
sources, 179
structure, 179
toxicity testing, 180
types of, 179, 180
Gelling agents
agar, 13, 17
alginates, 13
blends of, 182
carrageenans, 13, 66, 230
curdlan, 103
egg, 121, 123–124
furcellaran, 64
gelatin, 4, 164–168
gellan gum, 179, 184, 185
gums, 5
konjac mannan, 217
microcrystalline cellulose, 81
mixtures, 182
pectin, 221, 230
phycocolloids, 13
and thickening agents, 182
xanthan gum, 272
Gelose, 2
Gel press extraction, 59
Gel strength, (defined), 17, 20, 26, 37–39, 57, 60, 62–63, 67, 84, 87, 102–103, 115, 162–164, 175, 181, 185, 209, 225, 227, 230, 251, 253, 274
Generally recognized as safe. *See* GRAS
Gigartina spp., 58
Ginger,
chopped, 263
fresh, 262, 264, 265, 280, 281
grated, 260
juice, 261, 262
pickled, 90

Ginkgo nuts, 130, 297
Glazes
  agar, 19
  gelatin, 166
  gum arabic, 198
  gums, 5
  pectins, 229, 235–236
  starch, 255
Glucomannans, 208, 274, 275; *See also* Konjac mannan
Glucose, 2, 19, 20, 40, 77, 101, 147, 153–154, 179, 2058, 211, 222, 245, 252, 271, 272, 297, 298
Glucose, 2, 19–20, 31, 40, 77, 82, 101, 117, 147–148, 153–154, 179, 208, 211, 218, 222, 241, 245, 251–252, 271–272, 284, 297–298, 305
D-Glucosyl, 101
Glucuronic acid, 179, 196, 272
Glue, gelatin, 161
Glutamic acid, 141, 272
Gluten free, 86, 153, 277, 283
Gluten production, 244
Glycan, 2, 142
Glycerol
  in agar gel, 31
  in alginate films, 42, 50
  beverage stabilization, 198
  in coatings, 8, 41, 184
  in xanthan gum solutions, 272, 283
Glycerol monostearate, 93, 247, 298
Glycine, 141, 161, 272
Glycols, 50
GMP (good manufacturing practice), 60
Glycoproteins, 196
Gracilaria spp., 13, 14, 15
Grape juice, 30, 185
GRAS
  agar, 16, 18
  alginates, 36
  carrageenan, 60
  guar gum, 140
  konjac mannan, 217
  locust bean gum, 140
Grateloupiaceae, 58
Gravy,
  brown, 202–203
  thick like sauce, 6
  viscous, 146
Green, H.C., 36
Guar gum
  applications, 42, 85, 86, 141–142, 144–147, 182; *See also specific foods*
  with carrageenan, 65, 66, 67, 145
  cooking tips, 158
  food grade, 140
  gel creation, 143–144
  industrial grade, 140
  with locust bean gum, 145
  molecular weight, 142
  processing, 140
  recipes, 45–46, 150, 151, 152, 294
  regulatory status, 140
  solution properties, 142–143, 144
  sources, 140
  structure, 141
  substitution ratios, 142
  world market, 3
  -xanthan gum interactions, 273, 274, 276
L-Guluronic acid, 36
Gum acacia, 1, 195
Gum arabic
  and agar gels, 7
  applications, 9, 10, 197–199, 238, 253
  common names, 1, 195
  composition of, 195
  cooking tips, 203
  and gelatin, 167
  grades of, 196
  history of, 1, 195
  production of, 195–196
  properties of, 196
  recipes with, 45–46, 153–154, 199–203
  solution viscosity, 197, 203
  sources of, 3, 195
  "wattle blossom" model, 196
  world market, 3
  and xanthan gum, 275
Gumdrops, 248, 249, 253
Gum karaya, 3, 8, 19, 199, 238
Gum tragacanth, 3, 5, 8, 43, 197, 199–200, 238

H

Ham products, 67–68, 103, 166, 168
Hand mixer, 44, 47, 49, 88–89, 95, 134, 149, 211, 278–279, 282, 291, 294, 298
Harada, Tokuya, 101
Heat coagulation, eggs, 121
Heavy metals
  agar, 16
  cellulose products, 78, 80, 84–85
  gelatins, 164
Hemicellulose, 77
Herring, 41, 42
Herter, J.R., 36
High-acyl (HA) gellan gum, 180, 181, 182, 183, 189, 190, 191

High-density lipoproteins (HDLs), egg yolk, 120, 122
High-methoxy pectin (HMP)
  applications, 38, 229–230
  degree of esterification, 225
  extraction times, 224
  -fructose gels, 227
  gels, 221, 226–227
  -sucrose gels, 227
High-shear mixing (stirring), 37, 40, 61, 226, 274
Hilba, 141, 156
Honey, 104, 150
Hoso-kanten, 15
HPC. *See* Hydroxypropylcellulose
Hydrocolloids
  classification, 3
  cost of, 4
  and crystallization, 9
  emulsion stabilization, 8
  food applications, 5–10
  gelation, 6–7
  importance of, 1
  microparticles, 21
  mixtures
    benefits of, 182, 287
    cooking tips, 295
    food applications, 288–289
    recipes with, 290–294
  nonfood applications, 1, 4
  regulatory aspects, 10
  solution preparation, 37–38
  sources, 1
  synergistic combinations, 287
  terminology, 1–3
  viscosity formation, 5–6
  world market, 3–4
Hydrogen bonds, 9, 77, 102–103, 226, 246, 272
Hydrophobic interactions, 103, 227
Hydroxyethyl starch, 252
Hydroxyproline, 161
Hydroxypropylated starches, 268
Hydroxypropylcellulose (HPC)
  manufacture of, 78
  precipitation temperature, 79
  properties of, 79–80
  solubility, 78, 79, 96
  solution viscosity, 79
  uses, 80, 84, 86, 95, 96
Hypercholesterolemia, 141
Hysteresis
  agar, 17
  defined, 63
  gellan gum gel, 181–182

I

Ice-cream maker, 93–94, 150, 294, 298
Ice-cream stabilization
  alginates, 35, 41, 85
  blends for, 85
  carboxymethylcellulose, 65, 85, 96
  carrageenans, 64, 65, 85, 145
  curdlan, 103, 104
  galactomannans, 144–145, 230
  gelatin, 85, 165, 175
  guar gum, 65, 85, 144–145, 158, 276
  gums, 4, 5
  locust bean gum, 65, 85, 144–145, 276
  pectins, 230
  starches, 252
  xanthan gum, 276
Ice crystallization inhibition
  Cellulose derivatives, 81, 85
  galactomannans, 144–145
  gelatin, 165, 175
  gums, 5
  hydrocolloids, 4, 9
  pectins, 230
  and xanthan gum, 275
Icing stabilization
  agar, 8, 18, 19, 31, 166
  alginates, 43, 166
  carrageenan, 8
  furcellaran, 64
  gelatin, 8, 166
  guar gum, 158
  gum arabic, 198
  gums, 5
  starches, 250
Infant formulas, 64, 65, 249
Inorganic salts, and agar gelation, 18
Insulin resistance,
  and fenugreek gum, 141, 147
  and gellan gum, 179
Internal setting, alginate gelation, 39, 40
Inulin, 43
Iodine, and agar gelation, 18
Iota-carrageenan. *See* ι-Carrageenan
Iron, in eggs, 120, 121
Ito-kanten, 15
Ivory nut mannan, 141

J

Jackfruit seed starch, 255
Jams
  agar, 20
  alginates in, 41
  furcellaran in, 64

galactomannans, 146
gellan gum in, 184
gums, 5
pectin in, 221, 222, 227–229
thin-layered curdlan gels, 104
Japanese broth, 89–90, 125–126, 128–129, 261–262, 280–281, 297–298
Japanese pepper, 125–126, 298
Jellies
agar in, 13, 19–20, 185
alginates in, 41, 43
carrageenans in, 67
curdlan in, 103
furcellaran in, 64
gelatin, 164
gellan gum in, 184, 185, 191
konjac mannan in, 209, 210
pectin in, 221, 225, 227–228, 230
starch, 249, 253
Jelly candies, 5, 19, 253, 277
Judas ear, 127–128, 198
Junction zones, 102–104, 143–144, 226

K

Kaku-kanten, 14, 15, 21, 23, 27
Kamaboko, 103, 106–107, 128–129
Kami, 1, 35
Kanten, 13
Kappa-carrageenan. *See* κ-Carrageenan
Kelco Company, 271
Kelp, 35; *See also* Seaweed
Kelzan, 271
Ketchup, 47, 50, 91–92, 104, 146, 172, 202
Kirin Kyowa Foods Co., Ltd., 101
Koch, Robert, 13
Konjac mannan/flour
applications, 209–211; *See also specific foods*
-carrageenan gels, 295
composition of, 208, 209
cooking tips, 217–218
and gellan gum, 183, 184
gels, 218
health benefits, 211
history of, 207
manufacture of, 208
percentage in konjac tuber, 207
-potato starch gels, 211, 254
quality of, 209
recipes with, 211–217, 291
regulatory status, 217
solution viscosity, 209, 217
structure, 208–209
xanthan gum interactions, 144, 275, 288
Konjac plant, 207
Konjac tuber, 207
Kon-nyaku, 207, 209, 211
Koori tofu, 104
Krefting, A., 35

L

β-Lactoglobulin, with carrageenan, 289
Lambda-carrageenan. See λ-Carrageenan
Laminaria cloustoni, 35
Laminaria digitata, 35
Laminaria hyperborea, 35
Laminaria saccharina, 35
Laminaria spp, 35, 36, 298
Larch gum, 2
Latex, 1
LBG. *See* Locust bean gum
Lead, 16, 141; *See also* Pb.
Leatherhead Food Research Association (LFRA) texture analyzer, 228
Lecithin, 65, 137, 146, 294, 298
LeGloahec, V.C.E., 36
Lemon juice powders, 199
Lignins, 77
Lily roots, 130, 298
Lintnerized starch, 244
Lipids
egg yolk, 120
gum arabic, 197
konjac, 207
pectin, 225, 230
Lipoproteins
egg yolk, 120, 122
galactmannan, 147
metabolism, and pectins, 225
Liquid pectin, 224
Liqueur Jelly, 188–189
Locust bean gum (LBG)
and agar, 7, 18, 19, 20, 31
amino acids in, 141
applications, 8, 19, 65, 144–147; *See also specific foods*
block structure, 141
carrageenan interactions, 4, 7, 8, 63, 65–67, 144, 288, 295
cellulose derivatives, 85
cooking tips, 31, 73, 158, 295
freeze-thaw gels, 143
and gellan gum, 183
manufacture of, 139–140
molecular weight, 142
and pectin, 7
regulatory status, 140
recipes, 147–149, 290, 292, 294

solution viscosity, 142, 273
sources, 139
substitution ratios, 142
world market, 3
-xanthan gum interactions, 144, 182, 274, 275, 276, 288
Low-acyl (LA) gellan gum, 180–184, 186, 187, 188, 191
Low-calorie products. *See* Reduced-calorie products
Low-density lipoproteins (LDLs), 120, 122, 123, 147
Low-methoxy pectins (LMP)
applications, 228–230
degree of esterification, 225, 228
egg-box model, 227
extraction, 221–224
gellan gum, 185
gels, 221, 224–225, 227–228, 238
Lozenges, 198
Lumps, 44, 70, 83,148, 195, 203, 213, 225, 259, 291
Lupine, 121
Lysozyme, 120, 122–123

M

Mackerel, 41, 42, 108, 266–268
Macrocystis pyrifera, 35, 36
Mannitol, 201, 298
D-Mannopyranosyl, 141, 288
Mannose
in fenugreek, 141
in galactomannans, 141
in konjac, 208
in locust bean gum, 288
in tara gum, 140
in xanthan gum, 272, 274
D-Mannuronic acid, 36
Mark-Houwink equation, 62
Marmalade, 64, 233
Marshmallows, 7, 164, 165, 173–175
Mashed potatoes, 67, 202
Mashiko, K., 207
Mayonnaise, 5, 90–91, 104, 145, 255, 279
MC. *See* Methylcellulose
MCC. *See* Microcrystalline cellulose
Meat-alginate oligosaccharide conjugate, 42
Meat products
agar in, 20
alginate coatings, 42
carrageenans in, 67–68, 73
cellulose derivateds, 87–88,
curdlan in, 103–104
furcellaran in, 64
gelatin in, 164, 165–166
gellan gum in, 183, 184
guar gum in, 146, 158
gums, 7
konjac mannan in, 209
locust bean gum in, 146
pectin, 230
starches in, 254
Melting point, 17, 59
Meringues, 7, 19, 119, 134–135, 175, 250
Methylcellulose (MC)
and agar, 19
applications, 9, 86, 87, 88
gelation, 79
manufacture of, 77–78
properties of, 78–79
recipes, 88–89
solution viscosity, 78–79
Methylhydroxypropylcellulose (MHPC)
applications, 86, 87
gelation, 79
manufacture of, 77, 78
properties of, 78–79
recipes, 91–92
solution viscosity, 78–79
MHPC. *See* Methylhydroxypropylcellulose
Microbial polysaccharide, 101, 272
Microcrystalline cellulose (MCC)
applications, 80, 81, 87
colloidal, 80–81
manufacture of, 78
properties of, 80–81, 87
types of, 80
Microencapsulation. *See* Encapsulating agents
Microparticles, hydrocolloid, 21
Microwaveable foods
alginate-based products, 43
cellulose derivatives and, 87
and xanthan gum, 273
Milk products, *See also* Dairy products and *individual products*
agar in, 20
alginates in, 41, 43
carboxymethylcellulose in, 65, 86, 96
carrageenans in, 63–64, 65–66, 289, 295
furcellaran in, 64, 65, 66, 73
galactomannans in, 144, 145
gelatin in, 164, 165–166
guar gum in, 65
locust bean gum in, 65
pectins in, 43, 221, 228, 229, 230
starch in, 250, 253255
xanthan gum in, 276
Milk proteins, 8, 63, 64, 86, 184, 197, 229 289; *See also* Caseins; Whey proteins

Milk pudding, 5, 41, 64, 66, 68, 73
Mirin, 128–129, 280–281, 298
Mitsu-mame jelly cubes, 23, 25, 185
Mizu yokan, 21–24,185
Modified starches, 68, 145, 185, 197, 198, 249, 250–252, 253, 255
Moisture retention. *See* Water-binding ability
Molding agents
  gum arabic, 198
  gums, 5
  starch, 249, 253
  xanthan gum, 276
Molds, 225
Mucilage(s), 2, 36

N

Nakajima, T., 207
Nongelling agents, blends of, 7, 182, 245
Noodles
  citrus pectin, 230
  curdlan, 103, 104, 105–106
  gellan gum, 185
  konjac, 210
Nutmeg, 91, 108, 298

O

Oil absorption reduction, cellulose derivatives, 86–87
Oil-in-water emulsion stabilizers, *See also* Salad dressing stabilizers
  alginates, 43
  carrageenans, 67
  egg whites, 122
  gelatin, 167
  gellan gum, 184, 185
  guar gum, 145, 158
  gum arabic, 197, 238
  gum karaya, 238
  gum tragacanth, 8, 43
  pectin, 238
  propylene glycol alginate, 8
  protein-polysaccharides, 287–288
  starch, 8, 43
  xanthan gum, 43, 67, 197–198, 283
Oils, and curdlan gels, 104, 112
Okara, 214–217, 298
Okara konjac recipes, 214–217
Oleic acid, 121
Onion, 156, 170, 262–263, 280–281, 299
  cut, 92, 264
  green, 89–90, 106, 210
  long, 127–128, 266–268
  powder, 71–72, 279
  rings, 47–48
  spring, 260, 265
Organic solvents, 60, 78–79, 82
Ovalbumin, 120, 122, 123
Ovomucoid, 120, 123
Ovotransferrin, 120, 123
Oxidized starches, 252

P

Papaya pieces, 41, 184
Paprika oleoresin emulsions, 198–199
Pasta, konjac mannan, 209, 210
Pastilles, 198
Pb, 16, 141; *See also* lead
Pastries, 119, 133, 198, 244
Pastry fillings, 43, 86
Peach
  alginate gels, 40
  canned, 256
  jelly, 114, 115
  juice, 114, 115
  pastry fillings, 43
  puree, 40
  sauce, 256
Pea starch, 42, 254
Pectate, 225
Pectic acid, 222, 224
Pectic substances, defined, 221–222
Pectin(s)
  amidated, 224–225, 227
  applications, 5, 10, 19, 38, 40, 42, 43, 66, 67, 184, 185, 199, 221, 228–230; *See also specific foods*
  classification, 3
  cooking tips, 238
  de-esterification, 223, 224
  defined, 222
  degree of amidation (DA), 225, 227
  degree of esterification (DE), 223, 225, 226, 227, 229
  extraction methods, 221, 222, 223–224
  gel enhancers, 7–8
  gel types and properties, 7, 225, 226–228, 238
  health benefits of, 225
  high-methoxy (HMP), 38, 221, 224–225, 226–227, 229–230
  history of, 221
  and locust bean gum, 7
  low-methoxy (LMP), 221, 225, 227–230, 238
  manufacture, 223–224
  nomenclature, 221–222
  properties of, 221, 222–223, 226

purity, 225
recipes with, 231–238
regulatory status, 225
solution properties, 225–226, 228
sources, 221, 222
structure, 222
world market, 3–4
Pectina de Mexico, 223
Pectinase, 273
Pectin casings, with sodium alginate, 42
Pectinesterase, 224
Pectinic acids, 222
Pectinmethylesterase, 224
Peelability, and cellulose derivatives, 87
Peptization, microcrystalline cellulose, 81
Peschardt, W.J.S., 40
Pet foods
alginates in, 183
carrageenans in, 67, 183
cellulose derivatives, 85
gellan gum in, 183–184
guar gum in, 158
gums in, 5–6
locust bean gum, 146, 274
starch, 274
xanthan gum, 274
PGA. *See* Propylene glycol alginate
Pharmaceutical tableting, microcrystalline cellulose, 80
Phase separation, 79, 145, 146, 184, 230, 274, 287, 288
Phospholipids, egg yolk, 120
Phosphorus, in eggs, 121
Phycocolle, 2
Phycocolloids, 2, 13
Pie fillings
agar in, 19
alginates in, 43
carboxymethylcellulose in, 84, 86, 96
carrageenans in, 64, 67
gellan gum in, 183, 184
gums in, 5, 7
starches in, 243, 245, 253, 254, 255
Piment d'Espelette, 298
Pineapple
canned, 24, 93
cut, 93
flavoring, 93
fresh, 94
gel, 190
ice cream, 93
jelly, 186
juice, 185
sauce, 257
syrup, 93
Pizza sauce, 104
Plants
cellulose, 77
guar gum, 140
gums, 3, 4
konjac, 207, 209
pectins, 221, 222, 224, 225
starches, 244, 245
Plasticizer, 7, 8, 42, 198
Polymorphic forms, 103
Polysacolloid, 2
Pork, alginate coatings, 42–43
"Portable gelatin," 164
Potassium alginates, 37
Potassium chloride, 58–59, 62–63, 67, 73, 79
Potassium sorbate, 79, 187, 298
Potato starch
applications, 244, 254
cooking tips, 268
dispersion properties, 246–247, 268
effect of food ingredients, 248
flavor of, 247
fractionation, 250–251
gelatinization, 244, 246, 248
-konjac flour gels, 211, 254
market for, 243
properties of, 244, 249, 268
recipes, 91–92, 107–108, 127, 215–216, 260–268
structure, 245
texture of, 251
Poultry
agar coatings, 20
alginate coatings, 42
carrageenans and, 68
cellulose derivatives, 88
gellan gum, 183
Precipitation temperature, hydroxypropylcellulose, 79
Pregelatinized starches, 51, 250, 253, 254, 255, 268
Proline, 161
Propylene glycol alginate (PGA)
development of, 35
emulsion stabilization, 43
pH and, 37–38, 50
production, 36
recipe, 49
regulatory status, 37
uses, 5, 8, 37, 43, 85
Proteins
with carrageenans, 67, 68, 73, 289
egg, 120, 121, 122, 123, 124
foaming properties, 8, 122
galactomannans, 145

with gelatin, 161, 162, 163, 167
with gellan gum, 185
with um arabic, 197, 199
-polysaccharide interactions, 287–288
with sodium alginate, 40
vegetable, 8, 86, 121
in xanthan gum gels, 274–275, 276
Protein stabilization
carboxymethylcellulose, 84, 85, 86
carrageenans, 63–64
pectins, 229, 230
Protopectin, 222
Pseudomonas elodea, 179
Pseudoplasticity
carrageenan, 61
cellulose, 78, 79, 83
galactomannans, 143
gellan gum, 182
gum arabic, 197
pectins, 226
sodium alginate, 182
xanthan gum, 272–273, 275, 276
Puddings
agar in, 20
alginates in, 40
furcellaran in, 64, 66, 73
gellan gum in, 183, 184
gums in, 5
konjac mannan in, 209
starches in, 243, 244, 248, 250, 253

R

Red algae
agar from, 14, 16
carrageenan from, 35, 58
gum types in, 58
Red bean paste, 21–23, 298
Reduced-calorie foods
agar in, 20
alginates in, 43
carrageenans in, 67
cellulose derivatives in, 80, 85, 86, 88
galactomannans, 20, 146
gelatin in, 175
gellan gum in, 184, 185
konjac mannan in, 209, 211, 254
starches in, 20, 244, 254, 268
Reducing agents, 14
Resins, 1, 3
Resistant starch-rich powder (RSRP), 244
Retorted products, and curdlan gel, 103, 104
Retrogradation, 243, 247, 248, 252, 255, 275, 277
Rhamnose, 179, 196, 222, 226
Rice, agar coatings, 20
Rice cakes
curdlan in, 103
galactomannans in, 146
Rice starch
manufacture of, 244
properties of, 249
structure, 245
Rum, 131, 256

S

SAG method, 225, 228
Sago starch, 244–245, 268
Sake, 128–129, 215–216, 262–266, 298
Salad dressing stabilizers, *See also* Oil-in-water emulsion stabilizers
alginates, 43
carboxymethylcellulose, 96
carrageenans, 67
galactomannans, 146
guar gum, 145
gums, 5, 8
pectins, 230
starches, 8, 243, 245, 248, 250, 254–255, 268
xanthan gum, 230, 274, 275, 283
Salmon,
eggs, 45–46
meat, 42
smoked, 155
Salmonella, 164, 225
Salami, 146
Salt
and agar gel strength, 18
and starch gelatinization, 248, 268
Saponins, 141
Sardine, 108, 259–262
Sauces
alginates in, 5, 43
carrageenans in, 67
cellulose derivatives in, 79, 83, 85, 87
curdlan in, 104
galactomannans in, 145–146
gums in, 5, 6
starch in, 243, 248, 250, 255
xanthan gum in, 274, 275, 276, 283
Sausage casings
alginates, 41–43
cellulose derivatives in, 87
collagen in, 166
gelatin in, 42, 166
gums in, 5
pectin in, 43

Sausages
alginates in, 43
carrageenans in, 67
curdlan in, 103–104, 108–111
gelatin in, 166
gums, 5
locust bean gum in, 146
Schmidt, C., 57
Seaweed
history of, 1–2, 13–14, 35–37, 57–59
rigid-types, 14
soft, 14
uses, 1–4, 18, 67
Seaweed gums, 2
Seaweed mucilage, 2
Seiko, 208, 211, 213, 214, 291
Sequestering agents
alginates and, 37, 38, 39, 41, 50
gellan gum and, 180, 181, 182
Serine, 141
Sesame oil, 263, 264, 265
Shape retention
curdlan, 103
gums, 5
Shear-thinning, 146, 182, 185
Shellfish, in alginate jelly, 42
Shiitake mushrooms, 128–129, 263–264, 267–268, 298
Shiratama rice flour, 24–25, 298
Shortening, 19, 91–92, 94–95, 152, 250, 298
Silicone tube, 29–30
Simulated fruit, 19, 40
"Skin" formation, 248
Slab fixation, 10
Sodium alginate
in agar gels, 18
applications, 8, 41, 42, 43, 199
cooking tips, 50
gel, 39–40
and pectin gels, 8
sales of, 37
solution properties, 35, 37–38, 182
recipes, 44, 45–46, 47
and xanthan gum, 50, 273
Sodium-alginate cornstarch meat coating, 42
Sodium bisulfite, 14
Sodium carbonate, 17, 36, 37, 218
Sodium carboxymethylcellulose. *See* Carboxymethylcellulose
Sodium monoglutamate, 298
Sodium nitrate, 249
Sodium sulfate, 249
Soft cheese spreads, 41, 158
Soft drinks, 73, 96
Solid-state condition, and gelation, 6, 7
Soluble soybean polysaccharide, 229
Sorbet, 67
Sorbitol, 298
Sorbitan esters, 19
Sorghum, 298
Sorghum starch, 152, 243, 244, 247, 298
Souffle, 281–282
Soups
gums, 5, 10
agar in, 20
cellulose derivatives in, 85, 87
curdlan in, 104
guar gum in, 146
gelatin in, 167
gum arabic in, 198
starch in, 244, 250, 254
xanthan gum in, 274, 276
Soup stock, 127–129, 170–172, 202, 216–217, 258–259, 261, 263–266, 291, 297, 299
Soy milk gel, noodle-shaped, 104
Soy protein
in foams, 8
and gum arabic, 198
uses, 7, 68, 91–92, 145
Soy sauce, 104, 125, 127–129, 215–216, 260–266, 280–281, 299
Soy yoghurt, 276
Spans, 197
Sphingomonas elodea, 179
Spray drying
flavor-fixation, 9–10
gum arabic, 196, 198–199
Stabilizer, 4–5, 8–9, 18–19, 35, 37, 41, 43, 63–65, 80–81, 85, 87, 102, 145, 147, 165, 217, 221, 230, 245, 249, 250, 252, 254–255, 275–277, 283, 289
Stanford, E.C.C., 3536
Staphylococcus, 88, 225
Starch(es)
applications, 5, 8, 243, 245, 246
X-ray diffraction 247, 250, 252–255; *See also specific foods*
and agar, 20
and slginate, 42, 43, 51
and carrageenans, 65, 66, 67, 68, 73
and cellulose derivatives, 80, 81, 8687
classification, 2
composition, 245
converted, 251
cooking tips, 268
cross-linking reactions, 251
and curdlan, 103
derivatives, 247, 252, 253
dry, 244, 246, 249–250, 251, 255

effect of food ingredients, 248–249
fractionation, 250
and galactomannans, 145, 146
and gelatin, 166
gelatinization, 243, 244, 246, 247, 248, 249, 250, 252, 253, 254, 268
and gellan gum, 183, 185
and gum arabic, 197, 198
and konjac, 211, 217
lintnerized, 244
modified, 243, 244, 246, 249, 250–252, 253, 254, 255
oxidized, 252, 253
pastes, 243, 244, 245, 246, 247, 248, 251, 252
and pectin, 230
pregelatinized, 51, 244, 250, 253, 254, 255, 268
properties of, 249–252
raw, 246
recipes with, 47, 91–92, 93, 106–108, 127, 152, 173–174, 215–216, 255–268, 280–282
retrogradation, 243, 247, 248, 252, 255, 275, 277
sources, 243
structure, 245
suspension properties, 246
thermophile-free, 254
varieties of, 243–245; *See also individual starches*
world market, 3–4
in xanthan gum gels, 274, 275, 276, 277
Starch acetates, 252
Starch phosphates, 253, 255
Stevens-LFRA texture analyzer, 163
St. John's bread, 139
"St. Patrick's soup," 57
Stractan, 3
Succinoglucan, 101
Sucrose, 39, 40, 66, 85, 87, 103, 122, 123, 1444, 145, 164, 165, 225, 227, 248, 250, 271
Sugar
and agar gel strength, 18
crystallization inhibition, 9, 139, 254
hydration inhabitation, 181
and gel properties, 7, 8, 63, 191, 209, 222, 227, 253
and starch gelatinization, 247, 248
uses, 10, 19, 40, 60, 65, 66, 67, 83, 144, 145, 158, 164, 165
and xanthan gum, 273
Sugarbeet waste, 223
Sunflower oil, 41, 184
Surface active agent, 19
Surface tension, 79, 82, 122, 145, 165, 184, 199
Surfactants, 8, 19, 80, 122, 167, 249
Surimi, 105
cellulose gels in, 87
curdlan gels in, 104
fish gelatin in, 166
Suspension stabilizers
carboxymethylcellulose, 85
carrageenans, 8, 64, 65
gums, 5
xanthan gum, 275, 276
Sweet potato starch, 257, 258, 268
Syneresis
agar gels, 14, 18
carrageenans, 59, 62, 145, 295
curdlan, 103, 115
defined, 7
factors influencing, 18
gellan gum gels, 181, 184
pectin, 38, 229
semi-automated methods, 14
starch gels, 248
Syneresis inhibitors
alginate, 38, 39, 41, 43
carboxymethylcellulose, 84
carrageenans, 65, 73
galactomannans, 143, 144, 145, 146, 158, 274, 295
gelatin, 166
gums, 5, 10
xanthan gum, 274, 276
Szechuan pepper, 264–265, 299

T

Tablespoon, 299
Tableting, microcrystalline cellulose, 80
Tackiness, 2, 80, 249
Tannic acid, 31
Tapioca starch
applications, 252–255
and carrageenans, 67
composition of, 245
cooking, 247, 268
and guar gum, 146
properties of, 244, 245, 247
reciepes, 152,281–282
texture of, 251
Tara gum, 139, 140–141, 153–154, 155, 274
Tarozaemon Minoya, 13
Tasti-Jax, 42
Teaspoon, 299
Tempura, 280–281
Tertiary structures, 62, 102, 123

Texturization
  alginates, 8, 40, 41, 42, 43
  carrageenans, 65, 66, 67, 68, 73, 288
  cellulose derivatives, 8, 78, 81, 82, 83, 85, 87
  curdlan, 102, 103, 104
  eggs, 121
  furcellaran, 64
  galactomannans, 7, 145, 146, 158, 288
  gelatin, 164, 165, 167, 175
  gellan gum, 183, 185, 191
  gum arabic, 198
  konjac mannan, 209, 217
  pectins, 221, 226, 228, 230
  soy protein, 7
  starches, 244, 246, 248-255, 268
  xanthan gum, 275–277
Texturized fruits, 20, 40
Thermoplasticity
  gum arabic, 199
  hydroxypropylcellulose, 80
Thermostability
  agar, 8, 18
  alginate, 39, 41
  carrageenans, 8
  cellulose derivatives, 81
  curdlan gels, 103
  gelatin, 8
  gellangum, 185
  pectin gels, 221, 228
  xanthan gum, 273, 276
  xanthan gum–guar gum, 274
Thermoirreversible, 103–104
Thickening agents
  added to gelling agents, 182
  alginates, 37, 41
  carrageenans, 64, 66
  cellulose derivatives, 80–81, 86–87
  cost of, 4
  eggs, 121
  furcellaran, 64
  galactomannans, 139, 143
  gelatin, 20
  gum arabic, 195
  gums, 5, 6
  konjac mannan, 209, 217
  starches, 245, 246–247, 250, 252–255, 268
  xanthan gum, 271, 272, 276, 277, 283
Thixotropy
  alginates, 38
  carboxymethylcellulose, 83
  carrageenans, 60
  gellan gum, 182
  microcrystalline cellulose, 80, 81, 87
  pectins, 226, 228, 230
Toban jan, 264–265, 299
Tobiko, 208
Tofu, 103, 104, 112–113, 264–265
Tokoroten, 185
Tomato
  cherry, 170, 172
  coating, 198
  green mature, 198
  ketchup, 50, 104
  paste, 87
  pectic substances from, 222
  purée, 50, 156, 299
  sauce, 153
Trefoil, 130, 299
Triglycerides, egg yolk, 120
Triglycerides, serum, and konjac mannan, 211
Triple helices, 102, 272
Trisodium citrate, 181, 186–187, 189–191, 200, 299
Tuna
  canned, in agar jelly, 20
  and curdlan, 104
  gelatin source, 166
Turkey gum, 1, 195
"Turkish delight," 253
Tweens, 197

U

Unilever, extrusion system, 40
Unipectine, 223
Universal testing machine (UTM), 39, 62, 228

V

Vanilla flavoring, 27, 69, 111, 112, 130, 151, 172, 173, 174, 256, 257, 292, 29, 294
Vanilla beans, 131
Vanilla pod, 130, 131
Vanilla powdered, 133
Vegetable gelatin, 2
Vegetable oils, 8, 37, 43, 198
Vegetable proteins, 8, 80, 121
Vegetarian, 20, 65, 92
Vinegar, and starch gelatinization, 248
Vinyl polymers, 3
Viscosity-forming agents
  alginates, 19, 35
  carrageenans, 19, 57, 60
  cellulose derivatives, 78, 82, 83, 85, 87
  galactomannans, 19, 142, 143, 144, 146
  gelatin, 165
  gellan gum, 182

gum karaya, 19
hydrocolloids, 5–6, 19
konjac mannan, 209
pectin, 19
starch, 4, 252
types of, 19
Vitamin A, 121
Vitamin B complex, 121
Vitamin D, 121
Volatile oils, fenugreek gum, 141, 198

W

Water, surface tension, 79
Water-binding/absorption/holding/retention ability
alginate, 41
carrageenans, 67
cellulose derivatives, 78, 79, 80, 84, 85, 86, 88
egg white gels, 123, 124
galactomannans, 147
gelatin, 165, 166, 167, 168
gellan gum, 184, 185
gums, 4
konjac mannan, 209
starches, 246, 248, 253–254
xanthan gum, 275
Water-soluble gums, 1–3, 250
"Wattle blossom" model, 196
Wax coatings, 145
Waxy cornstarch
with alginate, 43
applications, 252–255
composition of, 245
flavor of, 247
gelatinization, 246
production of, 245
suspension properties, 246–247
texture of, 251
Waxy maize, 245, 246
Weeping
sausages, 146
starch gels, 248
Weight-reduction diets
citrus pectin noodles, 230
eggs in, 121
Wheat starch, 243, 244, 247, 252, 268, 280–281
Whey-off prevention, 65, 73, 85, 295
Whey proteins
carboxymethylcellulose, 85, 86
carrageenans, 66
pectin, 230
-xanthan gum gels, 274, 276
Whipped cream/toppings
alginates in, 41, 43
carrageenans in, 8, 9, 64, 65
cellulose derivatives in, 79, 80, 81, 86, 95–96
gum karaya in, 8
gums in, 5, 8
locust bean gum in, 8
Whisk, 88–89, 94, 124, 147, 149, 153–155, 211–213, 234, 290, 292, 294, 299
White dextrins, 251
Williamson etherification reaction, 78, 82
Wine, fining agents
agar, 20
alginates, 43
gellan gum, 183
gum arabic, 199
gums, 5

X

Xanthan gum
and alginate, 43, 50
applications, 275–277, 283; *See also specific foods*
and carrageenans, 50, 67, 68
and cellulose drivatives, 80–81, 85
conformations, 272
cooking tips, 283
and eggs, 121, 122
and gellan gum, 183, 185
grades of, 271–272
and guar gum, 146
and gum arabic, 197
interactions, 274–275
and konjac mannan, 209, 288
locust bean gum interactions, 144–145, 182, 274, 275, 276, 288, 295
manufacture of, 271–272
and pectin, 230
recipes with, 277–283, 290–291, 293–294
regulatory status, 271, 277
solutions, 182, 272–274, 283
sources, 179, 271
stability of, 273, 274, 275–276, 283
and starch, 146, 255
structure, 272
toxicity testing, 277
world market, 3–4
Xanthomonas campestris, 271
Xìngrén Jelly, 28
X-ray diffraction, 62, 103
Xylose, 222

Y

Yeast, 43, 225
Yellow cake, 121
Yellow dextrins, 251–252
Yield stress, 63
Yoghurt
    carrageenans in, 8
    curdlan in, 115
    furcellaran in, 64
    gelatin in, 165–166
    konjac mannan in, 209
    xanthan gum in, 276
Yokan, 185
Yuzu, 128–129, 261–262, 299

Z

Zero-shear, 143

Recipe Index

Appetizers

Aspic jelly salad 170
Artificial *ikura* (salmon eggs) 45
Fish cake (*kamaboko*) 106
Ham 71
Hot savory jelly 291
Kamaboko (fish cake) 106
Kinugoshi tofu (soybean curd) 112
Soybean curd (*kinugoshi tofu*) 112
Tempura 280

Soups

Crab meat and egg drop soup 127
Creamy corn soup 258
Fenugreek soup (*sharbat hilba*) 156
Sardine meatball soup 259
Sharbat hilba (fenugreek soup) 156
Squid meatball soup 261

Pasta, noodles & rice

Japanese noodle (*udon*) 105
Parsley spaghetti 88
Udon (Japanese noodle) 105

Main courses

BBQ *okara* konjac 215
Deep-fried mackerel with vegetable 266
Fenugreek soup (*sharbat hilba*) 156
Fried *okara* konjac 216
Goo lou yok (sweet and sour pork) 262
Japanese rolled omelettes (*sushi* egg) 125
Mappo *tofu* (spicy *tofu*) 264
Plain omelette 124
Sausages 108
Sharbat hilba (fenugreek soup) 156
Soy burger patties 91
Spicy *tofu* (Mappo *tofu*) 264
Sushi egg (Japanese rolled omelets) 125
Sweet and sour pork (*goo lou yok*) 262
Tempura 280

Side dishes

BBQ *okara* konjac 215
Chawan-mushi (savory egg custard) 128
Fried *okara* konjac 216
Hot savory jelly 291
Okara (soybean fiber) konjac … 214
Onion rings 47
Savory egg custard (*Chawan-mushi*) 128
Soybean fiber (*okara*) konjac 214
Tempura 280

Breakfast, Sandwiches & Burgers

Doughnuts 111
Empanada dough 152
Japanese rolled omelettes (*sushi* egg) 125
Octopus dumpling (*takoyaki*) 89
Plain omelette 124
Soy burger patties 91
Sushi egg (Japanese rolled omelets) 125
Takoyaki (octopus dumpling) 89

Drinks

Chocolate drink 71
Frozen fruit drink 151
Fruit juice 199

Desserts

Agar jelly with egg white (*awayuki-kan*). 27
Agar jelly with red bean paste (*mizu yokan*) 21
Agar jelly with rice dumplings (*Mitsu-mame*) 23
Agar spaghetti 29
Awayuki-kan (agar jelly with egg white) 27
Banana and nuts ice cream 150
Bavarian cream (*Bavaroise*) 172
Bavaroise (Bavarian cream) 172
Blanc manger 255
Brown sugar sherbet … 147
Caramel custard 130
Chinese almond pudding (*xìngrén* jelly) 28
Chocolate *soufflé* … 281
Coffee jelly 290
Cream puffs 132
Crème brûlée 131
Doughnuts 111
Flan 69
Fruit gummy candy 169
Fruit jelly 168
Fruit-juice jelly 186
Guimauve (marshmallows) 173
Konjac jelly 213
Liqueur jelly 188

Loukoums ... 153
Low fat ice cream ... 293
Marshmallows (*guimauve*) ... 173
Meringues ... 134
Milk pudding ... 68
Mizu yokan (agar jelly with red bean paste) ... 21
Mitsu-mame (agar jelly with rice dumplings) ... 23
Multilayered jelly ... 114
Orange jelly ... 26
Pineapple ice cream ... 93
Pulp-suspension fluid gel ... 189
Sugar-free candy ... 200
Sweet potato starch jelly (*warabi-mochi*) ... 257
Warabi-mochi (sweet potato starch jelly) ... 257
Xìngrén jelly (Chinese almond pudding) ... 28
Yoghurt spheres ... 44

Sauces, dressings & jams

Apple jam ... 231
Brown gravy ... 202
Clear glaze (*nappage neutre*) ... 235
Creamy Italian dressing ... 279
Dulce de batata (sweet potato jam) ... 292
French dressing ... 48
Herb-suspension oil-free dressing ... 190
Italian dressing ... 277
Low-solids jam ... 187
Milk jam ... 234
Nappage neutre (clear glaze) ... 235
Orange marmalade ... 233
Pear and fenugreek jam ... 157
Raspberry sauce ... 148
Sweet potato jam (*dulce de batata*) ... 292
Yoghurt dressing ... 155

Others

Agar spaghetti ... 29
Artificial *ikura* (salmon eggs) ... 45
Dessert base ... 236
Fish cake (*kamaboko*) ... 106
Ham ... 71
Hot savory jelly ... 291
Kamaboko (fish cake) ... 106
Kinugoshi tofu (soybean curd) ... 112
Konjac (*kon-nyaku*) ... 211
Kon-nyaku (konjac) ... 211
Low-fat whipped cream ... 95
Meringues ... 134
Okara (soybean fiber) konjac ... 214
Octopus dumpling (*takoyaki*) ... 89
Parsley spaghetti ... 88

Sausages....108
Soybean curd (*kinugoshi tofu*) 112
Soybean fiber (o*kara*) konjac 214
Sugarpaste....94
Takoyaki (octopus dumpling) 89
Yoghurt spheres.... 44